PROBLEMS AND SOLUTIONS IN QUANTUM MECHANICS

This collection of solved problems corresponds to the standard topics covered in established undergraduate and graduate courses in quantum mechanics. Completely up-to-date problems are also included on topics of current interest that are absent from the existing literature.

Solutions are presented in considerable detail, to enable students to follow each step. The emphasis is on stressing the principles and methods used, allowing students to master new ways of thinking and problem-solving techniques. The problems themselves are longer than those usually encountered in textbooks and consist of a number of questions based around a central theme, highlighting properties and concepts of interest.

For undergraduate and graduate students, as well as those involved in teaching quantum mechanics, the book can be used as a supplementary text or as an independent self-study tool.

KYRIAKOS TAMVAKIS studied at the University of Athens and gained his Ph.D. at Brown University, Providence, Rhode Island, USA in 1978. Since then he has held several positions at CERN's Theory Division in Geneva, Switzerland. He has been Professor of Theoretical Physics at the University of Ioannina, Greece, since 1982.

Professor Tamvakis has published 90 articles on theoretical high-energy physics in various journals and has written two textbooks in Greek, on quantum mechanics and on classical electrodynamics. This book is based on more than 20 years' experience of teaching the subject.

D0024343

PROBLEMS AND SOLUTIONS IN QUANTUM MECHANICS

KYRIAKOS TAMVAKIS

University of Ioannina

CAMBRIDGE
UNIVERSITY PRESS

CAMBRIDGE UNIVERSITY PRESS
Cambridge, New York, Melbourne, Madrid, Cape Town, Singapore, São Paulo

Cambridge University Press
The Edinburgh Building, Cambridge CB2 2RU, UK

Published in the United States of America by Cambridge University Press, New York

www.cambridge.org
Information on this title: www.cambridge.org/9780521840873

First published 2005

Printed in the United Kingdom at the University Press, Cambridge

A catalogue record for this book is available from the British Library

Library of Congress Cataloguing in Publication data

ISBN-13   978-0-521-84087-3 hardback
ISBN-10   0-521-84087-2 hardback

ISBN-13   978-0-521-60057-6 paperback
ISBN-10   0-521-60057-X paperback

Contents

Preface

This collection of quantum mechanics problems has grown out of many years of teaching the subject to undergraduate and graduate students. It is addressed to both student and teacher and is intended to be used as an auxiliary tool in class or in self-study. The emphasis is on stressing the principles, physical concepts and methods rather than supplying information for immediate use. The problems have been designed primarily for their educational value but they are also used to point out certain properties and concepts worthy of interest; an additional aim is to condition the student to the atmosphere of change that will be encountered in the course of a career. They are usually long and consist of a number of related questions around a central theme. Solutions are presented in sufficient detail to enable the reader to follow every step. The degree of difficulty presented by the problems varies. This approach requires an investment of time, effort and concentration by the student and aims at making him or her fit to deal with analogous problems in different situations. Although problems and exercises are without exception useful, a collection of solved problems can be truly advantageous to the prospective student only if it is treated as a learning tool towards mastering ways of thinking and techniques to be used in addressing new problems rather than a solutions manual. The problems cover most of the subjects that are traditionally covered in undergraduate and graduate courses. In addition to this, the collection includes a number of problems corresponding to recent developments as well as topics that are normally encountered at a more advanced level.

1

Wave functions

Problem 1.1 Consider a particle and two normalized energy eigenfunctions $\psi_1(\mathbf{x})$ and $\psi_2(\mathbf{x})$ corresponding to the eigenvalues $E_1 \neq E_2$. Assume that the eigenfunctions vanish outside the two non-overlapping regions Ω_1 and Ω_2 respectively.

(a) Show that, if the particle is initially in region Ω_1 then it will stay there forever.
(b) If, initially, the particle is in the state with wave function

$$\psi(\mathbf{x}, 0) = \tfrac{1}{\sqrt{2}} [\psi_1(\mathbf{x}) + \psi_2(\mathbf{x})]$$

show that the probability density $|\psi(\mathbf{x}, t)|^2$ is independent of time.
(c) Now assume that the two regions Ω_1 and Ω_2 overlap partially. Starting with the initial wave function of case (b), show that the probability density is a periodic function of time.
(d) Starting with the same initial wave function and assuming that the two eigenfunctions are real and isotropic, take the two partially overlapping regions Ω_1 and Ω_2 to be two concentric spheres of radii $R_1 > R_2$. Compute the probability current that flows through Ω_1.

Solution

(a) Clearly $\psi(\mathbf{x}, t) = e^{-iEt/\hbar}\psi_1(\mathbf{x})$ implies that $|\psi(\mathbf{x}, t)|^2 = |\psi_1(\mathbf{x})|^2$, which vanishes outside Ω_1 at all times.

(b) If the two regions do not overlap, we have

$$\psi_1(\mathbf{x})\psi_2^*(\mathbf{x}) = 0$$

everywhere and, therefore,

$$|\psi(\mathbf{x}, t)|^2 = \tfrac{1}{2}[|\psi_1(\mathbf{x})|^2 + |\psi_2(\mathbf{x})|^2]$$

which is time independent.

(c) If the two regions overlap, the probability density will be

$$|\psi(\mathbf{x}, t)|^2 = \tfrac{1}{2}\left[|\psi_1(\mathbf{x})|^2 + |\psi_2(\mathbf{x})|^2\right]$$
$$+ |\psi_1(\mathbf{x})|\,|\psi_2(\mathbf{x})|\cos[\phi_1(\mathbf{x}) - \phi_2(\mathbf{x}) - \omega t]$$

where we have set $\psi_{1,2} = |\psi_{1,2}|e^{i\phi_{1,2}}$ and $E_1 - E_2 = \hbar\omega$. This is clearly a periodic function of time with period $T = 2\pi/\omega$.

(d) The current density is easily computed to be

$$\mathcal{J} = \hat{\mathbf{r}}\frac{\hbar}{2m}\sin\omega t\left[\psi_2'(r)\psi_1(r) - \psi_1'(r)\psi_2(r)\right]$$

and vanishes at R_1, since one or the other eigenfunction vanishes at that point. This can be seen through the continuity equation in the following alternative way:

$$I_{\Omega_1} = -\frac{d}{dt}\mathcal{P}_{\Omega_1} = \int_{S(\Omega_1)} d\mathbf{S}\cdot\mathcal{J} = \int_{\Omega_1} d^3x\,\nabla\cdot\mathcal{J} = -\int_{\Omega_1} d^3x\,\frac{\partial}{\partial t}|\psi(\mathbf{x}, t)|^2$$
$$= \omega\sin\omega t\int_{\Omega_1} d^3x\,\psi_1(r)\psi_2(r)$$

The last integral vanishes because of the orthogonality of the eigenfunctions.

Problem 1.2 Consider the one-dimensional normalized wave functions $\psi_0(x)$, $\psi_1(x)$ with the properties

$$\psi_0(-x) = \psi_0(x) = \psi_0^*(x), \qquad \psi_1(x) = N\frac{d\psi_0}{dx}$$

Consider also the linear combination

$$\psi(x) = c_1\psi_0(x) + c_2\psi_1(x)$$

with $|c_1|^2 + |c_2|^2 = 1$. The constants N, c_1, c_2 are considered as known.

(a) Show that ψ_0 and ψ_1 are orthogonal and that $\psi(x)$ is normalized.

(b) Compute the expectation values of x and p in the states ψ_0, ψ_1 and ψ.

(c) Compute the expectation value of the kinetic energy T in the state ψ_0 and demonstrate that

$$\langle\psi_0|T^2|\psi_0\rangle = \langle\psi_0|T|\psi_0\rangle\langle\psi_1|T|\psi_1\rangle$$

and that

$$\langle\psi_1|T|\psi_1\rangle \geq \langle\psi|T|\psi\rangle \geq \langle\psi_0|T|\psi_0\rangle$$

(d) Show that

$$\langle\psi_0|x^2|\psi_0\rangle\langle\psi_1|p^2|\psi_1\rangle \geq \frac{\hbar^2}{4}$$

(e) Calculate the matrix element of the commutator $[x^2, p^2]$ in the state ψ.

Solution

(a) We have

$$\langle\psi_0|\psi_1\rangle = N\int dx\,\psi_0^*\frac{d\psi_0}{dx} = N\int dx\,\psi_0\frac{d\psi_0}{dx}$$

$$= \frac{N}{2}\int dx\,\frac{d\psi_0^2}{dx} = \frac{N}{2}\left\{\psi_0^2(x)\right\}_{-\infty}^{+\infty} = 0$$

The normalization of $\psi(x)$ follows immediately from this and from the fact that $|c_1|^2 + |c_2|^2 = 1$.

(b) On the one hand the expectation value $\langle\psi_0|x|\psi_0\rangle$ vanishes because the integrand $x\psi_0^2(x)$ is odd. On the other hand, the momentum expectation value in this state is

$$\langle\psi_0|p|\psi_0\rangle = -i\hbar\int dx\,\psi_0(x)\psi_0'(x)$$

$$= -\frac{i\hbar}{N}\int dx\,\psi_0(x)\psi_1(x) = -\frac{i\hbar}{N}\langle\psi_0|\psi_1\rangle = 0$$

as we proved in the solution to (a). Similarly, owing to the oddness of the integrand $x\psi_1^2(x)$, the expectation value $\langle\psi_1|x|\psi_1\rangle$ vanishes. The momentum expectation value is

$$\langle\psi_1|p|\psi_1\rangle = -i\hbar\int dx\,\psi_1^*\psi_1' = -i\hbar\frac{N}{N^*}\int dx\,\psi_1\psi_1'$$

$$= -i\hbar\frac{N}{2N^*}\int dx\,\frac{d\psi_1^2}{dx} = -i\hbar\frac{N}{2N^*}\left\{\psi_1^2\right\}_{-\infty}^{+\infty} = 0$$

(c) The expectation value of the kinetic energy squared in the state ψ_0 is

$$\langle\psi_0|T^2|\psi_0\rangle = \frac{\hbar^4}{4m^2}\int dx\,\psi_0\psi_0'''' = -\frac{\hbar^4}{4m^2}\int dx\,\psi_0'\psi_0'''$$

$$= \frac{\hbar^2}{2m|N|^2}\langle\psi_1|T|\psi_1\rangle$$

Note however that

$$\langle\psi_0|T|\psi_0\rangle = -\frac{\hbar^2}{2m}\int dx\,\psi_0\psi_0'' = \frac{\hbar^2}{2m}\int dx\,\psi_0'\psi_0'$$

$$= \frac{\hbar^2}{2m|N|^2}\langle\psi_1|\psi_1\rangle = \frac{\hbar^2}{2m|N|^2}$$

Therefore, we have

$$\langle\psi_0|T^2|\psi_0\rangle = \langle\psi_0|T|\psi_0\rangle\langle\psi_1|T|\psi_1\rangle$$

Consider now the *Schwartz inequality*

$$|\langle \psi_0 | \psi_2 \rangle|^2 \leq \langle \psi_0 | \psi_0 \rangle \langle \psi_2 | \psi_2 \rangle = \langle \psi_2 | \psi_2 \rangle$$

where, by definition,

$$\psi_2(x) \equiv -\frac{\hbar^2}{2m} \psi_0''(x)$$

The right-hand side can be written as

$$\langle \psi_2 | \psi_2 \rangle = \langle \psi_0 | T^2 | \psi_0 \rangle = \langle \psi_0 | T | \psi_0 \rangle \langle \psi_1 | T | \psi_1 \rangle$$

Thus, the above Schwartz inequality reduces to

$$\langle \psi_0 | T | \psi_0 \rangle \leq \langle \psi_1 | T | \psi_1 \rangle$$

In order to prove the desired inequality let us consider the expectation value of the kinetic energy in the state ψ. It is

$$\langle \psi | T | \psi \rangle = |c_1|^2 \langle \psi_0 | T | \psi_0 \rangle + |c_2|^2 \langle \psi_1 | T | \psi_1 \rangle$$

The off-diagonal terms have vanished due to oddness. The right-hand side of this expression, owing to the inequality proved above, will obviously be smaller than

$$|c_1|^2 \langle \psi_1 | T | \psi_1 \rangle + |c_2|^2 \langle \psi_1 | T | \psi_1 \rangle = \langle \psi_1 | T | \psi_1 \rangle$$

Analogously, the same right-hand side will be larger than

$$|c_1|^2 \langle \psi_0 | T | \psi_0 \rangle + |c_2|^2 \langle \psi_0 | T | \psi_0 \rangle = \langle \psi_0 | T | \psi_0 \rangle$$

Thus, finally, we end up with the double inequality

$$\langle \psi_0 | T | \psi_0 \rangle \leq \langle \psi | T | \psi \rangle \leq \langle \psi_0 | T | \psi_0 \rangle$$

(d) Since the expectation values of position and momentum vanish in the states ψ_0 and ψ_1, the corresponding uncertainties will be just the expectation values of the squared operators, namely

$$(\Delta x)_0^2 = \langle \psi_0 | x^2 | \psi_0 \rangle, \qquad (\Delta p)_0^2 = \langle \psi_0 | p^2 | \psi_0 \rangle, \qquad (\Delta p)_1^2 = \langle \psi_1 | p^2 | \psi_1 \rangle$$

We now have

$$\langle \psi_0 | x^2 | \psi_0 \rangle \langle \psi_1 | p^2 | \psi_1 \rangle \geq \langle \psi_0 | x^2 | \psi_0 \rangle \langle \psi_0 | p^2 | \psi_0 \rangle = (\Delta x)_0^2 (\Delta p)_0^2 \geq \frac{\hbar^2}{4}$$

as required.

(e) Finally, it is straightforward to calculate the matrix element value of the commutator $[x^2, p^2]$ in the state ψ. It is

$$\langle\psi|[x^2, p^2]|\psi\rangle = 2i\hbar\langle\psi|(xp+px)|\psi\rangle = 2i\hbar\left(\langle\psi|xp|\psi\rangle + \langle\psi|xp|\psi\rangle^*\right)$$

which, apart from an imaginary coefficient, is just the real part of the term

$$\langle\psi|xp|\psi\rangle = -i\hbar\int dx\,\psi^*x\psi'$$

$$= |c_1|^2\int dx\,\psi_0 x\psi_0' - i\hbar|c_2|^2\int dx\,\psi_1^*x\psi_1'$$

where the mixed terms have vanished because the operator has odd parity. Note however that this is a purely imaginary number. Thus, its real part will vanish and so

$$\langle\psi|[x^2, p^2]|\psi\rangle = 0$$

Problem 1.3 Consider a system with a *real* Hamiltonian that occupies a state having a real wave function both at time $t = 0$ and at a later time $t = t_1$. Thus, we have

$$\psi^*(x, 0) = \psi(x, 0), \qquad \psi^*(x, t_1) = \psi(x, t_1)$$

Show that the system is *periodic*, namely, that there exists a time T for which

$$\psi(x, t) = \psi(x, t + T)$$

In addition, show that for such a system the eigenvalues of the energy have to be integer multiples of $2\pi\hbar/T$.

Solution

If we consider the complex conjugate of the evolution equation of the wave function for time t_1, we get

$$\psi(x, t_1) = e^{-it_1 H/\hbar}\psi(x, 0) \qquad \Longrightarrow \qquad \psi(x, t_1) = e^{it_1 H/\hbar}\psi(x, 0)$$

The inverse evolution equation reads

$$\psi(x, 0) = e^{it_1 H/\hbar}\psi(x, t_1) = e^{2it_1 H/\hbar}\psi(x, 0)$$

Also, owing to reality,

$$\psi(x, 0) = e^{-2it_1 H/\hbar}\psi(x, 0)$$

Thus, for any time t we can write

$$\psi(x, t) = e^{-it H/\hbar}\psi(x, 0) = e^{-it H/\hbar}e^{-2it_1 H/\hbar}\psi(x, 0) = \psi(x, t + 2t_1)$$

It is, therefore, clear that the system is periodic with period $T = 2t_1$.

Expanding the wave function in energy eigenstates, we obtain

$$\psi(x, t) = \sum_n C_n e^{-i E_n t/\hbar} \psi_n(x)$$

The periodicity of the system immediately implies that the exponentials $\exp(-iT E_n/\hbar)$ must be equal to unity. This is only possible if the eigenvalues E_n are integer multiples of $2\pi\hbar/T$.

Problem 1.4 Consider the following superposition of plane waves:

$$\psi_{k,\delta k}(x) \equiv \frac{1}{2\sqrt{\pi \delta k}} \int_{k-\delta k}^{k+\delta k} dq \, e^{iqx}$$

where the parameter δk is assumed to take values much smaller than the wave number k, i.e.

$$\delta k \ll k$$

(a) Prove that the wave functions $\psi_{k,\delta k}(x)$ are normalized and orthogonal to each other.
(b) For a free particle compute the expectation value of the momentum and the energy in such a state.

Solution

(a) The proof of normalization goes as follows:

$$\int_{-\infty}^{+\infty} dx \, |\psi_{k,\delta k}(x)|^2 = \frac{1}{4\pi \delta k} \int_{-\infty}^{+\infty} dx \int_{k-\delta k}^{k+\delta k} dq' \int_{k-\delta k}^{k+\delta k} dq'' \, e^{i(q'-q'')x}$$

$$= \frac{1}{2\delta k} \int_{k-\delta k}^{k+\delta k} dq' \int_{k-\delta k}^{k+\delta k} dq'' \, \delta(q' - q'')$$

$$= \frac{1}{2\delta k} \int_{k-\delta k}^{k+\delta k} dq' \, \Theta(k + \delta k - q')\Theta(q' - k + \delta k)$$

$$= \frac{1}{2\delta k} \int_{k-\delta k}^{k+\delta k} dq' = 1$$

The proof of orthogonality proceeds similarly ($|k - k'| > \delta k + \delta k'$):

$$\int_{-\infty}^{+\infty} dx \, \psi_{k,\delta k}^*(x)\psi_{k',\delta k'}(x) = \frac{1}{2\sqrt{\delta k \delta k'}} \int_{k-\delta k}^{k+\delta k} dq' \int_{k'-\delta k'}^{k'+\delta k'} dq'' \, \delta(q' - q'')$$

$$= \frac{1}{2\sqrt{\delta k \delta k'}} \int_{k-\delta k}^{k+\delta k} dq' \Theta(k'+\delta k'-q')\Theta(q'-k'+\delta k')=0$$

since there is no overlap between the range over which the theta functions are defined and the range of integration.

(b) Proceeding in a straightforward fashion, we have

$$\langle p \rangle = \frac{1}{4\pi\,\delta k} \int_{-\infty}^{+\infty} dx \int_{k-\delta k}^{k+\delta k} dq' \int_{k-\delta k}^{k+\delta k} dq''\, e^{-iq'x} (-i\hbar\partial_x) e^{iq''x}$$

$$= \frac{1}{2\delta k} \int_{k-\delta k}^{k+\delta k} dq' \int_{k-\delta k}^{k+\delta k} dq''\, \hbar q'' \delta(q' - q'')$$

$$= \frac{1}{2\delta k} \int_{k-\delta k}^{k+\delta k} dq'\, \hbar q' = \frac{\hbar}{4\delta k}[(k + \delta k)^2 - (k - \delta k)^2] = \hbar k + O(\delta k)$$

Similarly, we obtain

$$\left\langle \frac{p^2}{2m} \right\rangle = \frac{\hbar^2 k^2}{2m} + O(\delta k)$$

Problem 1.5 Consider a state characterized by a real wave function up to a multiplicative constant. For simplicity consider motion in one dimension. Convince yourself that such a wave function should correspond to a bound state by considering the probability current density. Show that this bound state is characterized by vanishing momentum, i.e. $\langle p \rangle_\psi = 0$. Consider now the state that results from the multiplication of the above wave function by an exponential factor, i.e. $\chi(x) = e^{ip_0 x/\hbar} \psi(x)$. Show that this state has momentum p_0. Study all the above in the momentum representation. Show that the corresponding momentum wave function $\tilde{\chi}(p)$ is translated in momentum, i.e. $\tilde{\chi}(p) = \tilde{\psi}(p - p_0)$.

Solution

The probability current density of such a wave function vanishes:

$$\mathcal{J} = \frac{\hbar}{2mi}[\psi^*\psi' - \psi(\psi^*)'] = 0$$

The vanishing of the probability current agrees with the interpretation of such a state as bound.

The momentum expectation value of such a state is

$$\langle \psi | p | \psi \rangle = -i\hbar \int_{-\infty}^{+\infty} dx\, \psi(x)\psi'(x)$$

$$= -\frac{i\hbar}{2} \int_{-\infty}^{+\infty} dx\, \frac{d}{dx}\psi^2(x) = -\frac{i\hbar}{2}[\psi^2(x)]_{\pm\infty} = 0$$

The wave function $\chi(x) = e^{ip_0 x/\hbar} \psi(x)$, however, has momentum

$$\langle \chi | p | \chi \rangle = -i\hbar \int_{-\infty}^{+\infty} dx\, e^{-ip_0 x/\hbar} \psi(x) \left[e^{ip_0 x/\hbar} \psi(x) \right]'$$

$$= -i\hbar \int_{-\infty}^{+\infty} dx\, \psi(x) \left[\frac{i}{\hbar} p_0\psi(x) + \psi'(x) \right] = \langle p \rangle_\psi + p_0 = p_0$$

The wave function has been assumed to be normalized.

The momentum wave function is

$$\tilde{\chi}(p) = \int \frac{dx}{\sqrt{2\pi\hbar}} e^{-ipx/\hbar} \chi(x) = \int \frac{dx}{\sqrt{2\pi\hbar}} e^{i(p-p_0)x/\hbar} \psi(x) = \tilde{\psi}(p - p_0)$$

Problem 1.6 The *propagator* of a particle is defined as

$$\mathcal{K}(\mathbf{x}, \mathbf{x}'; t - t_0) \equiv \langle \mathbf{x} | e^{-i(t-t_0)H/\hbar} | \mathbf{x}' \rangle$$

and corresponds to the probability amplitude for finding the particle at $\mathbf{x}$ at time t if initially (at time t_0) it is at $\mathbf{x}'$.

(a) Show that, when the system (i.e. the Hamiltonian) is invariant in space translations[1] $\mathbf{x} \to \mathbf{x} + \boldsymbol{\alpha}$, as for example in the case of a free particle, the propagator has the property

$$\mathcal{K}(\mathbf{x}, \mathbf{x}'; t - t_0) = \mathcal{K}(\mathbf{x} - \mathbf{x}'; t - t_0)$$

(b) Show that when the energy eigenfunctions are real, i.e. $\psi_E(\mathbf{x}) = \psi_E^*(\mathbf{x})$, as for example in the case of the harmonic oscillator, the propagator has the property

$$\mathcal{K}(\mathbf{x}, \mathbf{x}'; t - t_0) = \mathcal{K}(\mathbf{x}', \mathbf{x}; t - t_0)$$

(c) Show that when the energy eigenfunctions are also parity eigenfunctions, i.e. odd or even functions of the space coordinates, the propagator has the property

$$\mathcal{K}(\mathbf{x}, \mathbf{x}'; t - t_0) = \mathcal{K}(-\mathbf{x}, -\mathbf{x}'; t - t_0)$$

(d) Finally, show that we always have the property

$$\mathcal{K}(\mathbf{x}, \mathbf{x}'; t - t_0) = \mathcal{K}^*(\mathbf{x}', \mathbf{x}; -t + t_0)$$

Solution

(a) Space translations are expressed through the action of an operator as follows:

$$\langle \mathbf{x} | e^{i\boldsymbol{\alpha}\cdot\mathbf{p}/\hbar} = \langle \mathbf{x} + \boldsymbol{\alpha} |$$

Space-translation invariance holds if

$$[\mathbf{p}, H] = 0 \qquad \Longrightarrow \qquad e^{i\boldsymbol{\alpha}\cdot\mathbf{p}/\hbar} H e^{-i\boldsymbol{\alpha}\cdot\mathbf{p}/\hbar} = H$$

which also implies that

$$e^{i\boldsymbol{\alpha}\cdot\mathbf{p}/\hbar} e^{-i(t-t_0)H/\hbar} e^{-i\boldsymbol{\alpha}\cdot\mathbf{p}/\hbar} = e^{-i(t-t_0)H/\hbar}$$

Thus we have

$$\mathcal{K}(\mathbf{x}, \mathbf{x}'; t - t_0) = \langle \mathbf{x} + \boldsymbol{\alpha} | e^{-i(t-t_0)H/\hbar} | \mathbf{x}' + \boldsymbol{\alpha} \rangle = \mathcal{K}(\mathbf{x} + \boldsymbol{\alpha}, \mathbf{x}' + \boldsymbol{\alpha}; t - t_0)$$

---

[1] The operator that can effect a space translation on a state is $e^{-i\mathbf{p}\cdot\boldsymbol{\alpha}/\hbar}$, since it acts on any function of $\mathbf{x}$ as the Taylor expansion operator:

$$\langle \mathbf{x} | e^{-i\mathbf{p}\cdot\boldsymbol{\alpha}/\hbar} = e^{\boldsymbol{\alpha}\cdot\nabla} \langle \mathbf{x} | = \sum_{n=0}^{\infty} \frac{1}{n!} (\boldsymbol{\alpha} \cdot \nabla)^n \langle \mathbf{x} | = \langle \mathbf{x} + \boldsymbol{\alpha} |$$

which clearly implies that the propagator can only be a function of the difference $\mathbf{x} - \mathbf{x}'$.

(b) Inserting a complete set of energy eigenstates, we obtain the propagator in the form

$$\mathcal{K}(\mathbf{x}, \mathbf{x}'; t - t_0) = \sum_E \psi_E(\mathbf{x}) e^{-i(t-t_0)E/\hbar} \psi_E^*(\mathbf{x}')$$

Reality of the energy eigenfunctions immediately implies the desired property.

(c) Clearly

$$\mathcal{K}(-\mathbf{x}, -\mathbf{x}'; t - t_0) = \sum_E \psi_E(-\mathbf{x}) e^{-i(t-t_0)E/\hbar} \psi_E^*(-\mathbf{x}')$$

$$= \sum_E (\pm)^2 \psi_E(\mathbf{x}) e^{-i(t-t_0)E/\hbar} \psi_E^*(\mathbf{x}') = \mathcal{K}(\mathbf{x}, \mathbf{x}'; t - t_0)$$

(d) In the same way,

$$\mathcal{K}(\mathbf{x}, \mathbf{x}'; t - t_0) = \sum_E \psi_E(\mathbf{x}) e^{-i(t-t_0)E/\hbar} \psi_E^*(\mathbf{x}')$$

$$= \left[\sum_E \psi_E^*(\mathbf{x}) e^{-i(t_0-t)E/\hbar} \psi_E(\mathbf{x}') \right]^* = \mathcal{K}^*(\mathbf{x}', \mathbf{x}; t_0 - t)$$

Problem 1.7 Calculate the propagator of a free particle that moves in three dimensions. Show that it is proportional to the exponential of the *classical action* $\mathcal{S} \equiv \int dt\, L$, defined as the integral of the Lagrangian for a free classical particle starting from the point $\mathbf{x}$ at time t_0 and ending at the point $\mathbf{x}'$ at time t. For a free particle the Lagrangian coincides with the kinetic energy. Verify also that in the limit $t \to t_0$ we have

$$\mathcal{K}_0(\mathbf{x} - \mathbf{x}'; 0) = \delta(\mathbf{x} - \mathbf{x}')$$

Solution
Inserting the plane-wave energy eigenfunctions of the free particle into the general expression, we get

$$\mathcal{K}_0(\mathbf{x}', \mathbf{x}; t - t_0) = \int \frac{d^3 p}{(2\pi)^3} e^{i\mathbf{p}\cdot\mathbf{x}/\hbar} \exp\left[-i \frac{p^2}{2m\hbar}(t - t_0)\right] e^{-i\mathbf{p}\cdot\mathbf{x}'/\hbar}$$

$$= \prod_{i=x,y,z} \int \frac{dp_i}{2\pi} \exp\left[-\frac{i}{\hbar}(x_i - x_i')p_i - \frac{ip_i^2}{2m\hbar}(t - t_0)\right]$$

$$= \left[\frac{m\hbar}{2\pi i(t - t_0)}\right]^{3/2} \exp\left[i \frac{m(\mathbf{x} - \mathbf{x}')^2}{2\hbar(t - t_0)}\right]$$

The exponent is obviously equal to $i/\hbar$ times the classical action

$$S = \int_{t_0}^{t} dt\, \frac{mv^2}{2} = (t - t_0)\frac{m}{2}\left(\frac{\mathbf{x} - \mathbf{x}'}{t - t_0}\right)^2 = \frac{m(\mathbf{x} - \mathbf{x}')^2}{2(t - t_0)}$$

In order to consider the limit $t \to t_0$, it is helpful to insert a small imaginary part into the time variable, according to

$$t \to t - i\epsilon$$

Then, we can safely take $t = t_0$ and consider the limit $\epsilon \to 0$. We get

$$K_0(\mathbf{x} - \mathbf{x}', 0) = \lim_{\epsilon \to 0}\left\{\left(\frac{m\hbar}{2\pi\epsilon}\right)^{3/2}\exp\left[-\frac{m(\mathbf{x} - \mathbf{x}')^2}{2\hbar\epsilon}\right]\right\} = \delta(\mathbf{x} - \mathbf{x}')$$

For the last step we needed the delta function representation

$$\delta(x) = \lim_{\epsilon \to 0}\left[(\epsilon\pi)^{-1/2}e^{-x^2/\epsilon}\right]$$

Problem 1.8 A particle starts at time t_0 with the initial wave function $\psi_i(x) = \psi(x, t_0)$. At a later time $t \geq t_0$ its state is represented by the wave function $\psi_f(x) = \psi(x, t)$. The two wave functions are related in terms of the propagator as follows:

$$\psi_f(x) = \int dx'\, K(x, x'; t - t_0)\psi_i(x')$$

(a) Prove that

$$\psi_i^*(x) = \int dx'\, K(x', x; t - t_0)\psi_f^*(x')$$

(b) Consider the case of a free particle initially in the plane-wave state

$$\psi_i(x) = (2\pi)^{-1/2}\exp\left(ikx - i\frac{\hbar k^2}{2m}t_0\right)$$

and, using the known expression for the free propagator,[2] verify the integral expressions explicitly. Comment on the reversibility of the motion.

Solution

(a) We can always write down the inverse evolution equation

$$\psi(x, t_0) = \int dx'\, K(x, x'; t_0 - t)\psi(x', t)$$

or

$$\psi_i(x) = \int dx'\, K(x, x'; t_0 - t)\psi_f(x')$$

---

[2] The expresssion is

$$K_0(x, x'; t - t_0) = \sqrt{\frac{m\hbar}{2\pi i(t - t_0)}}\,\exp\left[i\frac{m(x - x')^2}{2\hbar(t - t_0)}\right]$$

Taking the complex conjugate and using relation (d) of problem 1.6, we get

$$\psi_i^*(x) = \int dx' \mathcal{K}^*(x, x'; t_0 - t)\psi_f^*(x') = \int dx' \mathcal{K}(x', x; t - t_0)\psi_f^*(x')$$

(b) Introducing the expression for $\psi_i(x)$, an analogous expression for the evolved wave function $\psi_f(x) = (2\pi)^{-1/2} \exp(ikx - i\hbar k^2 t/2m)$ and the given expression for $\mathcal{K}_0(x - x'; t - t_0)$, we can perform a Gaussian integration of the type

$$\int_{-\infty}^{+\infty} dx' \exp[ia(x - x')^2 + ikx'] = \sqrt{\frac{\pi}{a}} \exp\left(ikx - \frac{ik^2}{4a}\right)$$

and so arrive at the required identity.

The reversibility of the motion corresponds to the fact that, in addition to the evolution of a free particle of momentum $\hbar k$ from a time t_0 to a time t, an alternative way to see the motion is as that of a free particle with momentum $-\hbar k$ that evolves from time t to time t_0.

Problem 1.9 Consider a normalized wave function $\psi(x)$. Assume that the system is in the state described by the wave function

$$\Psi(x) = C_1\psi(x) + C_2\psi^*(x)$$

where C_1 and C_2 are two known complex numbers.

(a) Write down the condition for the normalization of Ψ in terms of the complex integral $\int_{-\infty}^{+\infty} dx \, \psi^2(x) = D$, assumed to be known.
(b) Obtain an expression for the probability current density $\mathcal{J}(x)$ for the state $\Psi(x)$. Use the polar relation $\psi(x) = f(x)e^{i\theta(x)}$.
(c) Calculate the expectation value $\langle p \rangle$ of the momentum and show that

$$\langle \Psi | p | \Psi \rangle = m \int_{-\infty}^{+\infty} dx \, \mathcal{J}(x)$$

Show that both the probability current and the momentum vanish if $|C_1| = |C_2|$.

Solution
(a) The normalization condition is

$$|C_1|^2 + |C_2|^2 + C_1^* C_2 D^* + C_1 C_2^* D = 1$$

(b) From the defining expression of the probability current density we arrive at

$$\mathcal{J}(x) = \frac{\hbar}{m}(|C_1|^2 - |C_2|^2)\theta'(x)f^2(x)$$

(c) The expectation value of the momentum in the state $\Psi(x)$ is[3]

$$\langle \Psi | p | \Psi \rangle = -i\hbar \int_{-\infty}^{+\infty} dx \, \Psi^*(x)\Psi'(x)$$

$$= \hbar \left(|C_1|^2 - |C_2|^2 \right) \int_{-\infty}^{+\infty} dx \, \theta'(x) f^2(x) = m \int dx \, \mathcal{J}(x)$$

Obviously, both the current and the momentum vanish if $|C_1| = |C_2|$.

Problem 1.10 Consider the complete orthonormal set of eigenfunctions $\psi_\alpha(\mathbf{x})$ of a Hamiltonian H. An arbitrary wave function $\psi(\mathbf{x})$ can always be expanded as

$$\psi(\mathbf{x}) = \sum_\alpha C_\alpha \psi_\alpha(\mathbf{x})$$

(a) Show that an alternative expansion of the wave function $\psi(\mathbf{x})$ is that in terms of the complex conjugate wave functions, namely

$$\psi(\mathbf{x}) = \sum_\alpha C'_\alpha \psi^*_\alpha(\mathbf{x})$$

Determine the coefficients C'_α.

(b) Show that the time-evolved wave function

$$\tilde{\psi}(\mathbf{x}, t) = \sum_\alpha C'_\alpha \psi^*_\alpha(\mathbf{x}) e^{-i E_\alpha t / \hbar}$$

does not satisfy Schroedinger's equation in general, but only in the case where the Hamiltonian is a *real* operator ($H^* = H$).

(c) Assume that the Hamiltonian is real and show that

$$\mathcal{K}(\mathbf{x}, \mathbf{x}'; t - t_0) = \mathcal{K}(\mathbf{x}', \mathbf{x}; t - t_0) = \mathcal{K}^*(\mathbf{x}, \mathbf{x}'; t_0 - t)$$

Solution

(a) Both the orthonormality and the completeness requirements are satisfied by the alternative set $\psi^*_\alpha(\mathbf{x})$ as well:

$$\int d^3x \, \psi^*_\alpha(\mathbf{x})\psi_\beta(\mathbf{x}) = \delta_{\alpha\beta} = \left[\int d^3x \, \psi^*_\alpha(\mathbf{x})\psi_\beta(\mathbf{x}) \right]^*$$

$$= \int d^3x \, \psi_\alpha(\mathbf{x})\psi^*_\beta(\mathbf{x})$$

$$\sum_\alpha \psi_\alpha(\mathbf{x})\psi^*_\alpha(\mathbf{x}') = \delta(\mathbf{x} - \mathbf{x}') = \sum_\alpha \psi^*_\alpha(\mathbf{x})\psi_\alpha(\mathbf{x}')$$

---

[3] Note the vanishing of the integrals of the type

$$\int_{-\infty}^{+\infty} dx \, ff' = \frac{1}{2} \int_{-\infty}^{+\infty} dx \, \frac{d}{dx}[f^2(x)] = \frac{1}{2}[f^2(x)]_{-\infty}^{+\infty} = 0$$

for a function that vanishes at infinity.

The coefficients of the standard expansion are immediately obtained as

$$C_\alpha = \int d^3x \, \psi_\alpha^*(\mathbf{x})\psi(\mathbf{x})$$

while those of the alternative (or complex-conjugate) expansion are

$$C_\alpha' = \int d^3x \, \psi_\alpha(\mathbf{x})\psi(\mathbf{x})$$

(b) As can be seen by substitution, the wave function $\tilde{\psi}$ does not satisfy the Schroedinger equation, since

$$H\psi_\alpha^*(\mathbf{x}) \neq E_\alpha \psi_\alpha^*(\mathbf{x})$$

This is true however in the case of a *real* Hamiltonian, i.e. one for which $H^* = H$.

(c) From the definition of the propagator using the reality of the Hamiltonian, we have

$$\mathcal{K}(\mathbf{x}, \mathbf{x}'; t - t_0) \equiv \langle \mathbf{x}|e^{-i(t-t_0)H/\hbar}|\mathbf{x}'\rangle = \left(\langle \mathbf{x}|e^{i(t-t_0)H/\hbar}|\mathbf{x}'\rangle\right)^*$$
$$= \mathcal{K}^*(\mathbf{x}, \mathbf{x}'; t_0 - t)$$

Also, using hermiticity,

$$\mathcal{K}(\mathbf{x}, \mathbf{x}'; t - t_0) \equiv \langle \mathbf{x}|e^{-i(t-t_0)H/\hbar}|\mathbf{x}'\rangle = \left(\langle \mathbf{x}'|e^{i(t-t_0)H/\hbar}|\mathbf{x}\rangle\right)^*$$
$$= \langle \mathbf{x}'|e^{-i(t-t_0)H/\hbar}|\mathbf{x}\rangle = \mathcal{K}(\mathbf{x}', \mathbf{x}; t - t_0)$$

Problem 1.11 A particle has the wave function

$$\psi(r) = Ne^{-\alpha r}$$

where N is a normalization factor and α is a known real parameter.

(a) Calculate the factor N.
(b) Calculate the expectation values

$$\langle \mathbf{x}\rangle, \qquad \langle r\rangle, \qquad \langle r^2\rangle$$

in this state.
(c) Calculate the uncertainties $(\Delta \mathbf{x})^2$ and $(\Delta r)^2$.
(d) Calculate the probability of finding the particle in the region

$$r > \Delta r$$

(e) What is the momentum-space wave function $\tilde{\psi}(\mathbf{k}, t)$ at any time $t > 0$?
(f) Calculate the uncertainty $(\Delta \mathbf{p})^2$.
(g) Show that the wave function is at all times *isotropic*, i.e.

$$\psi(\mathbf{x}, t) = \psi(r, t)$$

What is the expectation value $\langle \mathbf{x}\rangle_t$?

Solution

(a) The normalization factor is determined from the normalization condition

$$1 = \int d^3r \, |\psi(r)|^2 = 4\pi N^2 \int_0^\infty dr \, r^2 e^{-2\alpha r} = \frac{\pi N^2}{\alpha^3}$$

which gives

$$N = \sqrt{\frac{\alpha^3}{\pi}}$$

We have used the integral ($n \geq 0$)

$$\int_0^\infty dx \, x^n e^{-x} = \Gamma(n+1) = n!$$

(b) The expectation value $\langle \mathbf{x} \rangle$ vanishes owing to spherical symmetry. For example,

$$\langle x \rangle = N^2 \int d^3r \, x e^{-2\alpha r} = N^2 \int_0^\infty dr \, r^3 \, e^{-2\alpha r} \int_{-1}^1 d\cos\theta \sin\theta \int_0^{2\pi} d\phi \, \cos\phi$$
$$= 0$$

The expectation value of the radius is

$$\langle r \rangle = N^2 \int d^3r \, r e^{-2\alpha r} = 4\pi N^2 \int_0^\infty dr \, r^3 e^{-2\alpha r} = \frac{3}{2\alpha}$$

The radius-squared expectation value is

$$\langle \mathbf{x}^2 \rangle = \langle r^2 \rangle = N^2 \int d^3r \, r^2 e^{-2\alpha r} = 4\pi N^2 \int_0^\infty dr \, r^4 e^{-2\alpha r} = \frac{3}{\alpha^2}$$

(c) For the uncertainties, we have

$$(\Delta \mathbf{x})^2 \equiv \langle r^2 \rangle - \langle \mathbf{x} \rangle^2 = \langle r^2 \rangle = \frac{3}{\alpha^2}$$

and

$$(\Delta r)^2 \equiv \langle r^2 \rangle - \langle r \rangle^2 = \frac{3}{\alpha^2} - \left(\frac{3}{2\alpha}\right)^2 = \frac{3}{4\alpha^2}$$

(d) The probability of finding the particle in the region $\Delta r < r < \infty$ is

$$\int_{\Delta r}^\infty d^3r \, |\psi(r)|^2 = 4\pi N^2 \int_{\sqrt{3}/2\alpha}^\infty dr \, r^2 e^{-2\alpha r} = \frac{1}{2} \int_{\sqrt{3}}^\infty dy \, y^2 e^{-y}$$
$$= \frac{1}{2}(5 + 2\sqrt{3})e^{-\sqrt{3}} \sim 0.7487$$

(e) From the Fourier transform

$$\tilde{\psi}(\mathbf{k}) = \int \frac{d^3 r}{(2\pi)^{3/2}} e^{-i\mathbf{k}\cdot\mathbf{x}} \psi(r)$$

we obtain

$$\tilde{\psi}(k) = \frac{N}{\sqrt{2\pi}} \int_0^\infty dr\, r^2 e^{-\alpha r} \int_{-1}^1 d\cos\theta\, e^{ikr\cos\theta}$$

$$= \frac{iN}{k\sqrt{2\pi}} \int_0^\infty dr\, r\, e^{-\alpha r}(e^{-ikr} - e^{ikr}) = \frac{4N\alpha}{\sqrt{2\pi}} \frac{1}{(\alpha^2 + k^2)^2}$$

Designating as $t = 0$ the moment at which the particle has the wave function $Ne^{-\alpha r}$, we obtain at time $t > 0$ the evolved momentum-space wave function

$$\tilde{\psi}(k, t) = \frac{4N\alpha}{\sqrt{2\pi}} \frac{e^{-i\hbar k^2 t/2m}}{(\alpha^2 + k^2)^2}$$

(f) Owing to the spherical symmetry of the momentum distribution, we have $\langle \mathbf{p} \rangle = 0$. The uncertainty squared is

$$(\Delta \mathbf{p})^2 = \langle \mathbf{p}^2 \rangle = \frac{16N^2\alpha^2}{2\pi} \int d^3 k \frac{\hbar^2 k^2}{(k^2 + \alpha^2)^4}$$

$$= \frac{32\alpha^5}{\pi} \int_0^\infty dk \left[\frac{1}{(k^2 + \alpha^2)^2} - \frac{2\alpha^2}{(k^2 + \alpha^2)^3} + \frac{\alpha^4}{(k^2 + \alpha^2)^4} \right]$$

$$= \frac{32\alpha^5}{\pi} \left[-\frac{\partial}{\partial\alpha^2} - \alpha^2 \left(\frac{\partial}{\partial\alpha^2} \right)^2 - \frac{\alpha^4}{6} \left(\frac{\partial}{\partial\alpha^2} \right)^3 \right] \mathcal{J}$$

where

$$\mathcal{J} = \int_0^\infty dk \frac{1}{k^2 + \alpha^2} = \frac{\pi}{2\alpha}$$

For the last step we have used the integral

$$\int dx \frac{1}{1 + x^2} = \arctan x$$

Thus, we end up with

$$(\Delta \mathbf{p})^2 = \hbar^2 \alpha^2$$

(g) From the Fourier transform we get

$$\psi(\mathbf{x}, t) = \int \frac{d^3k}{(2\pi)^{3/2}} e^{i\mathbf{k}\cdot\mathbf{x}} \tilde{\psi}(k, t)$$

$$= \frac{1}{\sqrt{2\pi}} \int_0^\infty dk\, k^2\, \tilde{\psi}(k, t) \int_{-1}^1 d\cos\theta\, e^{ikr\cos\theta} = \psi(r, t)$$

Consequently, the expectation value $\langle \mathbf{x} \rangle_t$ will vanish at all times owing to spherical symmetry.

2

The free particle

Problem 2.1 A free particle is initially (at $t = 0$) in a state described by the wave function

$$\psi(\mathbf{x}, 0) = N e^{-\alpha r}$$

where α is a real parameter.

(a) Compute the normalization factor N.
(b) Show that the probability density of finding the particle with momentum $\hbar \mathbf{k}$ is *isotropic*, i.e. it does not depend on the direction of the momentum.
(c) Show that the spatial probability density

$$\mathcal{P}(\mathbf{x}, t) = |\psi(\mathbf{x}, t)|^2$$

is also isotropic.
(d) Calculate the expectation values

$$\langle \mathbf{p} \rangle_t, \quad \langle \mathbf{x} \rangle_t$$

(e) Show that the expectation value of r^2 increases with time, i.e. it satisfies the inequality

$$\langle r^2 \rangle_t \geq \langle r^2 \rangle_0$$

(f) Modify the initial wave function, assuming that initially the particle is in a state described by

$$\psi(\mathbf{x}, 0) = N e^{-\alpha r} e^{i \mathbf{k}_0 \cdot \mathbf{x}}$$

Calculate the expectation values $\langle \mathbf{p} \rangle_t$, $\langle \mathbf{x} \rangle_t$ for this case.

Solution

(a) $N = (\alpha^3 / \pi)^{3/2}$.

(b) The momentum wave function will be

$$\tilde{\psi}(k) = N \int \frac{d^3 x}{(2\pi)^{3/2}} e^{-\alpha r - i \mathbf{k} \cdot \mathbf{x}} \propto \frac{1}{k} \int_0^\infty dr \, r e^{-\alpha r} \left(e^{ikr} - e^{-ikr} \right) = \tilde{\psi}(k)$$

17

where we have taken the z-axis of the integration variables to coincide with the momentum direction (so that $\mathbf{k} \cdot \mathbf{x} = kr \cos \theta$). Thus,

$$\Pi(\mathbf{k}) = |\tilde{\psi}(k)|^2 = \Pi(k)$$

Note that, since we have a free particle, its momentum wave function will also be an energy eigenfunction and will evolve in time in a trivial way:

$$\tilde{\psi}(\mathbf{k}, t) = \tilde{\psi}(k)e^{-i\hbar k^2 t/2m}$$

Its corresponding probability density $\Pi(k)$ will be time independent.

(c) The evolved wave function will be

$$\psi(\mathbf{x}, t) = \int \frac{d^3 k}{(2\pi)^{3/2}} \, \tilde{\psi}(k)e^{-i\hbar k^2 t/2m} \, e^{i\mathbf{k}\cdot\mathbf{x}}$$

Taking the $\hat{\mathbf{z}}$-axis of the integration variables to coincide with the direction of $\mathbf{x}$, we obtain

$$\psi(\mathbf{x}, t) \propto \frac{1}{r} \int_0^\infty dk \, k\tilde{\psi}(k)e^{-i\hbar k^2 t/2m} \left(e^{ikr} - e^{-ikr} \right) \propto \psi(r, t)$$

Thus, the probability density

$$\mathcal{P}(\mathbf{x}, t) = |\psi(r, t)|^2 = \mathcal{P}(r, t)$$

will be isotropic at all times, i.e. it will not depend on angle.

(d) The momentum expectation value will clearly not depend on time:

$$\langle \mathbf{p} \rangle = \int d^3 k \, \hbar\mathbf{k}|\tilde{\psi}(k)|^2$$

Note also that isotropy implies the vanishing of this integral. An easy way to see this is to apply the parity transformation to the integration variable by taking $\mathbf{k} \to -\mathbf{k}$, which leads to

$$\langle \mathbf{p} \rangle = -\langle \mathbf{p} \rangle = 0$$

The same argument applies to the position expectation value, which also vanishes at all times:

$$\langle \mathbf{x} \rangle_t = 0$$

(e) The expectation value of the position squared can be expressed in terms of the momentum wave function as

$$\langle r^2 \rangle_t = -\int d^3 k \, \tilde{\psi}(k, 0)e^{iEt/\hbar} \, \nabla_{\mathbf{k}}^2 \left[\tilde{\psi}(k, 0)e^{-iEt/\hbar} \right]$$

Note that, in the case that we are considering, the momentum wave function is not only isotropic but also real. We have

$$\nabla_{\mathbf{k}}\left[\tilde{\psi}(k,0)e^{-iEt/\hbar}\right] = -\frac{i\hbar t}{m}\mathbf{k}\tilde{\psi}e^{-iEt/\hbar} + \hat{\mathbf{k}}\tilde{\psi}'e^{-iEt/\hbar}$$

$$\nabla_{\mathbf{k}}^2\left[\tilde{\psi}(k,0)e^{-iEt/\hbar}\right] = -\frac{3i\hbar t}{m}\tilde{\psi}e^{-iEt/\hbar} - \frac{2i\hbar kt}{m}\tilde{\psi}'e^{-iEt/\hbar}$$

$$-\frac{\hbar^2 k^2 t^2}{m^2}\tilde{\psi}e^{-iEt/\hbar} + \frac{2}{k}\tilde{\psi}'e^{-iEt/\hbar} + \tilde{\psi}''e^{-iEt/\hbar}$$

The expectation value can be written as

$$\langle r^2\rangle_t = \langle r^2\rangle_0 + \frac{\hbar^2 t^2}{m^2}\int d^3k\, k^2\tilde{\psi}^2 + \frac{2i\hbar t}{m}\int d^3k\, k\tilde{\psi}\tilde{\psi}' + \frac{3i\hbar t}{m}\int d^3k\, \tilde{\psi}^2$$

The terms linear in time vanish since

$$\int d^3k\, k\tilde{\psi}\tilde{\psi}' = 2\pi\int_0^\infty dk\, k^3(\tilde{\psi}^2)' = -2\pi\int_0^\infty dk(k^3)'\tilde{\psi}^2 = -\frac{3}{2}\int d^3k\,\tilde{\psi}^2 = -\frac{3}{2}$$

Note that we used the normalization $\int d^3k\,\tilde{\psi}^2 = 1$. Finally, we have

$$\langle r^2\rangle_t = \langle r^2\rangle_0 + \frac{\hbar^2 t^2}{m^2}\int d^3k\, k^2\tilde{\psi}^2$$

which demonstrates the validity of the inequality $\langle r^2\rangle_t \geq \langle r^2\rangle_0$. Note that this inequality corresponds to the general fact that, for a free particle, the *uncertainty* $(\Delta x)^2$ always increases in time.

(f) It is not difficult to see that in this case

$$\tilde{\psi}(\mathbf{k},0) = \tilde{\psi}(|\mathbf{k}-\mathbf{k}_0|,0)$$

and that

$$\psi(\mathbf{x},t) = e^{i\mathbf{k}_0\cdot\mathbf{x}}f(|\mathbf{x}-\hbar t\mathbf{k}_0/m|)$$

Thus we have

$$\langle\mathbf{p}\rangle_t = \hbar\int d^3k\,\mathbf{k}\,\tilde{\psi}^2(|\mathbf{k}-\mathbf{k}_0|,0) = \hbar\int d^3q\,\mathbf{q}\,\tilde{\psi}(q,0) + \hbar\mathbf{k}_0\int d^3k\,\tilde{\psi}^2 = \hbar\mathbf{k}_0$$

and

$$\langle\mathbf{x}\rangle_t = \int d^3r\,\mathbf{x}\left|\psi\left(\left|\mathbf{x}-\frac{\hbar t}{m}\mathbf{k}_0\right|,t\right)\right|^2$$

$$= \int d^3\rho\,\boldsymbol{\rho}|\psi(\rho,t)|^2 + \frac{\hbar t}{m}\mathbf{k}_0\int d^3r\,|\psi|^2 = \frac{\hbar t}{m}\mathbf{k}_0$$

Problem 2.2 Show that the 'spherical waves'

$$\psi_\pm(r, t) = \frac{N}{r} e^{\pm ikr - i\hbar k^2 t/2m}$$

satisfy the Schroedinger equation for a free particle of mass m except at the origin $r = 0$. Show also that, in contrast with a plane wave, which satisfies the continuity equation everywhere, the above spherical waves do not satisfy the continuity equation at the origin. Give a physical interpretation of this non-conservation of probability. Does the probability interpretation of $\psi_\pm$ break down at the origin? Find a linear combination of the above spherical waves $\psi_\pm$ that is finite at the origin and reexamine the validity of the continuity equation everywhere.

Solution

The current density corresponding to $\psi_\pm$ is

$$\mathcal{J}_\pm = \pm \frac{\hbar k}{m} |N|^2 \frac{\hat{\mathbf{r}}}{r} = \mp \frac{\hbar k}{m} |N|^2 \nabla \left(\frac{1}{r} \right)$$

The probability density is $\mathcal{P}_\pm = |N|^2/r^2$ and it is independent of time. Thus, we have

$$\nabla \cdot \mathcal{J} + \dot{\mathcal{P}} = \mp \frac{\hbar k}{m} |N|^2 \nabla^2 \left(\frac{1}{r} \right) = \pm 4\pi |N|^2 \left(\frac{\hbar k}{m} \right) \delta(\mathbf{x})$$

The physical interpretation of this non-zero probability density rate, in the framework of a statistical ensemble of identical systems, is *the number of particles created or destroyed per unit volume per unit time*. The non-conservation of probability arises here from the fact that the $\psi_\pm$ are not acceptable wave functions since they diverge at the origin.

In contrast, the spherical wave

$$\psi_0(r) = \frac{1}{2i} [\psi_+(r) - \psi_-(r)] = N \frac{\sin kr}{r} e^{-i\hbar k^2 t/2m}$$

is finite at the origin and satisfies everywhere the free Schroedinger equation and the continuity equation. In fact, we get $\mathcal{J} = 0$ and $\dot{\mathcal{P}} = 0$.

Problem 2.3 A free particle is initially (at $t = 0$) in a state corresponding to the wave function

$$\psi(r) = \left(\frac{\gamma}{\pi} \right)^{3/4} e^{-\gamma r^2/2}$$

(a) Calculate the probability density of finding the particle with momentum $\hbar \mathbf{k}$ at any time t. Is it isotropic?

(b) What is the probability of finding the particle with energy E?

(c) Examine whether the particle is in an eigenstate of the square of the angular momentum $\mathbf{L}^2$, and of its z-component L_z, for any time t.

Solution

(a) The momentum wave function derived from $\psi(r)$ is[1]

$$\tilde{\psi}(k) = \int \frac{d^3x}{(2\pi)^{3/2}} \, \psi(r) \, e^{-i\mathbf{k}\cdot\mathbf{x}} = (\gamma\pi)^{-3/4} \, e^{-k^2/2\gamma}$$

Since the particle is free, the momentum wave function will also be an eigenfunction of energy and will evolve trivially with a time phase:

$$\tilde{\psi}(k, t) = (\gamma\pi)^{-3/4} \, e^{-k^2/2\gamma} \, e^{-i\hbar k^2 t/2m}$$

The corresponding momentum probability density is obviously constant and isotropic.

(b) If we denote by $\mathcal{P}(E)$ the probability density for the particle to have energy $E = \hbar^2 k^2/2m$, we shall have

$$1 = \int_0^\infty dE \, \mathcal{P}(E) = \frac{\hbar^2}{m} \int_0^\infty dk \, k\mathcal{P}(E)$$

Comparing this formula with

$$1 = \int d^3k \, |\tilde{\psi}(k)|^2 = 4\pi \int_0^\infty dk \, k^2 \, (\gamma\pi)^{-3/2} \, e^{-k^2/\gamma}$$

we can conclude that

$$\mathcal{P}(E) = \frac{m}{\hbar^2} \, 4\pi k(\gamma\pi)^{-3/2} \, e^{-k^2/\gamma}$$

(c) Since the initial wave function is spherically symmetric, it will be an eigenfunction of angular momentum with vanishing eigenvalues. Moreover, since the Hamiltonian of the free particle is spherically symmetric or, equivalently, it commutes with the angular momentum operators, the time-evolved wave function will continue to be an angular momentum eigenfunction with the same eigenvalue.

Problem 2.4 Consider a free particle that moves in one dimension. Its initial ($t = 0$) wave function is

$$\psi(x, 0) = \left(\frac{\alpha}{\pi}\right)^{1/4} e^{ik_0 x - \alpha x^2/2}$$

where α and k_0 are real parameters.

---

[1] We can use the Gaussian integral

$$\int d^3x \, e^{-ar^2} \, e^{-i\mathbf{q}\cdot\mathbf{x}} = \left(\frac{\pi}{a}\right)^{3/2} e^{-q^2/4a}$$

with $\mathrm{Re}(a) > 0$.

(a) Calculate the momentum wave function $\tilde{\psi}(k, t)$ at all times $t > 0$ and the corresponding momentum probability density $\Pi(k)$. What is the most probable momentum?

(b) Compute the wave function $\psi(x, t)$ at all times $t > 0$ and the corresponding probability density $\mathcal{P}(x, t)$. How does the most probable position evolve in time? Consider the limit $t \to \infty$ and comment on its position dependence.

(c) Calculate the expectation values of the position $\langle x \rangle_t$ and momentum $\langle p \rangle_t$. Show that they satisfy the classical equations of motion

$$\langle p \rangle_t = m \frac{d \langle x \rangle_t}{dt}, \qquad \frac{d \langle p \rangle_t}{dt} = m \frac{d^2 \langle x \rangle_t}{dt^2} = 0$$

(d) Calculate the probability current density $\mathcal{J}(x, t)$. What is the probability current density in the limit $t \to \infty$? Verify explicitly the continuity equation in this limit.

(e) Compute the expectation values $\langle p^2 \rangle_t$ and $\langle x^2 \rangle_t$. Determine the uncertainties $(\Delta x)_t^2$ and $(\Delta p)_t^2$. Verify the validity of the uncertainty relation

$$(\Delta x)_t^2 (\Delta p)_t^2 \geq \frac{\hbar^2}{4}$$

(f) Calculate the uncertainty in the energy, given by

$$(\Delta E)^2 = \langle H^2 \rangle - \langle H \rangle^2$$

(g) Consider the quantity

$$\tau = \frac{(\Delta x)_t}{|d \langle x \rangle_t / dt|}$$

which has the dimensions of time. What is the physical meaning of τ? Show that it satisfies a time–energy uncertainty inequality,

$$\tau (\Delta E) \geq \frac{\hbar}{2}$$

Solution

(a) From $\tilde{\psi}(k, 0) = \int_{-\infty}^{+\infty} dx/\sqrt{2\pi} \, e^{-ikx} \, \psi(x, 0)$, we obtain

$$\tilde{\psi}(k, t) = \tilde{\psi}(k, 0) e^{-i\hbar k^2 t/2m} = \frac{1}{(\alpha\pi)^{1/4}} e^{-(k-k_0)^2/2\alpha} e^{-i\hbar k^2 t/2m}$$

and

$$\Pi(k) = |\tilde{\psi}(k, t)|^2 = \frac{1}{\sqrt{\alpha\pi}} e^{-(k-k_0)^2/\alpha}$$

The most probable momentum value is $\hbar k_0$.

(b) From $\psi(x, t) = \int_{-\infty}^{+\infty} dk/\sqrt{2\pi} \, e^{ikx} \, \tilde{\psi}(k, t)$, we obtain

$$\psi(x, t) = \left(\frac{\alpha}{\pi}\right)^{1/4} \frac{1}{|z|} \exp\left(-\frac{\alpha x^2}{2z} + \frac{ik_0 x}{z} - \frac{i\hbar k_0^2 t}{2mz}\right)$$

where

$$z \equiv 1 + i \left(\frac{\hbar t \alpha}{m} \right)$$

Note that $|z|^2 = 1 + (\hbar t \alpha / m)^2$. The corresponding probability density is

$$P(x, t) = |\psi(x, t)|^2 = \frac{1}{|z|} \sqrt{\frac{\alpha}{\pi}} \, \exp \left[-\frac{\alpha}{|z|^2} \left(x - \frac{\hbar k_0}{m} t \right)^2 \right]$$

In the limit $t \to \infty$, we get

$$\mathcal{P}_\infty = \sqrt{\frac{\alpha}{\pi}} \left(\frac{m}{\alpha \hbar t} \right) \exp \left(-\frac{k_0^2}{\alpha} \right)$$

which is position independent and tends to zero with time.

(c) It is straightforward to calculate

$$\langle x \rangle_t = \frac{\hbar k_0}{m} t, \qquad \langle p \rangle_t = \hbar k_0$$

Note that the momentum expectation value coincides with the most probable momentum value. It is trivial to verify that these expectation values satisfy the classical equation of Newton for the motion of a free particle.

(d) The probability current density can be calculated in a straightforward fashion. It is

$$\mathcal{J}(x, t) = \left(\frac{\hbar k_0}{m} \right) \frac{1}{|z|^3} \sqrt{\frac{\alpha}{\pi}} \, \exp \left[-\frac{\alpha}{|z|^2} \left(x - \frac{\hbar k_0}{m} t \right)^2 \right] \left(1 + \frac{\hbar t \alpha^2 x}{m k_0} \right)$$

$$= \left(\frac{\hbar k_0}{m} \right) \frac{1}{|z|^2} \left(1 + \frac{\hbar t \alpha^2 x}{m k_0} \right) \mathcal{P}(x, t)$$

In the limit $t \to \infty$ the current density is

$$\mathcal{J}_\infty = \sqrt{\frac{\alpha}{\pi}} \left(\frac{m}{\alpha \hbar} \right) \exp \left(-\frac{k_0^2}{\alpha} \right) \frac{x}{t^2} = -x \frac{d}{dt} \mathcal{P}_\infty$$

It is clear, then, that

$$\frac{d}{dx} \mathcal{J}_\infty = -\frac{d}{dt} \mathcal{P}_\infty$$

which is the continuity equation.

(e) We obtain[2]

$$\langle p^2 \rangle_t = \frac{\hbar^2 \alpha}{2} + \hbar^2 k_0^2$$

$$\langle x^2 \rangle_t = \left(\frac{\hbar k_0 t}{m} \right)^2 + \frac{1}{2\alpha} \left[1 + \left(\frac{\hbar t \alpha}{m} \right)^2 \right]$$

The corresponding uncertainties are

$$(\Delta x)_t^2 = \frac{1}{2\alpha} \left[1 + \left(\frac{\hbar t \alpha}{m} \right)^2 \right], \qquad (\Delta p)_t^2 = \frac{\hbar^2 \alpha}{2}$$

and their product satisfies Heisenberg's inequality:

$$(\Delta x)_t^2 (\Delta p)_t^2 = \frac{\hbar^2}{4} \left[1 + \left(\frac{\hbar t \alpha}{m} \right)^2 \right] \geq \frac{\hbar^2}{4}$$

(f) The average value of the energy is proportional to that of the square of the momentum, which has been computed previously. It is

$$\langle H \rangle = \frac{1}{2m} \langle p^2 \rangle = \hbar^2 \left(k_0^2 + \frac{\alpha}{2} \right)$$

The expectation value of the square of the energy is equal to the expectation value of the fourth power of the momentum operator divided by $4m^2$. Thus, we consider

$$\langle p^4 \rangle = \frac{\hbar^4}{\sqrt{\alpha\pi}} \int_{-\infty}^{+\infty} dk\, k^4 e^{-(k-k_0)^2/\alpha} = \frac{\hbar^4}{\sqrt{\alpha\pi}} \int_{-\infty}^{+\infty} dk \left(k^4 + k_0^4 + 6k^2 k_0^2 \right) e^{-k^2/\alpha}$$

$$= (\hbar k_0)^4 + \frac{6k_0^2}{\sqrt{\alpha\pi}} \int dk\, k^2 e^{-k^2/\alpha} + \frac{\hbar^4}{\sqrt{\alpha\pi}} \int dk\, k^4 e^{-k^2/\alpha}$$

The two integrals involved are found as follows:

$$\int dk\, k^2 e^{-k^2/\alpha} = -\frac{\partial}{\partial(1/\alpha)} \int dk\, e^{-k^2/\alpha} = \alpha^2 \frac{\partial}{\partial\alpha} \sqrt{\alpha\pi} = \frac{\alpha}{2}\sqrt{\alpha\pi}$$

$$\int dk\, k^4 e^{-k^2/\alpha} = -\frac{\partial}{\partial(1/\alpha)} \int dk\, k^2 e^{-k^2/\alpha} = \alpha^2 \frac{\partial}{\partial\alpha} \frac{\alpha}{2} \sqrt{\pi\alpha} = \frac{3\alpha^2}{4}\sqrt{\pi\alpha}$$

Substituting, we obtain

$$\langle p^4 \rangle = (\hbar k_0)^4 + 3(\hbar k_0)^2 (\hbar^2 \alpha) + \frac{3}{4}(\hbar^2 \alpha)^2$$

---

[2] Note that

$$\int_{-\infty}^{+\infty} dk\, e^{-k^2/\alpha} = \sqrt{\alpha\pi}, \qquad \int_{-\infty}^{+\infty} dk\, k^2 e^{-k^2/\alpha} = \frac{\alpha}{2}\sqrt{\alpha\pi}$$

The uncertainty in energy is thus

$$(\Delta E)^2 = \frac{\hbar^2 \alpha}{8m^2} \left[\hbar^2 \alpha + 4(\hbar k_0)^2 \right]$$

(g) The quantity τ is easily computed to be

$$\tau = \frac{m}{\hbar k_0 \sqrt{2\alpha}} \sqrt{1 + \left(\frac{\hbar t \alpha}{m} \right)^2}$$

Its physical meaning is that of the characteristic time scale in which the modification of the spatial distribution, or spreading of the wave packet, will become apparent relative to the overall motion of its centre. Taking the minimum value of the characteristic time τ,

$$\tau_0 = \frac{1}{\sqrt{2\alpha}} \left(\frac{m}{\hbar k_0} \right) = \frac{1}{\sqrt{2\alpha}} \left(\frac{1}{v_0} \right)$$

we are led to a time–energy uncertainty product,

$$\tau_0 \, (\Delta E) = \frac{\hbar}{2} \sqrt{1 + \frac{\alpha}{4k_0^2}}$$

that is always greater than $\hbar/2$.

Problem 2.5 Consider a free particle that is initially (at $t = 0$) extremely well localized at the origin $x = 0$ and has a Gaussian wave function,

$$\psi(x, 0) = \left(\frac{\alpha}{\pi} \right)^{1/4} e^{-\alpha x^2/2} e^{ik_0 x}$$

for very large values of the real parameter α ($\alpha \to \infty$); k_0 is also real.

(a) Write down the position probability density $\mathcal{P}(x, 0) = |\psi(x, 0)|^2$ and show that

$$\lim_{\alpha \to \infty} [\mathcal{P}(x, 0)] = \delta(x)$$

(b) Calculate the evolved wave function $\psi(x, t)$ at times $t > 0$, keeping the parameter α finite.

(c) Write down the evolved position probability density $\mathcal{P}(x, t)$. Take the limit $\alpha \to \infty$ and observe that even for infinitesimal values of time ($t \sim m/\hbar\alpha$) it becomes space independent. Give a physical argument for the contrast of this behaviour with the initial distribution.

Solution

(a) Using the well-known representation of the delta function $\lim_{\epsilon \to 0}[(\epsilon\pi)^{-1/2} e^{-x^2/\epsilon}]$, we immediately see that

$$\lim_{\alpha \to \infty} [\mathcal{P}(x, 0)] = \delta(x)$$

(b) The momentum wave function is easily obtained from the Fourier transform:

$$\tilde{\psi}(k) = \int \frac{dx}{\sqrt{2\pi}} e^{-ikx} \psi(x, 0) = (\alpha\pi)^{-1/4} e^{-(k-k_0)^2/2\alpha}$$

The time evolution, since we have a free particle, is trivial, namely

$$\tilde{\psi}(k, t) = (\alpha\pi)^{-1/4} e^{-(k-k_0)^2/2\alpha} e^{-i\hbar k^2 t/2m}$$

The evolved wave function can now be obtained in terms of the inverse Fourier transform as

$$\psi(x, t) = \int \frac{dx}{\sqrt{2\pi}} e^{ikx} \tilde{\psi}(k, t) = \frac{(\alpha\pi)^{-1/4}}{\sqrt{2\pi}} \int dk \, e^{ikx} e^{-(k-k_0)^2/2\alpha} e^{-i\hbar k^2 t/2m}$$

$$= \frac{(\alpha\pi)^{-1/4}}{\sqrt{2\pi}} e^{ik_0 x} \int dk \, e^{ikx} e^{-k^2/2\alpha} e^{-i\hbar(k+k_0)^2 t/2m}$$

$$= \frac{(\alpha\pi)^{-1/4}}{\sqrt{2\pi}} e^{ik_0 x} e^{-i\hbar k_0^2 t/2m} \int dk \, e^{ik(x-v_0 t)} e^{-zk^2/2\alpha}$$

where

$$z \equiv 1 + i\left(\frac{\hbar t \alpha}{m}\right), \qquad v_0 \equiv \frac{\hbar k_0}{m}$$

Finally, we get

$$\psi(x, t) = \frac{1}{\sqrt{z}} \left(\frac{\alpha}{\pi}\right)^{1/4} \exp\left(ik_0 x - \frac{i\hbar k_0^2 t}{2m}\right) \exp\left[-\frac{\alpha}{2z}(x - v_0 t)^2\right]$$

(c) The evolved position probability density is

$$\mathcal{P}(x, t) = \sqrt{\frac{\alpha}{\pi |z|^2}} \exp\left[-\frac{\alpha}{|z|^2}(x - v_0 t)^2\right]$$

Note that this distribution has the same form as the initial one apart from the (essential) replacement

$$\alpha \to \alpha(t) \equiv \frac{\alpha}{|z|^2} = \frac{\alpha}{1 + (\hbar t \alpha/m)^2}$$

In the limit $\alpha \to \infty$ the parameter $\alpha(t)$, *for any $t > m/\hbar\alpha$*, goes to zero and the spatial probability density becomes position independent, approaching zero:

$$\mathcal{P}(x, t) \sim \frac{1}{\sqrt{\pi\alpha}} \left(\frac{m}{\hbar t}\right) \exp\left[-\frac{1}{\alpha}\left(\frac{m}{\hbar t}\right)^2 (x - v_0 t)^2\right] \to 0$$

Note that this behaviour is a direct consequence of the extreme localization of the initial state, which according to the (inescapable) uncertainty principle is accompanied by $(\Delta p)_0 = \infty$. Thus, since the initial wave packet includes modes of infinite momentum, the particle reaches all space immediately.

Problem 2.6 Consider a free particle moving in one dimension. At time $t = 0$ its wave function is

$$\psi(x, 0) = Ne^{-\alpha(x+x_0)^2/2} + Ne^{-\alpha(x-x_0)^2/2}$$

where α and x_0 are known real parameters.

(a) Compute the normalization factor N and the momentum wave function $\tilde{\psi}(k)$.
(b) Find the evolved wave function $\psi(x, t)$ for any time $t > 0$.
(c) Write down the position probability density and discuss the physical interpretation of each term.
(d) Obtain the expression for the probability current density $\mathcal{J}(x, t)$.

  Solution
  (a) The normalization factor is

$$N = \frac{1}{\sqrt{2}} \left(\frac{\alpha}{\pi}\right)^{1/4} \frac{1}{\sqrt{1 + e^{-\alpha x_0^2}}}$$

The momentum wave function is given by

$$\tilde{\psi}(k, 0) = \frac{2N}{\sqrt{\alpha}} \cos kx_0 \, e^{-k^2/2\alpha}$$

and

$$\tilde{\psi}(k, t) = \frac{2N}{\sqrt{\alpha}} \cos kx_0 \, e^{-zk^2/2\alpha}$$

where

$$z \equiv 1 + i \left(\frac{\hbar t \alpha}{m}\right)$$

  (b) The evolved wave function is obtained from the Fourier transform

$$\psi(x, t) = \int \frac{dk}{\sqrt{2\pi}} e^{ikx} \tilde{\psi}(k, t)$$

It is

$$\psi(x, t) = \frac{N}{\sqrt{z}} \left[e^{-\alpha(x+x_0)^2/2z} + e^{-\alpha(x-x_0)^2/2z} \right]$$

  (c) The position probability density is

$$\mathcal{P}(x, t) = \frac{|N|^2}{|z|} \left[e^{-\alpha(x+x_0)^2/|z|^2} + e^{-\alpha(x-x_0)^2/|z|^2} + 2e^{-\alpha x^2/|z|^2} e^{-\alpha x_0^2/|z|^2} \cos 2qx \right]$$

where

$$q(t) \equiv \left(\frac{\hbar \alpha t}{m}\right) \frac{x_0 \alpha}{|z|^2}$$

The first two terms in $\mathcal{P}(x, t)$ correspond to wave packets localized at $\mp x_0$, while the oscillatory term corresponds to the interference of the two wave packets.

(d) The probability current density, after a straightforward but tedious calculation, turns out to be

$$\mathcal{J}(x, t) = \frac{\hbar |N|^2}{2m|z|^2} \left\{ \left(\frac{\hbar t \alpha^2}{m}\right) \left[(x + x_0)e^{-\alpha(x+x_0)^2/|z|^2} + (x - x_0)e^{-\alpha(x-x_0)^2/|z|^2}\right] \right.$$
$$\left. + 4\left[-\alpha x_0 \sin 2qx + \left(\frac{\hbar t \alpha^2}{m}\right) x \cos 2qx\right] e^{-\alpha(x^2+x_0^2)/|z|^2} \right\}$$

Problem 2.7 Consider the initial ($t = 0$) free-particle wave function

$$\psi(x, 0) = Ne^{ik_0 x}e^{-\alpha(x+x_0)^2/2} + Ne^{-ik_0 x}e^{-\alpha(x-x_0)^2/2}$$

(a) Calculate the normalization constant N and the expectation values $\langle x \rangle_0$, $\langle x^2 \rangle_0$.
(b) Calculate the momentum wave function $\tilde{\psi}(k, 0)$.
(c) Write down the momentum probability density. Find the expectation values $\langle p \rangle$, $\langle p^2 \rangle$, as well as the uncertainty $(\Delta p)^2$. Verify the Heisenberg uncertainty relation at time $t = 0$.
(d) Find the evolved wave function $\psi(x, t)$ at time $t > 0$. Write down the position probability density $\mathcal{P}(x, t)$. How does it behave at very late times?
(e) Consider the probability of finding the particle at the origin and discuss its dependence upon time.

Solution
(a) The normalization constant is

$$N = \frac{1}{\sqrt{2}} \left(\frac{\alpha}{\pi}\right)^{1/4} \frac{1}{\sqrt{1 + e^{-k_0^2/\alpha}e^{-\alpha x_0^2}}}$$

The initial position probability density is

$$\mathcal{P}(x, 0) = |N|^2 \left[e^{-\alpha(x+x_0)^2} + e^{-\alpha(x-x_0)^2} + 2e^{-\alpha x^2}e^{-\alpha x_0^2} \cos 2k_0 x\right]$$

Owing to the fact that this distribution is even, the expectation value of the position can immediately be seen to vanish. For the square of the position we have

$$\langle x^2 \rangle_0 = |N|^2 \left[J_+ + J_- + e^{-\alpha x_0^2}(J_0 + J_0^*)\right]$$

where

$$J_\pm = \int dx\, x^2 e^{-\alpha(x\pm x_0)^2} = \int dx\, (x \mp x_0)^2 e^{-\alpha x^2}$$

$$= \int dx\, (x^2 + x_0^2 \mp 2xx_0)\, e^{-\alpha x^2} = x_0^2 \sqrt{\frac{\pi}{\alpha}} - \frac{\partial}{\partial\alpha} \int dx\, e^{-\alpha x^2}$$

$$= \sqrt{\frac{\pi}{\alpha}} \left(x_0^2 + \frac{1}{2\alpha} \right)$$

However,

$$J_0 = J_0^* = -\frac{\partial}{\partial\alpha} \int dx\, e^{-\alpha x^2 + 2ik_0 x} = -\frac{\partial}{\partial\alpha} \int dx\, e^{-\alpha(x - ik_0/\alpha)^2}\, e^{-k_0^2/\alpha}$$

$$= -\frac{\partial}{\partial\alpha} \sqrt{\frac{\pi}{\alpha}}\, e^{-k_0^2/\alpha} = \sqrt{\frac{\pi}{\alpha}} \left(\frac{1}{2\alpha} - \frac{k_0^2}{\alpha^2} \right) e^{-k_0^2/\alpha}$$

Substituting, we obtain

$$\langle x^2 \rangle_0 = 2|N|^2 \sqrt{\frac{\pi}{\alpha}} \left[x_0^2 + \frac{1}{2\alpha} + e^{-\alpha x_0^2 - k_0^2/\alpha} \left(\frac{1}{2\alpha} - \frac{k_0^2}{\alpha^2} \right) \right]$$

or

$$\langle x^2 \rangle_0 = (\Delta x)_0^2 = \frac{1}{2\alpha} \left[\frac{1 + 2\alpha x_0^2 + (1 - 2k_0^2/\alpha)\, e^{-k_0^2/\alpha - \alpha x_0^2}}{1 + e^{-k_0^2/\alpha - \alpha x_0^2}} \right]$$

(b) The momentum wave function is

$$\tilde{\psi}(k, 0) = \frac{N}{\sqrt{\alpha}} \left[e^{i(k - k_0)x_0}\, e^{-(k - k_0)^2/2\alpha} + e^{-i(k + k_0)x_0}\, e^{-(k + k_0)^2/2\alpha} \right]$$

(c) The momentum probability density at all times is

$$|\tilde{\psi}(k, t)|^2 = \frac{|N|^2}{\alpha} \left[e^{-(k - k_0)^2/\alpha} + e^{-(k + k_0)^2/\alpha} + 2e^{-k^2/\alpha}\, e^{-k_0^2/\alpha}\, \cos 2kx_0 \right]$$

Since it is even in k, the expectation value of the momentum $\langle p \rangle$ will vanish. The expectation value of the square of the momentum is

$$(\Delta p)^2 = \langle p^2 \rangle = \frac{\hbar^2 |N|^2}{\alpha} \left[I_- + I_+ + e^{-k_0^2/\alpha}(I_0 + I_0^*) \right]$$

where we have the integrals

$$I_\pm = \int_{-\infty}^{+\infty} dk\, k^2 e^{-(k \pm k_0)^2/\alpha} = \int_{-\infty}^{+\infty} dk\, (k^2 + k_0^2 \mp 2kk_0) e^{-k^2/\alpha}$$

$$= \left(k_0^2 - \frac{\partial}{\partial(1/\alpha)} \right) \int_{-\infty}^{+\infty} dk\, e^{-k^2/\alpha} = \left(k_0^2 + \alpha^2 \frac{\partial}{\partial\alpha} \right) \sqrt{\pi\alpha}$$

$$= \sqrt{\pi\alpha} \left(k_0^2 + \frac{\alpha}{2} \right)$$

and

$$I_0 = I_0^* = \int_{-\infty}^{+\infty} dk\, k^2 e^{-k^2/\alpha + ikx_0} = \alpha^2 \frac{\partial}{\partial \alpha} \int_{-\infty}^{+\infty} e^{-k^2/\alpha + ikx_0}$$

$$= \alpha^2 \frac{\partial}{\partial \alpha} \sqrt{\pi \alpha}\, e^{-\alpha x_0^2} = \sqrt{\pi \alpha} \left(\frac{\alpha}{2} - \alpha^2 x_0^2 \right) e^{-\alpha x_0^2}$$

Finally, we get

$$(\Delta p)^2 = \frac{\hbar^2 \alpha}{2} \left[\frac{1 + 2k_0^2/\alpha + (1 - 2\alpha x_0^2)e^{-\alpha x_0^2}}{1 + e^{-k_0^2/\alpha - \alpha x_0^2}} \right]$$

The uncertainty product is

$$(\Delta x)_0^2 (\Delta p)_0^2 = \frac{\hbar^2}{4} f(\xi, \zeta)$$

where

$$f(\xi, \zeta) \equiv \frac{\left[1 + 2\xi + (1 - 2\zeta)e^{-\xi - \zeta}\right]\left[1 + 2\zeta + (1 - 2\xi)e^{-\xi - \zeta}\right]}{\left(1 + e^{-\xi - \zeta}\right)^2}$$

This is a symmetric function of $\xi \equiv \alpha x_0^2$ and $\zeta \equiv k_0^2/\alpha$ that is always greater than unity.

(d) We start from the Fourier transform

$$\tilde{\psi}(k, t) = \frac{N}{\sqrt{\alpha}} e^{ikx} \left[e^{i(k-k_0)x_0}\, e^{-(k-k_0)^2/2\alpha} + e^{-i(k+k_0)x_0}\, e^{-(k+k_0)^2/2\alpha} \right] e^{-i\hbar k^2 t/2m}$$

Introducing $z = 1 + i\hbar t\alpha/m$ as well as $v_0 = \hbar k_0/m$ and changing variables by setting $k \to k \pm k_0$, we obtain

$$\psi(x, t) = \frac{N}{\sqrt{2\pi\alpha}} e^{-i\hbar k_0^2 t/2m} \left\{ e^{ik_0 x} \int dk\, \exp\left[i(x + x_0 - v_0 t)k\right] e^{-zk^2/2\alpha} \right.$$

$$\left. + e^{-ik_0 x} \int dk\, \exp\left[i(x - x_0 + v_0 t)k\right] e^{-zk^2/2\alpha} \right\}$$

or

$$\psi(x, t) = \frac{N}{\sqrt{z}} e^{-i\hbar k_0^2 t/2m} \left\{ e^{ik_0 x} \exp\left[-\frac{\alpha}{2z}(x + x_0 - v_0 t)^2 \right] \right.$$

$$\left. + e^{-ik_0 x} \exp\left[-\frac{\alpha}{2z}(x - x_0 + v_0 t)^2 \right] \right\}$$

The corresponding probability density is

$$\mathcal{P}(x,t) = \frac{|N|^2}{|z|} \left\{ \exp\left[-\frac{\alpha}{|z|^2}(x + x_0 - v_0 t)^2 \right] + \exp\left[-\frac{\alpha}{|z|^2}(x - x_0 + v_0 t)^2 \right] \right.$$

$$\left. + 2 \exp\left[-\frac{\alpha}{|z|^2}\left(x^2 + (x_0 - v_0 t)^2 \right) \right] \cos 2\tilde{q}x \right\}$$

where

$$\tilde{q}(t) = \frac{k_0 + \hbar\alpha^2 x_0 t}{|z|^2}$$

Note that $|z|^2 = 1 + (\hbar t \alpha / m)^2$.

In the limit $t \to \infty$ we have

$$\mathcal{P} = \frac{1}{t} \left(\frac{4m|N|^2 e^{-k_0^2/\alpha}}{\hbar\alpha} \right)$$

which is position independent.

(e) We can easily obtain

$$\mathcal{P}(0, t) = \frac{\mathcal{P}(0, 0)}{|z|^2} \exp\left[-\frac{\alpha}{|z|^2}(v_0^2 t^2 - 2v_0 x_0 t) \right]$$

The ratio $\mathcal{P}(0, t)/\mathcal{P}(0, 0)$ starts from unity with a positive slope, reaches a maximum value at some point determined by x_0/v_0 and then decreases. At large values of time it decreases as $\propto 1/t^2$.

3

Simple potentials

Problem 3.1 Consider a particle incident on an infinite planar surface separating empty space and an infinite region with constant potential energy V. The energy of the particle is $E > V$ (Fig. 1). Choose coordinates such that

$$V(x) = \begin{cases} 0, & x < 0 \\ V, & x > 0 \end{cases}$$

The incident particle is represented by a wave function of plane-wave form, $\psi_i(\mathbf{x}, t) = A\, e^{i(\mathbf{k}\cdot\mathbf{x} - \omega t)}$. The reflected and transmitted *wave functions* are

$$\psi_r(\mathbf{x}, t) = B e^{i(\mathbf{k}'\cdot\mathbf{x} - \omega t)}, \qquad \psi_t(\mathbf{x}, t) = C e^{i(\mathbf{q}\cdot\mathbf{x} - \omega t)}$$

The incident, reflected and transmitted *wave vectors* are

$$\mathbf{k} = k(\hat{\mathbf{x}}\cos\theta + \hat{\mathbf{y}}\sin\theta), \qquad \mathbf{k}' = k'(-\hat{\mathbf{x}}\cos\theta' + \hat{\mathbf{y}}\sin\theta')$$
$$\mathbf{q} = q(\hat{\mathbf{x}}\cos\theta'' + \hat{\mathbf{y}}\sin\theta'')$$

(a) Show that the angle of reflection equals the angle of incidence, i.e. $\theta' = \theta$. Show the validity of *Snell's law*,

$$\frac{\sin\theta''}{\sin\theta} = n$$

where $n = v_1/v_2$ is the *index of refraction* of the $x > 0$ region.

(b) Compute the coefficient ratios B/A and C/A.

(c) Calculate the incident, reflected and transmitted probability current densities $\mathcal{J}_i$, $\mathcal{J}_r$, $\mathcal{J}_t$.

(d) Demonstrate that the component of current perpendicular to the $x = 0$ plane is conserved, i.e.

$$\mathcal{J}_{i,x} + \mathcal{J}_{r,x} = \mathcal{J}_{t,x}$$

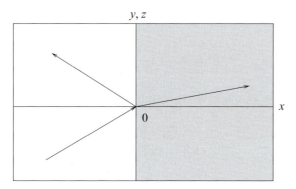

Fig. 1 Reflection and transmission at a three-dimensional potential step with $E > V$.

(e) Compute the *transmission* and *reflection coefficients*, defined as

$$\mathcal{T} = \frac{\mathcal{J}_{t,x}}{\mathcal{J}_{i,x}}, \qquad \mathcal{R} = \frac{|\mathcal{J}_{r,x}|}{\mathcal{J}_{i,x}}$$

in terms of the incidence and refraction angles θ and θ'' and check explicitly that

$$\mathcal{T} + \mathcal{R} = 1$$

Solution

(a) From Schroedinger's equation we immediately get

$$E = \hbar\omega = \frac{\hbar^2 k^2}{2m} = \frac{\hbar^2 k'^2}{2m} = \frac{\hbar^2 q^2}{2m} + V$$

which implies that

$$k' = k = \sqrt{\frac{2mE}{\hbar^2}}, \qquad q = k\sqrt{1 - \frac{V}{E}}$$

The refraction index is

$$n = \frac{v_1}{v_2} = \frac{k}{q} = \frac{1}{\sqrt{1 - V/E}}$$

In order to maintain continuity at every point $\mathbf{x}_0 = y\hat{\mathbf{y}} + z\hat{\mathbf{z}}$ of the $x = 0$ plane, the phases of the incident, reflected and transmitted components must have the same value, i.e.

$$\mathbf{k} \cdot \mathbf{x}_0 = \mathbf{k}' \cdot \mathbf{x}_0 = \mathbf{q} \cdot \mathbf{x}_0$$

which implies that

$$k\sin\theta = k\sin\theta' = q\sin\theta''$$

or, equivalently,

$$\theta' = \theta, \qquad \frac{\sin \theta''}{\sin \theta} = \frac{k}{q}$$

(b) The continuity of the wave function and of its derivative imply further that

$$A + B = C$$
$$i\mathbf{k}A + i\mathbf{k}'B = i\mathbf{q}\,C$$

These relations can be solved to give

$$\frac{B}{A} = \frac{\tan \theta'' - \tan \theta}{\tan \theta'' + \tan \theta}$$
$$\frac{C}{A} = \frac{2 \tan \theta''}{\tan \theta'' + \tan \theta}$$

These can be reexpressed as

$$\frac{B}{A} = \frac{k \cos \theta - q\sqrt{1 - (k^2/q^2)\sin^2 \theta}}{k \cos \theta + q\sqrt{1 - (k^2/q^2)\sin^2 \theta}}$$

$$\frac{C}{A} = \frac{2k \cos \theta}{k \cos \theta + q\sqrt{1 - (k^2/q^2)\sin^2 \theta}}$$

Note that for perpendicular incidence ($\theta = 0$) the above reduce to the well-known one-dimensional expressions

$$\frac{B}{A} = \frac{k - q}{k + q}, \qquad \frac{C}{A} = \frac{2k}{k + q}$$

(c) The probability current densities are easily calculated to be

$$\mathcal{J}_{\mathrm{i}} = \frac{\hbar \mathbf{k}}{m}|A|^2 = \frac{\hbar k}{m}|A|^2(\hat{\mathbf{x}} \cos \theta + \hat{\mathbf{y}} \sin \theta)$$

$$\mathcal{J}_{\mathrm{r}} = \frac{\hbar \mathbf{k}'}{m}|B|^2 = \frac{\hbar k}{m}|B|^2(-\hat{\mathbf{x}} \cos \theta + \hat{\mathbf{y}} \sin \theta)$$

$$\mathcal{J}_{\mathrm{t}} = \frac{\hbar \mathbf{q}}{m}|C|^2 = \frac{\hbar q}{m}|C|^2(\hat{\mathbf{x}} \cos \theta'' + \hat{\mathbf{y}} \sin \theta'')$$

(d) From the continuity equation, which in our case takes the static form $\nabla \cdot \mathcal{J} = 0$, considering its integral over a volume Ω we get

$$\int_{\Omega} d^3x\, \nabla \cdot \mathcal{J} = \oint_{S(\Omega)} d\mathbf{S} \cdot \mathcal{J} = 0$$

Taking as the volume Ω a cylinder of infinitesimal width L and arbitrary base $S = \pi R^2$, we arrive at

$$-\hat{\mathbf{x}} \cdot \mathcal{J}_- S + \hat{\mathbf{n}} \cdot \mathcal{J} \, 2\pi RL + \hat{\mathbf{x}} \cdot \mathcal{J}_+ S = 0$$

which in the limit $L \to 0$ gives

$$-\mathcal{J}_{i,x} - \mathcal{J}_{r,x} + \mathcal{J}_{t,x} = 0$$

(e) The transmission and reflection coefficients are easily obtained as

$$\mathcal{T} = \frac{4 \tan\theta \tan\theta''}{(\tan\theta'' + \tan\theta)^2}, \qquad \mathcal{R} = \left(\frac{\tan\theta'' - \tan\theta}{\tan\theta'' + \tan\theta} \right)^2$$

and they satisfy $\mathcal{R} + \mathcal{T} = 1$.

Problem 3.2 Again, consider a particle incident on an infinite planar surface separating empty space and an infinite region with constant potential energy V. Now, though, take the energy of the particle to be $E < V$ (Fig. 2). Again choose coordinates such that

$$V(\mathbf{x}) = \begin{cases} 0, & x < 0 \\ V, & x > 0 \end{cases}$$

As before, the incident particle is represented by a wave function of plane-wave form,

$$\psi_i(\mathbf{x}, t) = A e^{i(\mathbf{k} \cdot \mathbf{x} - \omega t)}$$

the *reflected* wave function is

$$\psi_r(\mathbf{x}, t) = B e^{i(\mathbf{k}' \cdot \mathbf{x} - \omega t)}$$

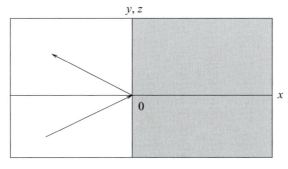

Fig. 2 A three-dimensional potential step with $V > E$.

while the wave function in the $x > 0$ region is

$$\psi_t(\mathbf{x}, t) = C e^{-\kappa x - iqy - i\omega t}$$

The *incident* and *reflected wave vectors* are

$$\mathbf{k} = k(\hat{\mathbf{x}} \cos \theta + \hat{\mathbf{y}} \sin \theta), \qquad \mathbf{k}' = k'(-\hat{\mathbf{x}} \cos \theta' + \hat{\mathbf{y}} \sin \theta')$$

(a) Show that $k' = k$, $\theta = \theta'$ and $q = k \sin \theta$, the analogue of Snell's law.

(b) Compute the coefficient ratios B/A and C/A.

(c) Calculate the incident, reflected and transmitted probability current densities

$$\mathcal{J}_i, \qquad \mathcal{J}_r, \qquad \mathcal{J}_t$$

(d) Demonstrate that the component of current perpendicular to the $x = 0$ plane is conserved, i.e.

$$\mathcal{J}_{i,x} + \mathcal{J}_{r,x} = \mathcal{J}_{t,x} = 0$$

Compute explicitly the *reflection coefficient*, defined as

$$\mathcal{R} = \frac{|\mathcal{J}_{r,x}|}{\mathcal{J}_{i,x}}$$

and verify that it equals unity.

Solution

(a) From Schroedinger's equation we have

$$E = \hbar\omega = \frac{\hbar^2 k^2}{2m} = \frac{\hbar^2 k'^2}{2m} = -\frac{\hbar^2 \kappa^2}{2m} + \frac{\hbar^2 q^2}{2m} + V$$

In order to be able to implement continuity at every point of the $x = 0$ plane, we must have the same value for the phases of the incident, reflected and 'transmitted' plane waves. This implies that

$$k \sin \theta = k' \sin \theta' = q$$

i.e.

$$\theta = \theta', \qquad q = k \sin \theta$$

(b) The continuity of the wave function and of its derivative give

$$A + B = C$$
$$ik_x A + ik'_x B = -\kappa C$$
$$ik_y A + ik'_y B = iqC$$

These relations amount to

$$\frac{C}{A} = \frac{2}{1 + i\kappa/k}$$

$$\frac{B}{A} = \frac{1 - i\kappa/k}{1 + i\kappa/k} = \exp\left[-2i\arctan(\kappa/k)\right]$$

(c) The probability current densities are

$$\boldsymbol{\mathcal{J}}_i = \frac{\hbar k}{m}|A|^2(\hat{\mathbf{x}}\cos\theta + \hat{\mathbf{y}}\sin\theta)$$

$$\boldsymbol{\mathcal{J}}_r = \frac{\hbar k}{m}|B|^2(-\hat{\mathbf{x}}\cos\theta + \hat{\mathbf{y}}\sin\theta)$$

$$\boldsymbol{\mathcal{J}}_t = \frac{\hbar q}{m}|C|^2 e^{-2\kappa x}\hat{\mathbf{y}} = \frac{\hbar k}{m}|C|^2 e^{-2\kappa x}\sin\theta\,\hat{\mathbf{y}}$$

(d) The continuity equation in our case takes the static form

$$\nabla \cdot \boldsymbol{\mathcal{J}} = 0$$

Considering its integral over a volume Ω, we get

$$\int_\Omega d^3x\,\nabla \cdot \boldsymbol{\mathcal{J}} = \oint_{S(\Omega)} d\mathbf{S} \cdot \boldsymbol{\mathcal{J}} = 0$$

Taking as the volume Ω the cylinder of infinitesimal width L and arbitrary base $S = \pi R^2$, we arrive at

$$-\hat{\mathbf{x}} \cdot \boldsymbol{\mathcal{J}}_- S + \hat{\mathbf{n}} \cdot \boldsymbol{\mathcal{J}}\,2\pi RL + \hat{\mathbf{x}} \cdot \boldsymbol{\mathcal{J}}_+ S = 0$$

which in the limit $L \to 0$ gives

$$-\mathcal{J}_{i,x} - \mathcal{J}_{r,x} + \mathcal{J}_{t,x} = 0$$

and, since $\mathcal{J}_{t,x} = 0$,

$$\mathcal{J}_{i,x} + \mathcal{J}_{r,x} = 0$$

The reflection coefficient is

$$\mathcal{R} = \frac{|\mathcal{J}_{r,x}|}{\mathcal{J}_{i,x}} = \frac{|B|^2}{|A|^2} = 1$$

Problem 3.3 Consider an infinite square well of width $2L$ with a particle of mass m moving in it ($-L < x < L$). The particle is in the lowest-energy state, so that

$$E_1 = \frac{\hbar^2\pi^2}{8mL^2}, \qquad \psi_1(x) = \frac{1}{\sqrt{L}}\cos\frac{\pi x}{2L}$$

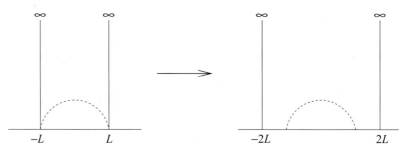

Fig. 3 Instantaneously expanding infinite square well.

Assume now that at $t = 0$ the walls of the well move instantaneously so that its width doubles $(-2L < x < 2L)$; see Fig. 3. This change does not affect the state of the particle, which is the same before and immediately after the change.

(a) Write down the wave function of the particle at times $t > 0$. Calculate the probability $\mathcal{P}_n$ of finding the particle in an arbitrary eigenstate of the modified system. What is the probability of finding the particle in an odd eigenfunction?

(b) Calculate[1] the expectation value of the energy at any time $t > 0$.

(c) If we assume instead that the walls move outwards with a finite speed u, our assumptions should still hold provided that this velocity is much larger than the characteristic velocity of the system, i.e. $u \gg v_0$. What is v_0?

Solution

(a) The new eigenvalues and eigenfunctions of the modified system are

$$\overline{E}_n = \frac{\hbar^2 n^2 \pi^2}{32 m L^2} = \frac{1}{4} E_n$$

$$\overline{\psi}_n(x) = \frac{1}{2\sqrt{2L}} \left[e^{in\pi x/4L} + (-1)^{n+1} e^{-in\pi x/4L} \right] = \frac{1}{\sqrt{2}} \psi_n\left(\frac{x}{2}\right)$$

where $n = 1, 2, \ldots$. The wave function of the particle at times $t > 0$ will be

$$\psi(x, t) = \sum_{n=1}^{\infty} C_n e^{-i\overline{E}_n t/\hbar}\, \overline{\psi}_n(x)$$

with coefficients

$$C_n = \int_{-2L}^{2L} dx\, \psi(x, 0)\, \overline{\psi}_n^*(x) = \frac{1}{\sqrt{L}} \int_{-L}^{L} dx \cos\frac{\pi x}{2L}\, \overline{\psi}_n^*(x)$$

[1] You can make use of the series

$$\sum_{v=0}^{\infty} \frac{(2v + 1)^2}{[(2v + 1)^2 - 4]^2} = \frac{\pi^2}{16}.$$

Substituting the initial wave function into the above integral expression for C_n, we obtain after performing some elementary exponential integrals

$$C_n = -\frac{4\sqrt{2}}{\pi}\frac{[1-(-1)^n]}{n^2-4}\cos\frac{n\pi}{4}, \qquad n = 0, 1, 2, \ldots$$

Clearly, the terms of even n, $n = 0, 2, \ldots$, corresponding to odd eigenfunctions, vanish.

The probability of finding the particle in the energy eigenstate $\overline{\psi}_n(x)$ is

$$\mathcal{P}_n = |C_n|^2 = \frac{32}{\pi^2}\frac{[1-(-1)^n]}{(n^2-4)^2}$$

Note that $\cos[(2\nu+1)\pi/4] = (-1)^{\nu+1}/\sqrt{2}$.

Since the Hamiltonian of the system is parity invariant, the initial-state parity is conserved at all times and the probability of finding the particle in an odd eigenstate is zero.

(b) The expectation value of the energy can be calculated in a straightforward fashion. It is

$$\langle H \rangle = \sum_{n=1}^{\infty} E_n|C_n|^2 = E_1\left\{\frac{16}{\pi^2}\sum_{\nu=0}^{\infty}\frac{(2\nu+1)^2}{[(2\nu+1)^2-4]^2}\right\} = E_1$$

which shows that the energy of the particle is conserved.

(c) The characteristic velocity of the system is

$$v_0 = \sqrt{\frac{2E_1}{m}} = \frac{\pi\hbar}{2mL}$$

and our results stay valid as long as

$$u \gg \frac{\pi\hbar}{2mL}$$

Problem 3.4 Consider particles incident on a one-dimensional step function potential $V(x) = V\Theta(x)$ with energy $E > V$. Calculate the reflection coefficient for either direction of incidence. Consider the limits $E \to V$ and $E \to \infty$.

Solution

Incidence from the left. In this case, the wave function is

$$\psi(x) = \begin{cases} e^{ikx} + Be^{-ikx}, & x < 0 \\ Ce^{iqx}, & x > 0 \end{cases}$$

where

$$k \equiv \sqrt{\frac{2mE}{\hbar^2}}, \qquad q \equiv \sqrt{\frac{2m(E-V)}{\hbar^2}}$$

Continuity of the wave function and of its derivative at $x = 0$ give

$$1 + B = C, \qquad k(1 - B) = qC$$

or, equivalently,

$$B = \frac{k - q}{k + q}, \qquad C = \frac{2k}{k + q}$$

The reflection coefficient is

$$\mathcal{R}_L = \frac{|J_r|}{J_i} = \frac{(\hbar k/m)|B|^2}{\hbar k/m} = |B|^2$$

or

$$\mathrm{R}_L = \left(\frac{k - q}{k + q}\right)^2$$

Incidence from the right. In this case, the wave function is of the form

$$\psi(x) = \begin{cases} De^{-ikx}, & x < 0 \\ e^{-iqx} + Fe^{iqx}, & x > 0 \end{cases}$$

Continuity at $x = 0$ gives

$$D = 1 + F, \qquad kD = q(1 - F)$$

or

$$F = \frac{q - k}{q + k}, \qquad D = \frac{2q}{q + k}$$

The reflection coefficient is

$$\mathcal{R}_R = |F|^2 = \left(\frac{q - k}{q + k}\right)^2$$

Thus, we see that the reflection coefficient is independent of the direction of incidence:

$$\mathcal{R}_L = \mathcal{R}_R$$

Expressed in terms of the energy, the reflection coefficient is

$$\mathcal{R} = \left(\frac{1 - \sqrt{1 - V/E}}{1 + \sqrt{1 - V/E}}\right)^2$$

In the limit $E \to V$, we approach the maximum value of the reflection coefficient, namely unity, and the barrier becomes impenetrable:

$$\lim_{E \to V+} (\mathcal{R}) \sim 1 - 4\sqrt{\frac{E}{V} - 1}$$

In the limit $E \to \infty$ the barrier should become irrelevant and the reflection coefficient should approach zero. Indeed,

$$\lim_{E \to \infty} (\mathcal{R}) \sim \frac{V^2}{16E^2}$$

Problem 3.5 Consider a step function potential $V(x) = V\Theta(x)$ and particles of energy $E > V$ incident on it from both sides simultaneously. The wave function is

$$\psi(x) = \begin{cases} Ae^{ikx} + Be^{-ikx}, & x < 0 \\ Ce^{iqx} + De^{-iqx}, & x > 0 \end{cases}$$

where $k \equiv \sqrt{2mE/\hbar^2}$ and $q \equiv \sqrt{2m(E - V)/\hbar^2}$.

(a) Determine two relations among the coefficients A, B, C and D from the continuity of the wave function and of its derivative at the point $x = 0$.

(b) Determine the matrix U defined by the relation

$$\begin{pmatrix} \sqrt{q}\, C \\ \sqrt{k}\, B \end{pmatrix} = \begin{pmatrix} U_{11} & U_{12} \\ U_{21} & U_{22} \end{pmatrix} \begin{pmatrix} \sqrt{k}\, A \\ \sqrt{q}\, D \end{pmatrix}$$

Show that U is a unitary matrix.

(c) Write down the probability current conservation and show that it is directly related to the unitarity of the matrix U.

Solution

(a) Continuity at the point $x = 0$ gives

$$A + B = C + D$$
$$ik(A - B) = iq(C - D)$$

These relations are equivalent to

$$\sqrt{q}\, C = -\left(\frac{1 - q/k}{1 + q/k}\right) \sqrt{q}\, D + \left(\frac{2\sqrt{q/k}}{1 + q/k}\right) \sqrt{k}\, A$$

$$\sqrt{k}\, B = \left(\frac{2\sqrt{q/k}}{1 + q/k}\right) \sqrt{q}\, D + \left(\frac{1 - q/k}{1 + q/k}\right) \sqrt{k}\, A$$

(b) From the last pair of relations we can conclude immediately that

$$U_{11} = U_{22} = \frac{2\sqrt{q/k}}{1 + q/k}$$

$$U_{12} = -U_{21} = -\frac{1 - q/k}{1 + q/k}$$

Unitarity requires that

$$|U_{11}|^2 + |U_{12}|^2 = |U_{22}|^2 + |U_{21}|^2 = 1$$

$$U_{11}U_{21}^* + U_{12}U_{22}^* = 0$$

These relations are easily seen to be satisfied and so U is unitary.

(c) The probability current densities in the left-hand region are

$$\mathcal{J}_i^{(-)} = \frac{\hbar k}{m}|A|^2, \qquad \mathcal{J}_r^{(-)} = -\frac{\hbar k}{m}|B|^2$$

Similarly, in the right-hand region, we have

$$\mathcal{J}_i^{(+)} = -\frac{\hbar q}{m}|D|^2, \qquad \mathcal{J}_r^{(+)} = \frac{\hbar q}{m}|C|^2$$

Current conservation is expressed as

$$\mathcal{J}_i^{(-)} + \mathcal{J}_r^{(-)} = \mathcal{J}_i^{(+)} + \mathcal{J}_r^{(+)}$$

Substituting the current expressions, this is equivalent to

$$q|C|^2 + k|B|^2 = k|A|^2 + q|D|^2$$

Expressing C and B in terms of the matrix U implies that

$$(|U_{11}|^2 + |U_{21}|^2)\,k|A|^2 + (|U_{12}|^2 + |U_{22}|^2)\,q|D|^2 + \sqrt{kq}\,AD^*(U_{11}^*U_{12} + U_{21}U_{22}^*)$$
$$+ \sqrt{kq}\,A^*D(U_{11}U_{12}^* + U_{22}U_{21}^*) = k|A|^2 + q|D|^2$$

This immediately implies the unitarity relations. Thus, current conservation is directly related to the unitarity of U.

Problem 3.6 Consider the standard one-dimensional infinite square well

$$V(x) = \begin{cases} +\infty, & x < -L \\ 0, & -L < x < L \\ +\infty, & x > L \end{cases}$$

(a) Write down the energy eigenfunctions and eigenvalues.
(b) Calculate the spatial uncertainty in an energy eigenstate.
(c) Calculate the matrix elements of the position operator,

$$x_{nn'} = \langle n|x|n'\rangle$$

and show that the matrix $x_{nn'}$ is Hermitian. In the $\{n\}$ representation the position operator is a purely imaginary antisymmetric matrix.
(d) Calculate the momentum matrix elements

$$p_{nn'} = \langle n|p|n'\rangle$$

and show that the matrix $p_{nn'}$ is a Hermitian, symmetric and real matrix. In addition, show that the square of this matrix divided by $2m$ gives the Hamiltonian matrix, i.e. a diagonal matrix with the energy eigenvalues as diagonal elements.[2]

Solution
(a) The eigenfunctions of positive parity are

$$\psi_{2v+1}(x) = \frac{1}{\sqrt{L}}\cos\frac{(2v+1)\pi x}{2L} \qquad (v = 0, 1, \ldots)$$

with eigenvalues

$$E_{2v+1} = \frac{\hbar^2(2v+1)^2\pi^2}{8mL^2}$$

The eigenvalues of negative parity are

$$\psi_{2v}(x) = \frac{i}{\sqrt{L}}\sin\frac{v\pi x}{L} \qquad (v = 1, 2, \ldots)$$

with eigenvalues

$$E_{2v} = \frac{\hbar^2 v^2\pi^2}{2mL^2}$$

(b) Since $|\psi_n(x)|^2$ is always even, the expectation value of position in an energy eigenstate will vanish owing to the oddness of the integrand. Thus $\langle x\rangle_n = 0$, and the uncertainty squared will be given by the expectation value:

$$(\Delta x)_n^2 = \langle n|x^2|n\rangle$$

---

[2] You may find useful the series

$$\sum_{j=0}^{\infty}\frac{(2j+1)^2}{[(2j+1)^2-n^2][(2j+1)^2-n'^2]} = \frac{\pi^2}{16}\delta_{nn'}$$

This can be computed in a straightforward fashion. For even $n = 2\nu$ we have

$$\langle x^2 \rangle_{2\nu} = \frac{1}{L} \int_{-L}^{L} dx\, x^2 \sin^2 \frac{\nu \pi x}{L} = \frac{1}{L} \int_{-L}^{L} dx\, x^2 \left(1 - \cos \frac{2\nu \pi x}{L} \right)$$

$$= \frac{L^2}{3} - \frac{L^2}{(2\nu \pi)^3} \int_{-2\nu \pi}^{2\nu \pi} dy\, y^2 \cos y$$

We now calculate the integral in the above expression:

$$\int_{-L}^{L} dy\, y^2 \cos y = \left[y^2 \sin y + 2y \cos y - 2 \sin y \right]_{-2\nu \pi}^{2\nu \pi} = 4\nu \pi$$

Thus we end up with

$$\langle x^2 \rangle_{2\nu} = \frac{L^2}{3} \left[1 - \frac{6}{(2\nu \pi)^2} \right]$$

An analogous result holds for the $n = 2\nu + 1$ case. Thus finally we have

$$(\Delta x)_n^2 = \langle x^2 \rangle_n = \frac{L^2}{3} \left[1 - \frac{6}{(n\pi)^2} \right]$$

(c) The position matrix elements between states of the same parity vanish due to the oddness of the integrand. Thus, the only non-vanishing matrix elements will be

$$x_{2\nu,\, 2\nu'+1} = \langle 2\nu | x | 2\nu' + 1 \rangle, \qquad x_{2\nu+1,\, 2\nu'} = \langle 2\nu + 1 | x | 2\nu' \rangle$$

For the first matrix element we have

$$x_{2\nu,\, 2\nu'+1} = -\frac{i}{L} \int_{-}^{L} dx\, x \sin \frac{\nu \pi x}{L} \cos \frac{(2\nu' + 1)\pi x}{2L}$$

$$= -\frac{i}{2L} \int_{-L}^{L} dx\, x \left[\sin \frac{(2\nu + 2\nu' + 1)\pi x}{2L} + \sin \frac{(2\nu - 2\nu' - 1)\pi x}{2L} \right]$$

Using the integral

$$\int_{-L}^{L} dx\, x \sin ax = \frac{2}{a^2} \sin aL - \frac{2L}{a} \cos aL$$

we eventually obtain

$$x_{2\nu,\, 2\nu'+1} = \frac{16i\, L}{\pi^2} (-1)^{\nu + \nu'} \frac{2\nu(2\nu' + 1)}{\left[(2\nu)^2 - (2\nu' + 1)^2 \right]^2}$$

and

$$x_{2v+1,\,2v'} = -\frac{16iL}{\pi^2}(-1)^{v+v'}\frac{2v'(2v+1)}{\left[(2v')^2 - (2v+1)^2\right]^2}$$

It is clear that

$$x_{nn'}^{\dagger} = x_{n'n}^{*} = x_{nn'}, \qquad x_{nn'}^{*} = -x_{nn'}$$

(d) In an analogous fashion we obtain the non-vanishing momentum matrix elements

$$p_{2v,\,2v'+1} = -\frac{2\hbar}{L}(-1)^{v+v'}\frac{2v(2v'+1)}{(2v)^2 - (2v'+1)^2}$$

$$p_{2v+1,\,2v'} = -\frac{2\hbar}{L}(-1)^{v+v'}\frac{2v'(2v+1)}{(2v')^2 - (2v+1)^2}$$

Clearly, we have

$$p_{nn'}^{\dagger} = p_{n'n}^{*} = p_{n'n} = p_{nn'}$$

The last question in (d) amounts to showing that

$$\frac{1}{2m}(p^2)_{nn'} = \frac{1}{2m}\sum_{n''}p_{nn''}p_{n''n'} = E_n\delta_{nn'}$$

Let us verify this explicitly in the case where n and n' are even. We have

$$\frac{1}{2m}\sum_{v''=0}^{\infty}p_{2v,\,2v''+1}\,p_{2v''+1,\,2v'}$$

$$= \frac{2\hbar^2}{mL^2}(-1)^{v+v'}4vv'\sum_{v''=0}^{\infty}\frac{(2v''+1)^2}{\left[(2v''+1)^2 - (2v)^2\right]\left[(2v''+1)^2 - (2v')^2\right]}$$

$$= \frac{2\hbar^2}{mL^2}(-1)^{v+v'}4vv'\frac{\pi^2}{16}\delta_{vv'} = \frac{\hbar^2(2v)^2\pi^2}{8mL^2}\delta_{vv'} = E_n\delta_{nn'}$$

The verification for n, n' odd proceeds in an identical fashion.

Problem 3.7 Consider a one-dimensional delta function potential $V(x) = g\delta(x)$ and the scattering of particles of energy $E > 0$ at it. Without loss of generality assume that the particles are incident from the left.

(a) Applying the appropriate continuity conditions at $x = 0$, find the wave function $\psi_E(x)$ and write it in the form

$$\psi_E(x) = e^{ikx} + Fe^{ik|x|}$$

where $k \equiv \sqrt{2mE/\hbar^2}$.

(b) Compute the probability current density and demonstrate that it is everywhere continuous.

(c) Consider the case of attractive coupling ($g < 0$) and solve the bound-state problem for $E < 0$. Find the bound-state wave function and compute the value of the bound-state energy.

(d) Show that the energy of the uniquely existing bound state corresponds to a pole of the coefficient F, previously calculated in (a).

Solution

(a) Starting from the usual type of ansatz,

$$\psi(x) = \begin{cases} e^{ikx} + Be^{-ikx}, & x < 0 \\ Ce^{ikx}, & x > 0 \end{cases}$$

we obtain

$$1 + B = C$$

from the continuity of the wave function at $x = 0$. The derivative of the wave function is discontinuous, with a finite jump given by

$$-\frac{\hbar^2}{2m}\left[\psi'(0+) - \psi'(0-)\right] + g\psi(0) = 0$$

This can be obtained by integrating Schroedinger's equation in an infinitesimal domain around the discontinuity point. This relation translates to

$$C + B - 1 = -i\frac{2mg}{k\hbar^2}C$$

Introducing

$$\tilde{g} \equiv \frac{2mg}{\hbar^2}$$

we obtain

$$C = \frac{2k}{2k + i\tilde{g}}, \qquad B = \frac{-i\tilde{g}}{2k + i\tilde{g}}$$

The wave function is

$$\psi_E(x) = \Theta(x)\left[e^{ikx} - \left(\frac{i\tilde{g}}{2k + i\tilde{g}}\right)e^{-ikx}\right] + \Theta(-x)\left(\frac{2k}{2k + i\tilde{g}}\right)e^{ikx}$$

It can be written in the compact form

$$\psi_E(x) = e^{ikx} + F(k)e^{ik|x|}$$

with

$$F(k) \equiv -\frac{i\tilde{g}}{2k + i\tilde{g}}$$

(b) The probability current density can be easily calculated to be

$$\mathcal{J}(x < 0) = \frac{\hbar k}{m} \left(1 - |B|^2\right) = \frac{\hbar k}{m} \left(\frac{4k^2}{4k^2 + \tilde{g}^2}\right)$$

$$\mathcal{J}(x > 0) = \frac{\hbar k}{m} |C|^2 = \frac{\hbar k}{m} \left(\frac{4k^2}{4k^2 + \tilde{g}^2}\right)$$

Thus, as a constant, it is conserved and continuous.

(c) It is easy to see that for an attractive potential ($g < 0$) there is a solution of negative energy

$$E \equiv -\frac{\hbar^2 \kappa^2}{2m} < 0$$

namely

$$\psi(x) = \sqrt{\kappa}\, e^{-\kappa|x|}$$

provided that

$$\kappa = -mg/\hbar^2$$

(d) Note that the previously obtained *scattering amplitude* $F(k)$ has a pole at

$$k = -i\frac{mg}{\hbar^2}$$

that corresponds to the negative energy

$$E = \frac{\hbar^2 k^2}{2m} = -\frac{mg^2}{2\hbar^2}$$

which coincides with the bound-state energy. This is a manifestation of the general property that the poles of the scattering amplitude correspond to the bound-state energies of the associated bound-state problem.

Problem 3.8 Consider a one-dimensional potential with a step function component and an attractive delta function component just at the edge (Fig. 4), namely

$$V(x) = V\Theta(x) - \frac{\hbar^2 g}{2m} \delta(x)$$

Compute the reflection coefficient for particles incident from the left with energy $E > V$. Consider the limit $E \to \infty$. Do you see any difference from the pure step

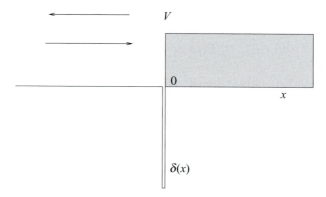

Fig. 4 Step function potential with an attractive delta function edge.

function case? Consider also the case $E < 0$ and determine the energy eigenvalues and eigenfunctions of any existing bound-state solutions.

Solution

The wave function will be of the form

$$\psi(x) = \begin{cases} e^{ikx} + Be^{-ikx}, & x < 0 \\ Ce^{iqx}, & x > 0 \end{cases}$$

with

$$k \equiv \sqrt{2mE/\hbar^2}, \qquad q \equiv \sqrt{2m(E-V)/\hbar^2}$$

Continuity of the wave function at $x = 0$ gives

$$1 + B = C$$

Integrating Schroedinger's equation over an infinitesimal interval around the origin gives

$$-\frac{\hbar^2}{2}\left[\psi'(0+) - \psi'(0-)\right] - \frac{\hbar^2 g}{2m}\psi(0) = 0$$

or

$$1 - B = -\frac{i}{k}(g + iq)C$$

From the two relationships between B and C we obtain

$$C = \frac{2}{1 + q/k - ig/k}$$

$$B = \frac{1 - q/k + ig/k}{1 + q/k - ig/k}$$

The reflection coefficient is

$$\mathcal{R} = \frac{|J_r|}{J_i} = \frac{(\hbar k/m)|B|^2}{\hbar k/m} = |B|^2$$

or

$$\mathcal{R} = \frac{(1 - q/k)^2 + g^2/k^2}{(1 + q/k)^2 + g^2/k^2}$$

In the high-energy limit we have

$$\lim_{E \to \infty}(\mathcal{R}) \sim \frac{g^2 \hbar^2}{8mE}$$

This is different from the case of a pure step barrier, where reflection drops off faster with incident energy, namely $\lim_{E \to \infty}(\mathcal{R}) \sim V^2/8E^2$.

In order to study the case of negative energy, $E < 0$, it is convenient to introduce the notation

$$\kappa_- \equiv \sqrt{\frac{2m|E|}{\hbar^2}}, \qquad \kappa_+ \equiv \sqrt{\frac{2m(V + |E|)}{\hbar^2}}$$

Then we can write the bound-state wave function as

$$\psi(x) = \begin{cases} Ae^{x\kappa_-}, & x < 0 \\ Ae^{-x\kappa_+}, & x > 0 \end{cases}$$

The discontinuity at the origin implies that

$$-\frac{\hbar^2}{2m}(-A\kappa_+ - A\kappa_-) - \frac{\hbar^2 g}{2m} A = 0$$

which is equivalent to the constraint

$$\kappa_- + \kappa_+ = g$$

Substituting into this relation the definitions of $\kappa_\pm$ and squaring, we are led to the unique bound-state energy eigenvalue

$$E = -\frac{m}{2\hbar^2 g^2}\left(\frac{\hbar^2 g^2}{2m} - V\right)^2$$

The corresponding wave number expressions are

$$\kappa_\pm^2 = \frac{m^2}{\hbar^4 g^2}\left(\frac{g^2 \hbar^2}{2m} \pm V\right)^2$$

The normalization constant of the bound-state wave function is easily computed to be $A = \sqrt{2\kappa_+ \kappa_-/g}$.

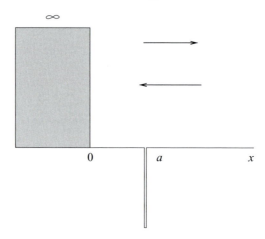

Fig. 5 Impenetrable potential wall with an attractive delta function.

Problem 3.9 Figure 5 shows an attractive delta function potential situated at a distance a from an impenetrable wall. Its form is

$$V(x) = \begin{cases} \infty, & x < 0 \\ -\dfrac{\hbar^2 g}{2m}\delta(x-a), & x > 0 \end{cases}$$

(a) Calculate the relative phase of the reflected waves for incident particles of energy $E > 0$. What is the behaviour of this phase for low and large energies?
(b) Study the $E < 0$ case. Are there any bound-state solutions?

Solution

(a) Since the wave function has to vanish at the wall, the only choice in the region $0 < x < a$ is $\sin kx$. Thus, the wave function is of the form

$$\psi(x) = \begin{cases} C\sin kx, & 0 < x < a \\ Ae^{-ikx} + Be^{ikx}, & x > a \end{cases}$$

Continuity of the wave function at $x = a$ gives

$$C\sin ka = Ae^{-ika} + Be^{ika}$$

However, the discontinuity of the derivative at that point gives

$$-ik(Ae^{-ika} - Be^{ika}) - kC\,\cos ka + gC\sin ka = 0$$

or

$$Ae^{-ika} - Be^{ika} = \frac{i}{k}(k\cos ka - g\sin ka)\,C$$

Solving for B and A, we obtain

$$A = \frac{C}{2}\left[i + \frac{g}{2k} - \frac{g}{2k}e^{2ika}\right]$$

$$B = \frac{C}{2}\left[-i + \frac{g}{2k} - \frac{g}{2k}e^{-2ika}\right]$$

The ratio of the reflected to the incident amplitude is a pure phase: namely, we have

$$\frac{B}{A} = \frac{-i + g/2k - (g/2k)e^{-2ika}}{i + g/2k - (g/2k)e^{2ika}} \equiv e^{2i\delta}$$

so that

$$\tan\delta \equiv \frac{-2k/g + \sin 2ka}{1 - \cos 2ka}$$

At very high energies ($k \to \infty$),

$$\tan\delta \sim \frac{k}{g\sin^2 ka} \qquad \Longrightarrow \qquad \delta \sim -\frac{\pi}{2}$$

At low energies ($k \to 0$),

$$\tan\delta \sim \frac{1}{ka}\left(1 - \frac{1}{ga}\right)$$

This gives either $\delta \sim \pi/2$ or $\delta \sim -\pi/2$, depending on how strong the delta-function potential is.

(b) Introducing $E = -\hbar^2\kappa^2/2m$, we write down the candidate bound-state wave function in the form

$$\psi(x) = \begin{cases} A\sinh\kappa x, & 0 < x < a \\ Be^{-\kappa x}, & x > a \end{cases}$$

The continuity of the wave function and the discontinuity of its derivative at $x = a$ give

$$A\sinh\kappa a = Be^{-\kappa a}$$

$$\kappa Be^{-\kappa a} + \kappa A\cosh\kappa a - gBe^{-\kappa a} = 0$$

From these we obtain the bound-state energy condition

$$\tanh\kappa a = \frac{1}{ga/\kappa a - 1}$$

which can be seen graphically to have one solution for $ga > 1$. The bound-state wave function is

$$\psi(x) = A\left[\Theta(x)\Theta(a - x)\sinh\kappa x + \Theta(x - a)\sinh\kappa a\, e^{-\kappa(x-a)}\right]$$

The normalization constant is given by

$$A^{-2} = \frac{a}{2}\left(\frac{1 + 2\kappa a - ga}{ga - 2\kappa a}\right)$$

Problem 3.10 Consider a standard one-dimensional square well,

$$V(x) = \begin{cases} 0, & |x| > a \\ -V_0, & |x| < a \end{cases}$$

(a) Particles of energy $E > 0$ are incident on it from the left. Calculate the transmission coefficient $\mathcal{T}$. How does $\mathcal{T}$ behave for very large energies? What is its low-energy limit?

(b) Are there any specific values of positive energy for which there is absolutely no reflection and the well is transparent? Verify explicitly that for these particular values the amplitude of the reflected wave vanishes.

(c) Consider now the bound-state problem ($E < 0$). Find the bound-state wave functions as well as the equation that determines the allowed energy eigenvalues. Assume that our square well is relatively shallow, so that

$$V_0 < \frac{\hbar^2 \pi^2}{8ma^2}$$

Show that in this case there is only one allowed bound state and that it corresponds to an even wave function.

(d) Assume that the depth of the square well increases while its width decreases in such a way that the product $V_0 \times 2a$ stays the same. This is equivalent to writing $V_0 \equiv g^2/2a$ and taking the limit $a \to 0$. Show that in this case there is a single bound state and determine the precise value of its energy eigenvalue.

Solution

(a) The wave function for particles incident from the left can be written as

$$\psi(x) = \begin{cases} e^{ikx} + Be^{-ikx}, & x < -a \\ Ce^{iqx} + De^{-iqx}, & -a < x < a \\ Fe^{ikx}, & x > a \end{cases}$$

where

$$k \equiv \sqrt{2mE/\hbar^2}, \qquad q \equiv \sqrt{2m(E + V_0)/\hbar^2}$$

Applying continuity of the wave function and of its first derivative at the points $x = -a$ and $x = a$ leads to the set of four equations

$$e^{-ika} + Be^{ika} = Ce^{-iqa} + De^{iqa}$$

$$e^{-ika} - Be^{ika} = \frac{q}{k}\left(Ce^{-iqa} - De^{iqa}\right)$$

$$Ce^{iqa} + De^{-iqa} = Fe^{ika}$$

$$Ce^{iqa} - De^{-iqa} = \frac{k}{q}Fe^{ika}$$

Solving for F, we obtain

$$F = -\frac{4qk}{(k-q)^2} \frac{e^{-2i(k+q)a}}{1 - e^{-4iqa}(k+q)^2/(k-q)^2}$$

The transmission coefficient is

$$\mathcal{T} = \frac{\mathcal{J}_t}{\mathcal{J}_i} = \frac{(\hbar k/m)|F|^2}{\hbar k/m} = |F|^2$$

and has the value

$$\mathcal{T} = \left(1 + \frac{V_0^2}{4E(E+V_0)} \sin^2 2qa\right)^{-1}$$

Its two extreme energy limits are

$$\lim_{E \to \infty} \{\mathcal{T}\} \sim 1 - \frac{V_0^2}{4E^2} \sin^2 2qa \to 1$$

$$\lim_{E \to 0} \{\mathcal{T}\} \sim \frac{4E}{V_0 \sin^2 \sqrt{8ma^2 V_0/\hbar^2}} \to 0$$

(b) If the values of the energy E are such that the sine in the denominator of the transmission coefficient vanishes, so that $2qa = n\pi$, the transmission coefficient becomes unity and no reflection occurs. These values are thus

$$E_n = -V_0 + \frac{n^2\hbar^2\pi^2}{8ma^2}$$

for $n = 1, 2, \ldots$. The physical reason for the overall vanishing of reflection is destructive interference between the waves reflected at $x = -a$ and those reflected at $x = a$; it is analogous to the phenomenon occurring in optics. It is immediately obvious from the expression for the reflected amplitude,

$$B = e^{-2ika} \frac{(k+q)(1 - e^{-4iqa})}{(k-q) - e^{-4iqa}(k+q)^2/(k-q)}$$

that for $2qa = n\pi$, $B = 0$.

(c) The wave function for an eigenfunction of negative energy is

$$\psi(x) = \begin{cases} Ae^{\kappa x}, & x < -a \\ B\cos qx + C\sin qx, & -a < x < a \\ De^{-\kappa x}, & x > a \end{cases}$$

with

$$\kappa \equiv \sqrt{2m|E|/\hbar^2}, \qquad q \equiv \sqrt{2m(V_0 - |E|)/\hbar^2}$$

The parity invariance of the potential allows us to choose the energy eigenstates to be simultaneously eigenstates of parity, i.e. odd and even functions. Thus, we have an even solution for

$$C = 0, \qquad D = A$$

and an odd solution for

$$B = 0, \qquad D = -A$$

The continuity conditions for the even solutions are

$$A = Be^{a\kappa} \cos qa, \qquad \tan qa = \kappa/q$$

and those for the odd solutions are

$$A = -Ce^{a\kappa} \sin qa, \qquad \tan qa = -q/\kappa$$

The energy eigenvalue conditions can be expressed in a more transparent form if we introduce

$$\xi \equiv qa, \qquad \beta^2 \equiv 2m V_0 a^2/\hbar^2$$

Then we have $\kappa a = \sqrt{\beta^2 - \xi^2}$ and the conditions are written as

$$\tan \xi = \sqrt{\frac{\beta^2}{\xi^2} - 1}, \qquad \tan \xi = -\frac{\xi}{\sqrt{\beta^2 - \xi^2}}$$

for the even and the odd solutions respectively. Plotting both sides of each equation, we can easily see in Figs. 6 and 7 that if the parameter β is smaller than $\pi/2$ then only the even condition can have a solution.

(d) The even eigenstates have eigenvalues that are solutions of the equation

$$\tan \xi = \sqrt{\frac{\beta^2}{\xi^2} - 1}$$

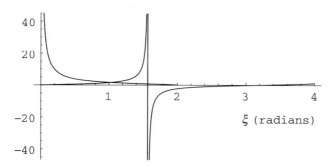

Fig. 6 Even solution with $\beta = 4$.

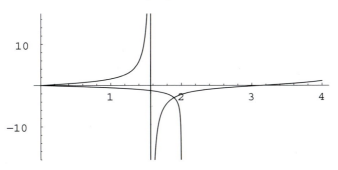

Fig. 7 Odd solution with $\beta = 4$.

Substituting the definitions of ξ and β given above and putting $V_0 = g^2/2a$, we get in the limit $a \to 0$

$$\xi = \sqrt{\frac{2m}{\hbar^2}\left(\frac{g^2a}{2} - |E|a^2\right)} \sim \sqrt{\frac{mg^2a}{\hbar^2}} \qquad \Longrightarrow \qquad \tan\xi \sim \xi \sim \sqrt{\frac{mg^2a}{\hbar^2}}$$

$$\sqrt{\frac{\beta^2}{\xi^2 - 1}} = \sqrt{\frac{|E|a^2}{g^2a/2 - |E|a^2}} \sim \sqrt{\frac{2a|E|}{g^2}}$$

The eigenvalue equation gives

$$|E| = \frac{mg^4}{2\hbar^2}$$

Taking the same limits in the odd bound-state eigenvalue equation

$$\tan\xi = -\frac{\xi}{\sqrt{\beta^2 - \xi^2}}$$

we get no solution. Thus, the above eigenvalue corresponds to the single bound state of the system in this limit.

Problem 3.11 Consider a one-dimensional infinite square well at the centre of which there is also a very-short-range attractive force, represented by a delta function potential. The potential energy of a particle moving in the well is

$$V(x) = \begin{cases} +\infty, & x < -L \\ -\dfrac{\hbar^2 g^2}{2m}\delta(x), & -L < x < L \\ +\infty, & x > L \end{cases}$$

(a) Find the positive energy eigenvalues $(E > 0)$.[3]

(b) Are there any negative energy eigenvalues $(E < 0)$?

(c) Consider the same problem with the sign of the coupling inverted, i.e. the case of a repulsive delta function in the centre of the infinite square well. What is the spectrum?

Solution

(a) The energy eigenfunctions and eigenvalues of the standard infinite square well, without the presence of the delta function potential, are known to be

$$\psi_n(x) = \frac{1}{2\sqrt{L}} \left[e^{in\pi x/2L} + (-1)^{n+1} e^{-in\pi x/2L} \right], \qquad E_n = \frac{\hbar^2 \pi^2 n^2}{8mL^2}$$

Schroedinger's equation for the complete problem reads, in the interval $-L < x < L$,

$$\psi''(x) + k^2 \psi(x) = -g^2 \psi(0)\delta(x)$$

where $E \equiv \hbar^2 k^2 / 2m$, the eigenfunctions $\psi(x)$ being subject to the boundary condition

$$\psi(-L) = \psi(L) = 0$$

Since, for the odd-parity eigenfunctions $\psi_{2n}(x)$ of the unperturbed problem we have $\psi_{2n}(0) = 0$, the delta function term will not contribute and we can conclude that all the infinite-square-well odd eigenfunctions (those for even n) will continue to be eigenfunctions of the problem at hand; thus

$$\psi_{2n} = \frac{i}{\sqrt{L}} \sin\frac{n\pi x}{L}, \qquad E_{2n} = \frac{\hbar^2 n^2 \pi^2}{2mL^2}$$

The sought-for even eigenfunctions of the present problem can always be expanded in terms of the even infinite-square-well eigenfunctions (those for odd n):

$$\psi_k(x) = \sum_{\nu=0}^{\infty} C_\nu \psi_{2\nu+1}(x)$$

Substituting in Schroedinger's equation, we obtain

$$\sum_{\nu=0}^{\infty} C_\nu \left[\frac{\pi^2 (2\nu+1)^2}{4L^2} - k^2 \right] \psi_{2\nu+1}(x) = g^2 \psi_k(0)\delta(x)$$

---

[3] You can use the mathematical formula

$$\sum_{n=0}^{\infty} \frac{1}{(2n+1)^2 - x^2} = \frac{\pi}{4x} \tan\frac{\pi x}{2}$$

Multiplying by $\psi^*_{2\nu'+1}(x)$, integrating over x, and using the orthonormality of the infinite-square-well eigenfunctions we get

$$C_{2\nu'+1} = \frac{g^2}{\pi^2(2\nu'+1)^2/4L^2 - k^2}\psi_k(0)\psi^*_{2\nu'+1}(0)$$

Going back to the expansion of $\psi_k(x)$, we obtain

$$\psi_k(x) = \psi_k(0)\,g^2 \sum_{\nu=0}^{\infty} \frac{\psi^*_{2\nu+1}(0)}{\pi^2(2\nu+1)^2/4L^2 - k^2}\psi_{2\nu+1}(x)$$

which can be true at the point $x = 0$ only if

$$\frac{1}{g^2} = \frac{4L^2}{\pi^2}\sum_{\nu=0}^{\infty} \frac{|\psi_{2\nu+1}(0)|^2}{(2\nu+1)^2 - (2Lk/\pi)^2}$$

Since we also know that $\psi_{2\nu+1}(0) = 1/\sqrt{L}$, we get

$$\frac{1}{g^2} = \frac{4L}{\pi^2}\sum_{\nu=0}^{\infty} \frac{1}{(2\nu+1)^2 - (2Lk/\pi)^2}$$

Summing the series, we arrive at

$$\frac{2}{g^2 L} = \frac{\tan kL}{kL}$$

There is an infinity of solutions $\tilde{E}_{2\nu+1}$ associated with the above equation:

$$\tilde{E}_{2\nu+1} \le E_{2\nu+1} = \frac{\hbar^2\pi^2(2\nu+1)^2}{8mL^2}$$

which coincide with $E_{2\nu+1}$ in the limit $g^2 \to 0$. These can be seen graphically if we plot $\tan x$ and the linear function $(2/g^2L)x$ on the same graph and look for points of intersection; see Fig. 8.

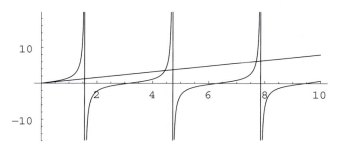

Fig. 8 Positive-energy eigenvalues for attractive coupling such that $g^2 L = 0.4$.

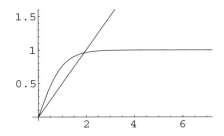

Fig. 9 Negative-energy eigenvalues for attractive coupling such that $g^2L = 5$.

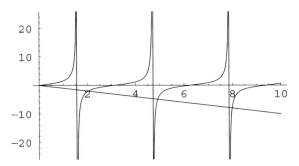

Fig. 10 Positive-energy eigenvalues for repulsive coupling such that $g^2L = 2$.

(b) Introducing $E = -\hbar^2\kappa^2/2m$, we write the candidate bound-state wave function as

$$\psi(x) = \begin{cases} A\,\sinh\kappa(x+L), & -L < x < 0 \\ -B\,\sinh\kappa(x-L), & 0 < x < L \end{cases}$$

The continuity and discontinuity conditions give

$$A = B, \qquad \tanh\kappa L = \frac{2}{g^2L}\,\kappa L$$

It is easily seen by plotting both sides of the latter equation that there is one solution provided that the coupling is strong enough, i.e. $g^2 > 2/L$. Thus, there can be only one even negative-energy solution; see Fig. 9.

(c) In this case there can be only positive-energy solutions. Obviously, the negative-parity (odd) states are still present since the delta function term in the potential is invisible to them. With regard to the positive-parity solutions, the discussion in (a) for the attractive delta function can be repeated in an analogous way. Thus we end up with the even-parity energy eigenvalue equation

$$\tan kL = -\frac{2}{g^2L}\,kL$$

Again plotting both sides of this equation (Fig. 10), we deduce that there is an infinity of solutions.

The corresponding energy eigenvalues are always larger than those we obtain with the delta function turned off, i.e.

$$\tilde{E}_{2v+1} \geq E_{2v+1} = \frac{\hbar^2 \pi^2 (2v+1)^2}{8mL^2}$$

Problem 3.12 Consider a potential consisting of two delta functions,

$$V(x) = \frac{\hbar^2 g_1}{2m} \delta(x+a) + \frac{\hbar^2 g_2}{2m} \delta(x-a)$$

(a) Find the bound-state spectrum in the case $g_1 = g_2 < 0$.
(b) Do the same in the case $g_1 = -g_2 > 0$.

Solution

(a) In the case of equal (attractive) couplings, i.e. $g_1 = g_2 \equiv -g^2 < 0$, the system is parity-even and we can exploit this symmetry to look for even and odd solutions.

An even candidate wave function is ($E = -\hbar^2 \kappa^2 / 2m$),

$$\psi_+(x) = \begin{cases} A e^{\kappa x}, & x < -a \\ B \cosh \kappa x, & -a < x < a \\ A e^{-\kappa x}, & x > a \end{cases}$$

The continuity and discontinuity conditions are

$$B = \frac{A e^{-\kappa a}}{\cosh \kappa a}$$

and

$$1 + \tanh \kappa a = \frac{g^2 a}{\kappa a}$$

We can read off easily from the plot of this equation (see Fig. 11) that there is always one (even) solution.

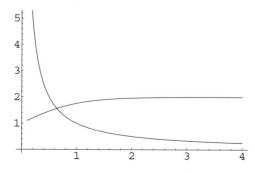

Fig. 11 Even solution for $g^2 a = 1$.

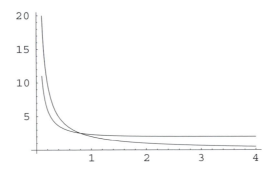

Fig. 12 Odd solution for $g^2a = 2$.

An odd candidate wave function is

$$\psi_-(x) = \begin{cases} -Ae^{\kappa x}, & x < -a \\ B\sinh \kappa x, & -a < x < a \\ Ae^{-\kappa x}, & x > a \end{cases}$$

The resulting conditions are

$$B = \frac{Ae^{-\kappa a}}{\sinh \kappa a}$$

and

$$1 + \coth \kappa a = \frac{g^2a}{\kappa a}$$

The graphical solution of the last condition is given in Fig. 12. It is clear that an odd solution does not always exist. In order to guarantee the presence of a bound state with odd-parity wave function we need a strong enough attractive coupling. It can be seen from the plot that $g^2 > 1/a$ is required.

(b) In the case of the asymmetric potential

$$V(x) = \frac{g^2\hbar^2}{2m} \left[\delta(x+a) - \delta(x-a)\right]$$

we can begin with the candidate bound-state wave function

$$\psi(x) = \begin{cases} Ae^{\kappa x}, & x < -a \\ Be^{\kappa x} + Ce^{-\kappa x}, & -a < x < a \\ De^{-\kappa x}, & x > a \end{cases}$$

Solving the continuity and discontinuity equation system, we get the energy eigenvalue condition

$$e^{4\kappa a} = \frac{g^4}{g^4 - 4\kappa^2}$$

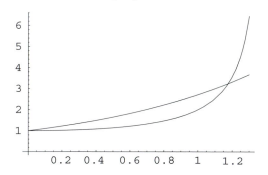

Fig. 13 Solution of the asymmetric case for $\lambda^2 = 2$.

which, in terms of $\xi \equiv 4ka$ and $\lambda^2 \equiv 4g^4a^2$, is

$$e^\xi = \frac{1}{1 - \xi^2/\lambda^2}$$

and can be solved graphically as in the previous cases to show that one bound state exists; see Fig. 13.

Problem 3.13 Consider the incidence of particles of energy $E > 0$ at a one-dimensional potential region consisting of two unequal delta functions:

$$V(x) = \frac{\hbar^2}{2m} \left[g_1 \delta(x + a) + g_2 \delta(x - a) \right]$$

(a) Calculate the transmission coefficient.
(b) Consider the case $g_1 = -g_2$ and show that there exist special values of the energy for which transmission is perfect and there is no reflection.
(c) Compare the low-energy behaviour ($E \to 0$) and high-energy behaviour ($E \to \infty$) in the three cases $g_1 = g_2$, $g_1 = -g_2$ and $g_1 \neq 0$, $g_2 = 0$.

Solution
(a) The wave function will be

$$\psi(x) = \begin{cases} e^{ikx} + Be^{-ikx}, & x < -a \\ Ce^{ikx} + De^{-ikx}, & -a < x < a \\ Fe^{ikx}, & x > a \end{cases}$$

The continuity conditions are

$$\lambda^{-1} + \lambda B = C\lambda^{-1} + D\lambda$$
$$F\lambda = C\lambda + D\lambda^{-1}$$

where we have defined

$$\lambda \equiv e^{ika}$$

The discontinuity conditions

$$\psi'(\pm a+) - \psi'(\pm a-) = g_{2,1}\psi(\pm a)$$

give

$$ikF\lambda - ikC\lambda + ikD\lambda^{-1} = g_2 F\lambda$$
$$ikC\lambda^{-1} - ikD\lambda - ik\lambda^{-1} + ikB\lambda = g_1(\lambda^{-1} + \lambda B)$$

From the second and third of the four continuity conditions we get

$$F = D\left(\frac{2ik}{g_2}\right)\lambda^{-2}, \qquad C = i\lambda^{-2}\left(\frac{2k + ig_2}{g_2}\right)D$$

Cancelling B from the other two equations gives

$$D = -2i\lambda^{-2}\frac{g_2 k}{g_1 g_2 + \lambda^{-4}(2k + ig_1)(2k + ig_2)}$$

and finally we obtain

$$F = \frac{4k^2}{\lambda^4 g_1 g_2 + (2k + ig_1)(2k + ig_2)}$$

The resulting transmission coefficient is

$$\mathcal{T} = |F|^2 = \left\{\left[1 - \frac{g_1 g_2}{4k^2}(1 - \cos 4ka)\right]^2 + \left(\frac{g_1 g_2}{4k^2}\sin 4ka + \frac{g_1 + g_2}{2k}\right)^2\right\}^{-1}$$

(b) In the case $g_1 = -g_2 \equiv g$, the expression simplifies to

$$\mathcal{T} = \left\{\left[1 + \frac{g^2}{4k^2}(1 - \cos 4ka)\right]^2 + \frac{g^4}{16k^4}\sin^2 4ka\right\}^{-1}$$

It is obvious that for $2ka = n\pi$ the transmission is perfect, i.e.

$$E_n = \frac{\hbar^2 n^2 \pi^2}{8ma^2} \quad (n = 1, 2, \ldots) \qquad \Longrightarrow \qquad \mathcal{T} = 1, \quad \mathcal{R} = 0$$

(c) The qualitative behaviour in all these three cases is the same. In detail, we have the following results.

For the case $g_1 = g_2 \equiv g$,

$$\lim_{k \to 0}\{\mathcal{T}\} \sim \frac{k^2}{g^2(1 + ag^2)}, \qquad \lim_{k \to \infty}\{\mathcal{T}\} \sim 1 - \frac{g^2}{k^2}\cos^2 2ka$$

For the case $g_1 = -g_2 \equiv g$,

$$\lim_{k \to 0}\{\mathcal{T}\} \sim \frac{k^2}{g^4 a^2}, \qquad \lim_{k \to \infty}\{\mathcal{T}\} \sim 1 - \frac{g^2}{k^2}\sin^2 2ka$$

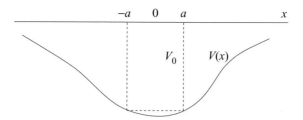

Fig. 14 Arbitrary one-dimensional well.

For the case $g_1 \equiv g \neq 0$, $g_2 = 0$,

$$\lim_{k \to 0} \{\mathcal{T}\} \sim \frac{4k^2}{g^2}, \qquad \lim_{k \to \infty} \{\mathcal{T}\} \sim 1 - \frac{g^2}{4k^2}$$

Problem 3.14 It is well known that the one-dimensional square well potential always has at least one negative-energy eigenstate. Consider an arbitrary one-dimensional potential that is always non-positive and bounded, i.e. $V(x) \leq 0$ and $|V(x)| < \infty$. You can prove that such a potential will always have at least one negative-energy eigenstate. In order to do this, first consider a square well $V_0(x) = -V_0\Theta(a - x)\Theta(x + a)$ inscribed in $V(x)$ (see Fig. 14). Then consider the ground state $\psi_0(x)$ of $V_0(x)$, with corresponding energy E_0. Prove that there is always a negative-energy eigenvalue $E \leq E_0$. *Hint: Take the matrix element of the Schroedinger operator between the state ψ_0 and the desired bound state $\psi_E(x)$.*

Solution

Let us consider the square-well potential $V_0(x)$ inscribed in $V(x)$. The time-independent Schroedinger equation for this square-well problem reads

$$-\frac{\hbar^2}{2m}\psi_0''(x) + V_0(x)\psi_0(x) = E_0\psi_0(x)$$

The corresponding eigenvalue equation for the potential $V(x)$ is

$$-\frac{\hbar^2}{2m}\psi_E''(x) + V(x)\psi_E(x) = E\psi_E(x)$$

Multiplying the first of these two equations by $\psi_E(x)$, the second by $\psi_0(x)$ and subtracting, we end up with

$$-\frac{\hbar^2}{2m}\left[\psi_E(x)\psi_0''(x) - \psi_0(x)\psi_E''(x)\right]$$
$$= -\left[V_0(x) - V(x)\right]\psi_0(x)\psi_E(x) + (E_0 - E)\psi_0(x)\psi_E(x)$$

Note that the left-hand side is proportional to the derivative

$$\left[\psi_E(x)\psi_0'(x) - \psi_0(x)\psi_E'(x)\right]'$$

Let us now integrate the equation over the whole interval $(-\infty, +\infty)$. The left-hand side will give a vanishing result and so we shall get

$$E_0 - E = \frac{\int_{-\infty}^{+\infty} dx[V_0(x) - V(x)]\psi_0(x)\psi_E(x)}{\int_{-\infty}^{+\infty} dx\, \psi_0(x)\psi_E(x)}$$

We have assumed implicitly that the candidate ground state $\psi_E(x)$ will not have any nodes, i.e. points where it vanishes. Thus the product $\psi_0(x)\psi_E(x)$ will be monotonic and will yield a non-vanishing integral by which it is legitimate to divide. Since, by construction, $V_0(x)$ is always larger than or equal to $V(x)$, the right-hand side of our last equation will be non-negative. Thus, we can conclude that

$$E \le E_0$$

and there will always be a negative-energy eigenvalue.

Problem 3.15 Consider the square well

$$V(x) = \begin{cases} 0, & |x| > a \\ -V_0, & |x| \le a \end{cases}$$

(a) What is the required value of the product $V_0 a^2$ for there to be four bound states?

(b) Consider a parabolic potential inscribed within the square well $V(x)$ (see Fig. 15):

$$\tilde{V}(x) = \begin{cases} 0, & |x| > a \\ V_0(x^2/a^2 - 1), & |x| \le a \end{cases}$$

Show that the parabolic potential will have at least one odd bound state and that it cannot have more than four bound states of either parity.

(c) Comment on the triangle potential

$$\overline{V} = \begin{cases} 0, & |x| > a \\ V_0(|x|/a - 1), & |x| \le a \end{cases}$$

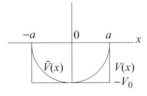

Fig. 15 Parabolic well $\tilde{V}$.

Solution

(a) The eigenvalue equations for the square well, written in terms of the variable $\xi = \sqrt{2ma^2(V_0 - |E|)/\hbar^2}$ and the area parameter $\beta^2 = 2ma^2V_0/\hbar^2$, are

$$\tan\xi = \sqrt{\frac{\beta^2}{\xi^2} - 1}, \qquad \tan\xi = -\frac{\xi}{\sqrt{\beta^2 - \xi^2}}$$

From these equations we can deduce graphically that we need β to be larger than $\pi/2$, $3\pi/2$, ... in order to have respectively two, three, etc. positive-parity bound states and one, two, etc. negative-parity bound states. Thus, for

$$\frac{3\pi}{2} < \beta < 2\pi \qquad \Longrightarrow \qquad \frac{9\hbar^2\pi^2}{8m} < V_0 a^2 < \frac{2\hbar^2\pi^2}{m}$$

we shall have four bound states.

(b) Since our potential is inscribed within a square well that has exactly four bound states, it is clear that we could not have more bound states than four. In order to find out the lowest number of allowed bound states in our potential, let us consider a second square well, characterized by depth V_1 and width $2a_1$, inscribed within the parabola. The fact that it is inscribed translates to the relation

$$\tilde{V}(a_1) = V_1 \qquad \Longrightarrow \qquad V_1 = V_0\left(\frac{a_1^2}{a^2} - 1\right)$$

Such a rectangular potential will have a maximum number of bound states if and only if its 'area' $V_1 a_1^2$ is maximal. Therefore, let us demand a maximum of the function $V_1(a_1)a_1^2$, i.e.

$$\frac{d}{da_1^2}[V_1(a_1)a_1^2] = V_0\frac{d}{da_1^2}\left(\frac{a_1^4}{a^2} - a_1^2\right) = V_0\left(\frac{2a_1^2}{a^2} - 1\right) = 0$$

or

$$a_1^2 = \frac{a^2}{2} \qquad \Longrightarrow \qquad \{a_1^2 V_1\}_{\max} = \frac{a^2 V_0}{4}$$

Thus, the corresponding parameter will be

$$\beta_1 \equiv \sqrt{\frac{2ma_1^2|V_1|}{\hbar^2}} = \frac{\beta}{2} \qquad \Longrightarrow \qquad \frac{3\pi}{4} < \beta_1 < \pi$$

and will correspond to two bound states. Thus, we can conclude that the lowest number of bound states that our parabolic potential can have is two: obviously, one is even (the ground state) and the other is odd.

(c) Applying exactly the same method as above to the triangular potential $\overline{V}$ (see Fig. 16), we conclude that the inscribed square well of maximal area has

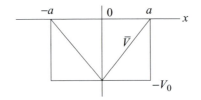

Fig. 16 Triangular well.

$a_1 = 2a/3$ and $(a_1^2|V_1|)_{\text{max}} = 4V_0a^2/27$. This corresponds to the area parameter value $\beta_1 = 2\beta/3\sqrt{3}$, so that

$$\frac{\pi}{\sqrt{3}} < \beta_1 < \frac{4\pi}{3\sqrt{3}}$$

For this range, again we can have two bound states, one even (the ground state) and the other odd.

Problem 3.16 A particle of mass m is bound in an attractive delta function potential

$$V(x) = -\frac{\hbar^2 g^2}{2m}\delta(x)$$

Its bound-state wave function is $\psi_0(x) = \sqrt{\kappa}\,e^{-\kappa|x|}$ and its energy is $E_0 = -\hbar^2\kappa^2/2m$ with $\kappa = g^2/2$. Assume that, instantaneously, a force acts on the particle, imparting momentum p_0 to it. Calculate the probability of finding the particle in its original state.

Solution

The initial state of the particle is characterized by vanishing momentum, i.e. $\langle\psi_0|p|\psi_0\rangle = 0$. If we act on the wave function with the operator $e^{-ip_0x/\hbar}$, which in momentum space clearly corresponds to a momentum-translation operator $e^{p_0\partial/\partial p}$, we obtain a state with wave function

$$\psi_{p_0}(x) = e^{ip_0x/\hbar}\,\psi_0(x)$$

This state has momentum expectation value

$$\langle\psi_{p_0}|p|\psi_{p_0}\rangle = \int_{-\infty}^{+\infty} dx\,\psi_0(x)\left[-i\hbar\psi_0'(x) + p_0\psi_0(x)\right] = p_0$$

The wave function of the system can be expanded in energy eigenfunctions of the unperturbed system as

$$\Psi(x, t) = C_0\psi_0(x)\,e^{-iE_0t/\hbar} + \int_0^\infty dE\,C(E)\psi_E(x)e^{-iEt/\hbar}$$

where the coefficients are the probability amplitudes for finding the particle in the corresponding states. Thus, the probability of finding the particle in the ground state immediately after we impart momentum p_0 to it is

$$\mathcal{P} = |C_0|^2 = \left| \int_{-\infty}^{+\infty} dx\, \psi_0(x)\psi_{p_0}(x) \right|^2 = \left| \int_{-\infty}^{+\infty} dx\, \psi_0^2(x)e^{ip_0x/\hbar} \right|^2$$

$$= \left| \frac{2\kappa}{2\kappa + ip_0/\hbar} \right|^2 = \frac{g^4\hbar^2}{p_0^2 + g^4\hbar^2}$$

Problem 3.17 Consider a particle moving in the three-dimensional attractive potential

$$V(x, y, z) = -\frac{\hbar^2}{2m} [\lambda_1\, \delta(x) + \lambda_2\, \delta(y) + \lambda_3\, \delta(z)]$$

(a) Find the energy and wave function of any existing bound state ($E < 0$).
(b) For the above state, compute the spatial and momentum uncertainties $(\Delta x)^2$, $(\Delta p)^2$ and check Heisenberg's inequality.

Solution
(a) The problem is separable. By setting

$$\psi(x, y, z) = X(x)Y(y)Z(z),$$

we obtain

$$\frac{X''(x) + \lambda_1 X(0)\delta(x)}{X(x)} + \frac{Y''(y) + \lambda_2 Y(0)\delta(y)}{Y(y)}$$

$$+ \frac{Z''(z) + \lambda_3 Z(0)\delta(z)}{Z(z)} = -\frac{2mE}{\hbar^2} = \kappa_1^2 + \kappa_2^2 + \kappa_3^2$$

which separates into three independent equations with solutions as shown:

$$X''(x) + \lambda_1 X(0)\delta(x) = \kappa_1^2 X(x) \quad \Longrightarrow \quad X(x) = X(0)e^{-\kappa_1|x|}$$
$$Y''(y) + \lambda_2 Y(0)\delta(y) = \kappa_2^2 Y(y) \quad \Longrightarrow \quad Y(y) = Y(0)e^{-\kappa_2|y|}$$
$$Z''(z) + \lambda_3 Z(0)\delta(z) = \kappa_3^2 Z(z) \quad \Longrightarrow \quad Z(z) = Z(0)e^{-\kappa_3|z|}$$

The discontinuity conditions at the origin give

$$X'(0^+) - X'(0^-) = -\lambda_1 X(0) \quad \Longrightarrow \quad \kappa_1 = \frac{\lambda_1}{2}$$

and, similarly,

$$\kappa_2 = \frac{\lambda_2}{2}, \qquad \kappa_3 = \frac{\lambda_3}{2}$$

Together, these relations lead to

$$\psi(x, y, z) = \left(\frac{\lambda_1 \lambda_2 \lambda_3}{8}\right)^{1/2} \exp\left(-\frac{\lambda_1}{2}|x| - \frac{\lambda_2}{2}|y| - \frac{\lambda_3}{2}|z|\right)$$

and

$$E = -\frac{\hbar^2}{8m}\left(\lambda_1^2 + \lambda_2^2 + \lambda_3^2\right)$$

(b) The expectation values $\langle x \rangle$, $\langle y \rangle$, $\langle z \rangle$ and $\langle p_x \rangle$, $\langle p_y \rangle$, $\langle p_z \rangle$ vanish owing to parity. Thus

$$(\Delta \mathbf{x})^2 = \langle x^2 + y^2 + z^2 \rangle$$

and

$$(\Delta \mathbf{p})^2 = \langle p_x^2 + p_y^2 + p_z^2 \rangle$$

We can easily compute

$$\langle \psi | x^2 | \psi \rangle = \langle X | x^2 | X \rangle \langle YZ | YZ \rangle = \frac{\lambda_1}{2} \int_{-\infty}^{+\infty} dx \, x^2 \, e^{-2\kappa_1 |x|}$$

$$= \frac{\lambda_1}{4} \frac{\partial^2}{\partial \kappa_1^2}\left(\int_0^\infty dx \, e^{-2\kappa_1 x}\right) = \frac{\lambda_1}{4} \frac{\partial^2}{\partial \kappa_1^2}\left(\frac{1}{2\kappa_1}\right) = \frac{\lambda_1}{4\kappa_1^3} = \frac{2}{\lambda_1^2}$$

Thus, finally,

$$(\Delta \mathbf{x})^2 = 2\left(\frac{1}{\lambda_1^2} + \frac{1}{\lambda_2^2} + \frac{1}{\lambda_3^2}\right)$$

In an analogous fashion, we get

$$\langle p_x^2 \rangle = \langle X | p_x^2 | X \rangle = -\frac{\lambda_1 \hbar^2}{2} \int_{-\infty}^{+\infty} dx \, e^{-\kappa_1 |x|} \frac{\partial^2}{\partial x^2} e^{-\kappa_1 |x|}$$

$$= -\frac{\lambda_1 \hbar^2}{2} \int_{-\infty}^{+\infty} dx \, e^{-\kappa_1 |x|} \frac{\partial}{\partial x}\left[-\kappa_1 \, \text{sign}(x) \, e^{-\kappa_1 |x|}\right]$$

$$= -\frac{\lambda_1 \hbar^2}{2} \int_{-\infty}^{+\infty} dx \, e^{-\kappa_1 |x|}\left[-2\kappa_1 \, \delta(x) + \kappa_1^2 \, e^{-\kappa_1 |x|}\right] = \frac{\lambda_1^2 \hbar^2}{4}$$

and, finally,

$$(\Delta \mathbf{p})^2 = \frac{\hbar^2}{4}\left(\lambda_1^2 + \lambda_2^2 + \lambda_3^2\right)$$

The uncertainty product gives

$$(\Delta \mathbf{x})^2 (\Delta \mathbf{p})^2 = \frac{\hbar^2}{2} \left(\lambda_1^2 + \lambda_2^2 + \lambda_3^2 \right) \left(\frac{1}{\lambda_1^2} + \frac{1}{\lambda_2^2} + \frac{1}{\lambda_3^2} \right)$$

which is always larger than $\hbar^2/4$, as required by Heisenberg's inequality.

Problem 3.18 Consider a three-dimensional square-well potential,

$$V(x, y, z) = -V_0 \, \Theta(a - |x|) \, \Theta(a - |y|) \, \Theta(a - |z|)$$

Find the eigenfunctions and energy eigenvalues corresponding to bound states ($E < 0$).

Solution
Introducing

$$E = -\frac{\hbar^2}{2m} \left(\kappa_x^2 + \kappa_y^2 + \kappa_z^2 \right), \qquad E + V_0 = \frac{\hbar^2}{2m} \left(q_x^2 + q_y^2 + q_z^2 \right)$$

we obtain the trial eigenfunctions

$$\psi(x, y, y) = X(x)Y(y)Z(z)$$

where each factor can be even, so that

$$X(x) = \Theta \left(a - |x| \right) A \, \cos q_x \, x + \Theta \left(|x| - a \right) B e^{-\kappa_x |x|}$$

or odd, so that

$$X(x) = \Theta \left(a - |x| \right) \overline{A} \, \sin q_x \, x + \Theta \left(|x| - a \right) \operatorname{sign}(x) \, \overline{B} e^{-\kappa_x |x|}$$

where A, B, $\overline{A}$, $\overline{B}$ are coefficients.

In an analogous fashion, replacing x by y and then z, we obtain the other wave function factors.

Continuity at $x = a$ gives

$$A \, \cos q_x a = B e^{-\kappa_x a}, \qquad - A q_x \, \sin q_x a = -B \kappa_x e^{-\kappa_x a}$$

for the even eigenfunctions and

$$\overline{A} \, \sin q_x a = \overline{B} e^{-\kappa_x a}, \qquad \overline{A} q_x \, \cos q_x a = -\overline{B} \kappa_x e^{-\kappa_x a}$$

for the odd eigenfunctions. Summarizing the energy eigenvalue equations, we have

$$\tan a q_i = \begin{cases} \kappa_i / q_i & \text{(even)} \\ -q_i / \kappa_i & \text{(odd)} \end{cases}$$

for $i = x, y, z$.

If we denote by

$$\varepsilon_1^{(+)}, \; \varepsilon_2^{(-)}, \; \varepsilon_3^{(+)}, \ldots$$

the energy values corresponding to the solutions of each of the above eigenvalue equations (we denote the corresponding parity of their wave functions by a superscript plus or minus), we shall have the following energy eigenstates and eigenvalues:

$$\psi_{111}^{(+++)} \;; \quad E_{111} = 3\varepsilon_1^{(+)}$$

$$\psi_{211}^{(-++)}, \; \psi_{121}^{(+-+)}, \; \psi_{112}^{(++-)} \;; \quad E_{211} = \varepsilon_2^{(-)} + 2\varepsilon_1^{(+)} = E_{121} = E_{211}$$

$$\psi_{221}^{(--+)}, \; \psi_{212}^{(-+-)}, \; \psi_{122}^{(+--)} \;; \quad E_{221} = 2\varepsilon_2^{(-)} + \varepsilon_1^{(+)} = E_{212} = E_{122}$$

$$\psi_{222}^{(---)} \;; \quad E_{222} = 3\varepsilon_2^{(-)}$$

and so on.

Problem 3.19 Consider a particle of mass m under the influence of an attractive delta function potential

$$V(x) = -\frac{\hbar^2 g^2}{2m}\delta(x)$$

The system is in its ground state, with energy $E_0 = -\hbar^2 g^4/8m$ and wave function[4] $\sqrt{g^2/2}\, e^{-g^2|x|/2}$. Since the only time dependence of this wave function comes from the phase factor $e^{-iEt/\hbar}$, the characteristic time of the system is $\tau \sim 8m/\hbar g^4$.

(a) In a time interval much shorter than the characteristic time of the system, the strength of the potential increases to a much larger value $\bar{g}^2 \gg g^2$, while the particle stays in the same state. What is the new ground state of the system? What are the new scattering states?

(b) What is the probability of finding the system in the (new) ground state?

(c) What is the probability density for finding the particle in an eigenstate of energy $E > 0$? What is the probability of finding the particle in any scattering state? What is its relation to the probability of part (b)?

(d) What is the energy required for the increase in strength of the potential?

Solution

(a) According to the assumptions, after the sudden increase in strength of the potential the wave function of the system will be the same at the instant after the

---

[4] The positive-energy eigenfunctions of the system are

$$\psi_k^{(\pm)}(x) = \frac{1}{\sqrt{2\pi}}\left(e^{\pm ikx} - \frac{g^2}{g^2 + 2ik}e^{ik|x|}\right)$$

with $k = \sqrt{2mE/\hbar^2} > 0$.

change.[5] It can be expanded in terms of the complete set of new energy eigenstates, given by

$$\overline{\psi}_0(x) = \sqrt{\frac{\overline{g}^2}{2}}\, e^{-\overline{g}^2|x|/2} \qquad \left(\overline{E}_0 = -\frac{\hbar^2 \overline{g}^4}{8m}\right)$$

$$\overline{\psi}_k^{(\pm)}(x) = \frac{1}{\sqrt{2\pi}}\left(e^{\pm ikx} - \frac{\overline{g}^2}{\overline{g}^2 + 2ik}\, e^{ik|x|}\right)$$

Thus, the wave function of the system will be

$$\Psi(x,t) = C_0 \overline{\psi}_0(x)\, e^{-i\overline{E}_0 t/\hbar} + \int_0^\infty dk\, C^{(+)}(k)\, e^{-iEt/\hbar}\, \overline{\psi}_k^{(+)}(x)$$
$$+ \int_0^\infty dk\, C^{(-)}(k)\, e^{-iEt/\hbar}\, \overline{\psi}_k^{(-)}(x)$$

At the instant after the change ($t = 0$) we have

$$\Psi(x,0) = \psi_0(x) = \sqrt{\frac{g^2}{2}}\, e^{-g^2|x|/2}$$
$$= C_0 \overline{\psi}_0(x) + \int_0^\infty dk\, C^{(+)}(k)\, \overline{\psi}_k^{(+)}(x) + \int_0^\infty dk\, C^{(-)}(k)\, \overline{\psi}_k^{(-)}(x)$$

(b) The probability of finding the system in the new ground state will be

$$P_0 = |C_0|^2 = \left|\int_{-\infty}^{+\infty} dx\, \psi_0(x)\overline{\psi}_0(x)\right|^2 = \frac{4g^2\overline{g}^2}{(g^2 + \overline{g}^2)^2}$$

(c) The probability density for the particle to be in a scattering eigenstate will be

$$\mathcal{P}_\pm(k) = |C^{(\pm)}(k)|^2$$

with

$$C^{(\pm)}(k) = \int_{-\infty}^{+\infty} dx\, \psi_0(x)\overline{\psi}_k^{(\pm)}(x)$$
$$= \frac{g}{2\sqrt{\pi}}\int_{-\infty}^{+\infty} dx\, e^{-g^2|x|/2}\left(e^{\pm ikx} - \frac{\overline{g}^2}{\overline{g}^2 + 2ik}\, e^{ik|x|}\right)$$
$$= \frac{g}{\sqrt{2}}\int_{-\infty}^{+\infty} dx\, e^{-g^2|x|/2}\left[\psi_k^{(\pm)}(x) + \frac{1}{\sqrt{2\pi}}\left(\frac{g^2}{g^2 + 2ik} - \frac{\overline{g}^2}{\overline{g}^2 + 2ik}\right)e^{ik|x|}\right]$$
$$= \frac{ikg}{\sqrt{\pi}}\frac{g^2 - \overline{g}^2}{(g^2 + 2ik)(\overline{g}^2 + 2ik)}\int_{-\infty}^{+\infty} dx\, e^{(ik - g^2/2)|x|}$$

---

[5] The energy of the system will change due to the change in potential, however.

The final integration gives

$$C^{(\pm)}(k) = \frac{4ikg}{\sqrt{\pi}} \frac{g^2 - \bar{g}^2}{(g^4 + 4k^2)(\bar{g}^2 + 2ik)}$$

and

$$\mathcal{P}_\pm(k) = \frac{16g^2 k^2}{\pi} \frac{(g^2 - \bar{g}^2)^2}{(g^4 + 4k^2)^2(\bar{g}^4 + 4k^2)}$$

The probability of finding the particle in any scattering state is[6]

$$P_\pm = \int_0^\infty dk\, \mathcal{P}_\pm(k) = \frac{1}{2}\left(\frac{g^2 - \bar{g}^2}{g^2 + \bar{g}^2}\right)^2$$

As expected,

$$P_0 + P_+ + P_- = 1$$

(d) The expectation value of the energy is

$$
\begin{aligned}
\langle H \rangle &= \frac{\hbar^2 g^2}{4m} \int_{-\infty}^{+\infty} dx\, e^{-g^2|x|/2} \left(-\frac{d^2}{dx^2} - \bar{g}^2 \delta(x)\right) e^{-g^2|x|/2} \\
&= \frac{\hbar^2 g^2}{4m} \int_{-\infty}^{+\infty} dx\, \left[\frac{g^4}{4}(|x|')^2 - \frac{g^2}{2}(|x|)'' + \bar{g}^2 \delta(x)\right] e^{-g^2|x|} \\
&= \frac{\hbar^2 g^2}{4m} \int_{-\infty}^{+\infty} dx\, \left[\frac{g^4}{4} - g^2 \delta(x) + \bar{g}^2 \delta(x)\right] e^{-g^2|x|}
\end{aligned}
$$

Finally, we can write

$$\langle H \rangle = E_0 \left(2\frac{\bar{g}^2}{g^2} - 1\right)$$

The difference in energy will be

$$\Delta W = E_0 - \langle H \rangle = -2E_0\left(\frac{\bar{g}^2}{g^2} - 1\right) = 2|E_0|\left(\frac{\bar{g}^2}{g^2} - 1\right)$$

6

$$
\begin{aligned}
P_\pm &= -\frac{16g^2}{\pi}(g^2 - \bar{g}^2)\frac{\partial}{\partial g^4} \int_0^\infty dk\, k^2 \left[\frac{1}{(\bar{g}^4 - g^4)}\left(\frac{1}{g^4 + 4k^2} - \frac{1}{\bar{g}^4 + 4k^2}\right)\right] \\
&= -\frac{2}{\pi}(g^2 - \bar{g}^2)^2 \frac{\partial}{\partial g^2} \int_0^\infty dk\, \frac{1}{(\bar{g}^4 - g^4)}\left(-\frac{g^4}{g^4 + 4k^2} + \frac{\bar{g}^4}{\bar{g}^4 + 4k^2}\right)
\end{aligned}
$$

Using the integral $\int_0^\infty dx\,(x^2 + a^2)^{-1} = \pi/2a$, we finally get

$$P_\pm = \frac{1}{2}\left(\frac{g^2 - \bar{g}^2}{g^2 + \bar{g}^2}\right)^2$$

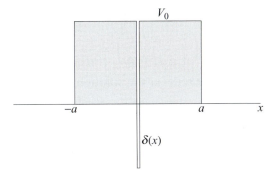

Fig. 17 Square barrier with an attractive delta function core.

Problem 3.20 A repulsive short-range potential with a strongly attractive core can be approximated by a square barrier with a delta function at its centre (Fig. 17), namely

$$V(x) = -\frac{\hbar^2 g^2}{2m}\delta(x) + V_0\Theta(a - |x|)$$

Show that there is a negative-energy eigenstate (the ground state). If E_0 is the ground-state energy of the delta function potential in the absence of the positive potential barrier, the ground-state energy of the present system obeys

$$E \geq E_0 + V_0$$

What is the particular value of V_0 for which we have the limiting case of a ground state with vanishing energy?

Solution

Let us define

$$\kappa^2 \equiv \frac{2m}{\hbar^2}|E|, \qquad q^2 = \frac{2m}{\hbar^2}(|E| + V_0), \qquad \beta^2 \equiv \frac{2m}{\hbar^2}V_0$$

The Schroedinger equation is

$$\psi'' = \kappa^2\psi \qquad (|x| > a)$$
$$\psi'' = q^2\psi \qquad (|x| < a)$$

The discontinuity at the origin gives

$$\psi'(+0) - \psi'(-0) = -g^2\psi(0)$$

Odd-parity solutions do not see the attractive delta function potential and, thus, cannot exist for $E < 0$. Even-parity solutions of the above equations have the form

$$\psi(x) = \begin{cases} Ae^{-\kappa|x|}, & |x| > a \\ Be^{q|x|} + Ce^{-q|x|}, & |x| < a \end{cases}$$

Continuity at a and 0 leads to the condition

$$e^{2qa}\left(\frac{1 - g^2/2q}{1 + g^2/2q}\right) = \frac{q - \kappa}{q + \kappa}$$

In the case of vanishing V_0, we recover the solution

$$E_0 = -\frac{\hbar^2}{2m}\left(\frac{g^2}{2}\right)^2$$

Since the right-hand side of the eigenvalue equation is always positive, we have necessarily

$$1 - \frac{g^2}{2q} > 0 \quad \Longrightarrow \quad \frac{2m}{\hbar^2}(-E + V_0) \geq \frac{g^4}{4}$$

or

$$E \leq V_0 - \frac{\hbar^2}{2m}\left(\frac{g^2}{2}\right)^2 = V_0 + E_0$$

We can see graphically that the above eigenvalue equation has only one solution, by defining

$$\xi \equiv qa, \qquad \lambda \equiv \frac{g^2 a}{2}, \qquad b \equiv \beta a$$

Then, we have

$$e^{2\xi}\left(\frac{\xi - \lambda}{\xi + \lambda}\right) = \frac{\xi - \sqrt{\xi^2 - b^2}}{\xi + \sqrt{\xi^2 - b^2}}$$

The solution exists provided that $\lambda \geq b$. In the limiting case $\lambda = b$, or, equivalently, $\beta = \sqrt{2m V_0/\hbar^2} = g^2/2$, we get a vanishing ground-state energy.

Problem 3.21 Consider a particle of mass m bound in the linear potential $V(x) = \lambda^2|x|$; see Fig. 18.

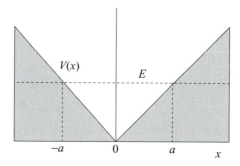

Fig. 18 Linear potential.

(a) Using dimensional analysis derive the dependence of the energy eigenvalues on the parameters of the system.

(b) Show the validity of the following approximate solution of the Schroedinger equation:

$$\psi(x) = \begin{cases} C\kappa^{-1/2} \exp\left[\int_x^a dx\, \kappa(x)\right], & x \gg a \\ C'k^{-1/2} \cos\left[\int_x^a dx\, k(x) - \pi/4\right], & x \ll a \end{cases}$$

where $k^2(x) = 2m[E - V(x)]/\hbar^2$, $\kappa^2(x) = 2m[V(x) - E]/\hbar^2$ and a is the *turning point* defined by $E = V(a)$. At what distance $L \gg a$ from the turning point does this solution become a good approximation?

(c) Find the energy eigenvalues.

Solution

(a) Since the only parameters that appear in the Schroedinger equation are λ^2 and $\hbar^2/m$, we must necessarily have

$$E = \left(\frac{\hbar^2}{m}\right)^\beta (\lambda^2)^\gamma \epsilon(n)$$

where β and γ are exponents to be determined and $\epsilon(n)$ is a dimensionless function of the quantum number n that determines the energy spectrum. Substituting the dimensions of the quantities involved (note that $[\lambda] = M^{1/2}L^{1/2}T^{-1}$), we obtain $\beta = 1/3$ and $\gamma = 2/3$. Thus

$$E = \left(\frac{\hbar^2}{m}\right)^{1/3} (\lambda^2)^{2/3} \epsilon(n)$$

(b) Let us consider the $x > a$ branch. Substituting into the Schroedinger equation, we obtain

$$\kappa^{3/2} - \kappa^{3/2} + \tfrac{1}{2}\kappa^{-1/2}\kappa' - \tfrac{1}{2}\kappa^{-1/2}\kappa' + \left[\tfrac{3}{4}\kappa^{-5/2}(\kappa')^2 - \tfrac{1}{2}\kappa^{-3/2}\kappa''\right] = 0$$

Thus, the above wave function is a valid approximate solution provided that the terms in the brackets are negligible. This amounts to

$$\kappa' \ll \kappa^2 \qquad \Longrightarrow \qquad (V - E)^{3/2} \gg \frac{\lambda^2}{2}\sqrt{\frac{\hbar^2}{2m}}$$

Setting $V(x) - E = \lambda^2(x - a) \equiv \lambda^2 L$, we obtain

$$L \gg \left(\frac{\hbar^2}{m\lambda^2}\right)^{1/3}$$

as expected on dimensional grounds.

(c) The wave function $\psi(x)$ should be odd or even under space reflection ($a \to -a$). Thus, we should have

$$\cos\left[\int_x^a dx\, k(x) - \frac{\pi}{4}\right] = \pm\cos\left[\int_{-a}^x dx\, k(x) - \frac{\pi}{4}\right]$$

or

$$\cos\left[-\int_x^a dx\, k(x) - \frac{\pi}{4}\right] = \cos\left[\int_x^a dx\, k(x) + \frac{\pi}{4}\right]$$

$$= \pm\cos\left[\int_{-a}^x dx\, k(x) + \frac{\pi}{4}\right]$$

This implies that

$$\int_x^a dx\, k(x) + \frac{\pi}{4} = -\int_{-a}^x dx\, k(x) - \frac{\pi}{4} + (n+1)\pi$$

with n a non-zero integer, or

$$\int_{-a}^a dx\, k(x) = \pi\left(n + \frac{1}{2}\right)$$

Calculating the integral, we have

$$\int_{-a}^a dx\, k(x) = 2\sqrt{\frac{2m\lambda^2}{\hbar^2}} \int_0^a dx\, (a-x)^{1/2} = \frac{4}{3}a^{3/2}\sqrt{\frac{2m\lambda^2}{\hbar^2}}$$

$$= \frac{4}{3}\left(\frac{E}{\lambda^2}\right)^{3/2}\sqrt{\frac{2m\lambda^2}{\hbar^2}}$$

we finally get

$$E_n = (\lambda^2)^{2/3}\left(\frac{\hbar^2}{m}\right)^{1/3}\left[\frac{3\pi}{4\sqrt{2}}\left(n + \frac{1}{2}\right)\right]^{2/3}$$

Problem 3.22 An electron moves under the influence of an electric field defined by

$$\mathcal{E} = \begin{cases} \mathcal{E}\hat{\mathbf{x}}, & x > a, \quad \forall y, z \\ 0, & -a < x < a, \quad \forall y, z \\ -\mathcal{E}\hat{\mathbf{x}}, & x < -a, \quad \forall y, z \end{cases}$$

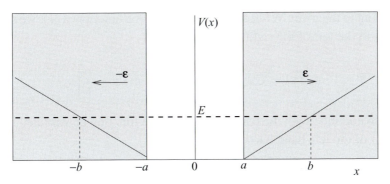

Fig. 19 Uniform electric field.

(see Fig. 19). In the WKB approximation the x-dependent part of its energy eigenfunctions in the $x > 0$ region is of the form

$$\psi(x) = \begin{cases} C\kappa^{-1/2} \exp\left[\int_x^b dx\,\kappa(x)\right], & x \gg b \\ C'k^{-1/2} \cos\left[\int_x^b dx\,k(x) - \pi/4\right], & x \ll b \end{cases}$$

where again $k^2(x) = 2m[E - V(x)]/\hbar^2$, $\kappa^2(x) = 2m[V(x) - E]/\hbar^2$ and b is the *turning point* defined by $E = V(b)$. Find the energy eigenvalues in this approximation.

Solution

The potential corresponding to the given electric field will be

$$V(x) = \begin{cases} -e\mathcal{E}x, & x > a \\ 0, & -a < x < a \\ e\mathcal{E}x, & x < -a \end{cases}$$

The Schroedinger equation can be reduced to one-dimensional equations by the separation of variables. The energy eigenfunctions are

$$\Psi(\mathbf{r}) = \frac{1}{2\pi} e^{ik_y y}\, e^{ik_z z} \psi(x)$$

where $\psi(x)$ solves the above one-dimensional potential with energy eigenvalues E. The total energy eigenvalues are

$$E_{\text{tot}} = E + \frac{\hbar^2 k_y^2}{2m} + \frac{\hbar^2 k_z^2}{2m}$$

According to the given WKB formula for the wave function $\psi(x)$, we shall have, for a point to the left of the right-hand barrier,

$$\psi_<(x) = C'k^{-1/2} \cos\left[\int_x^b dx\, k(x) - \frac{\pi}{4}\right]$$

This should coincide with the corresponding wave function constructed for a point to the right of the left-hand barrier, namely

$$\psi_>(x) = C'k^{-1/2} \cos\left[\int_{-b}^x dx\, k(x) - \frac{\pi}{4}\right] = C'k^{-1/2} \cos\left[\int_x^{-b} dx\, k(x) + \frac{\pi}{4}\right]$$

Thus, we must have

$$\psi_>(x) = \pm\psi_<(x) \qquad \Longrightarrow \qquad \int_{-b}^b dx\, k(x) - \frac{\pi}{2} = n\pi \quad (n = 1, 2, \ldots)$$

This is the WKB quantization condition. It gives

$$a + \int_a^b dx\, \sqrt{1 - \frac{x}{b}} = a + \frac{2b}{3}\left(1 - \frac{a}{b}\right)^{3/2} = \frac{\pi}{2\sqrt{E}}\left(n + \frac{1}{2}\right)\sqrt{\frac{\hbar^2}{2m}}$$

with

$$b = \frac{E}{|e|\mathcal{E}}$$

It is convenient to introduce the characteristic length scale of the potential,

$$L \equiv \left(\frac{\hbar^2}{m|e|\mathcal{E}}\right)^{1/3}$$

and the dimensionless energy variable

$$\xi = \sqrt{\frac{E}{a|e|\mathcal{E}}} - \sqrt{\frac{b}{a}}$$

Then the eigenvalue equation becomes

$$\xi + \frac{2\xi^3}{3}\left(1 - \frac{1}{\xi^2}\right)^{3/2} = \frac{\pi}{2\sqrt{2}}\left(n + \frac{1}{2}\right)\left(\frac{L}{a}\right)^{3/2}$$

Note that $\xi > 1$.

For very large energies or, equivalently, for $\xi \gg 1$, we get

$$\xi_0 = n^{1/3} \left(\frac{3\pi}{4\sqrt{2}}\right)^{1/3} \left(\frac{L}{a}\right)^{1/2}$$

and

$$E_n^{(0)} = n^{2/3} \left(\frac{9\pi^2 \hbar^2 e^2 \mathcal{E}^2}{32m}\right)^{1/3}$$

This is independent of a and coincides with the WKB result for the potential $V = -e\mathcal{E}|x|$.

In order to go beyond this approximation, we can expand the above equation in terms of ξ^{-2}. We obtain

$$\xi + \frac{2\xi^3}{3}\left(1 - \frac{3}{2\xi^2} + \frac{3}{8\xi^4} + \cdots\right) = \frac{\pi n}{2\sqrt{2}}\left(\frac{L}{a}\right)^{3/2}$$

or, keeping the next to leading term,

$$\frac{2\xi^3}{3} + \frac{1}{4\xi} = \frac{\pi n}{2\sqrt{2}}\left(\frac{L}{a}\right)^{3/2}$$

Now, we can substitute the trial solution

$$\xi = \xi_0 + \xi_1$$

and obtain

$$\frac{2\xi_0^3}{3}\left(1 + \frac{3\xi_1}{\xi_0} + \cdots\right) + \frac{1}{4\xi_0}\left(1 - \frac{\xi_1}{\xi_0} + \cdots\right) = \frac{\pi n}{2\sqrt{2}}\left(\frac{L}{a}\right)^{3/2} = \frac{2\xi_0^3}{3}$$

or

$$\xi_1 \approx -\frac{\xi_0^{-3}}{8}$$

Thus finally we get

$$E_n \approx n^{2/3}\left(\frac{9\hbar^2 \pi^2 e^2 \mathcal{E}^2}{32m}\right)^{1/3}\left[1 - \frac{1}{4n^{4/3}}\left(\frac{4\sqrt{2}}{3\pi}\right)^{4/3}\left(\frac{a}{L}\right)^2\right]$$

Problem 3.23 A particle in a one-dimensional crystal is free to move in the region $-a < x < a$ but it is subject in harmonic forces beyond this range. The potential energy of the particle can be stated in the form

$$V(x) = \begin{cases} \frac{1}{2}m\omega^2(x-a)^2, & x > a \\ 0, & -a < x < a \\ \frac{1}{2}m\omega^2(x+a)^2, & x < -a \end{cases}$$

where m is the mass of the particle and $m\omega^2$ is the spring constant. Find the approximate energy eigenvalues in the WKB approximation. How does the result behave in the two opposite limits of very small or very large a?

Solution

The WKB eigenvalue condition is

$$\int_{-b}^{b} dx \sqrt{\frac{2m}{\hbar^2}[E - V(x)]} = \pi\left(n + \frac{1}{2}\right)$$

The energy E is related to the turning point b by $E = \frac{1}{2}m\omega^2(b-a)^2$. The above integral becomes

$$\int_{-a}^{a} dx \sqrt{\frac{2m}{\hbar^2}E} + \int_{-b}^{-a} dx \sqrt{\frac{2m}{\hbar^2}\left[E - \frac{1}{2}m\omega^2(x+a)^2\right]}$$

$$+ \int_{a}^{b} dx \sqrt{\frac{2m}{\hbar^2}\left[E - \frac{m\omega^2}{2}(x-a)^2\right]}$$

or

$$2a\sqrt{\frac{2mE}{\hbar^2}} + 2\left(\frac{m\omega}{\hbar}\right)(b-a)^2 \int_{0}^{1} d\chi \sqrt{1 - \chi^2}$$

The last integral is just $\pi/4$. Thus, we get

$$\left(\frac{2m\omega}{\hbar}\right)\left[a(b-a) + \frac{\pi}{4}(b-a)^2\right] = \pi\left(n + \frac{1}{2}\right)$$

It is convenient to introduce the characteristic oscillator length $L \equiv \sqrt{\hbar/m\omega}$ and the dimensionless energy variable $\epsilon \equiv E/\hbar\omega$. In terms of them, the eigenvalue equation becomes

$$\epsilon + \left(\frac{2\sqrt{2}a}{\pi L}\right)\sqrt{\epsilon} - \left(n + \frac{1}{2}\right) = 0$$

Solving this equation, we obtain the energy eigenvalues as

$$E_n = \hbar\omega \left(\frac{2}{\pi^2}\right) \left(\frac{a}{L}\right)^2 \left[-1 + \sqrt{1 + \frac{\pi^2 L^2}{2a^2} \left(n + \frac{1}{2}\right)}\right]^2$$

In the limit $a \ll L$, we obtain, as expected,

$$E_n \approx \hbar\omega \left(n + \frac{1}{2}\right)$$

In the opposite limit, $a \gg L$, we obtain

$$E_n \approx \frac{\pi^2 \hbar^2 n^2}{8ma^2}$$

which coincides with the eigenvalues of an infinite square well.

4

The harmonic oscillator

Problem 4.1 Consider a one-dimensional harmonic oscillator with potential energy $V(x) = \frac{1}{2}m\omega^2 x^2$. The initial $(t = 0)$ wave function of the system is

$$\psi(x, 0) = \frac{1}{\sqrt{2}}\left(1 - \frac{x}{|x|}\right) f(x)$$

where $f(x)$ is a real $(f^*(x) = f(x))$ normalized function that is odd under space reflection $x \to -x$, i.e. $f(-x) = -f(x)$.

(a) Is $\psi(x, 0)$ normalized?
(b) What is the initial probability density at the point $x = 0$?
(c) What is the initial probability of finding the particle in the region $[0, +\infty]$? What is the initial probability for the region $[-\infty, 0]$?
(d) What is the parity of the initial wave function? What is the parity of the wave function at a later arbitrary time $t > 0$?
(e) Is there a time t_1 at which we can be certain that the particle will be in the region $x \geq 0$?
(f) Is there a time $T > 0$ at which we can be certain that the particle will be in the region $x \leq 0$?
(g) Calculate the current densities $\mathcal{J}(x, 0)$, $\mathcal{J}(x, t_1)$, $\mathcal{J}(x, T)$.
(h) Is there a time t_2 at which the probabilities of finding the particle in the regions $x > 0$ and $x < 0$ will be equal?

Solution
(a) The normalization condition is satisfied since

$$1 = \int_{-\infty}^{+\infty} dx\, |\psi(x, 0)|^2 = \int_{-\infty}^{0} dx\, 2[f(x)]^2$$

$$= \int_{-\infty}^{0} dx\, [f(x)]^2 - \int_{+\infty}^{0} d(-x)\,[-f(-x)]^2 = \int_{-\infty}^{+\infty} dx\,[f(x)]^2 = 1$$

82

(b) Owing to the odd parity of $f(x)$, we have $f(0) = -f(0) = 0$ and, therefore, $\mathcal{P}(0, 0) = 0$.

(c) Owing to the vanishing of the wave function in the $x > 0$ region, we have $P_+ \equiv \int_0^{+\infty} dx \, \mathcal{P}(x, 0) = 0$. Obviously

$$P_- \equiv \int_{-\infty}^0 dx \, \mathcal{P}(x, 0) = 1 - P_+ = 1$$

(d) The initial wave function does not have a definite parity. The time-evolved wave function will be a mixture of odd and even eigenfunctions and, therefore, will not have a definite parity either.

(e) The wave function at positive times will be

$$\psi(x, t) = e^{-i\omega t/2} \sum_{n=0}^{\infty} c_n e^{-in\omega t} \psi_n(x)$$

The coefficients c_n are given by

$$c_n = \sqrt{2} \int_{-\infty}^0 dx \, \psi_n(x) f(x)$$

Note that, owing to the reality of the eigenfunctions and the reality of $f(x)$, the c_n are also real.

It should be noted that the eigenfunctions have definite parity, $\psi_n(-x) = (-1)^n \psi_n(x)$. Then, it is clear that for

$$t_1 = \frac{\pi}{\omega}$$

we have

$$\psi(x, t_1) = -i \sum_n c_n \psi_n(-x) = -i\psi(-x, 0)$$

which implies that

$$\mathcal{P}(x, t_1) = \mathcal{P}(-x, 0)$$

or that at the time t_1 the particle has completely moved to the right with exactly the same distribution it had initially when it occupied the negative part of the x-axis.

(f) Similarly, we can see that at the time

$$T = \frac{2\pi}{\omega}$$

we have

$$\psi(x, T) = -\psi(x, 0)$$

and, thus, the particle is distributed exactly as initially, or

$$P(x, T) = P(x, 0)$$

(g) All three wave functions $\psi(x, 0)$, $\psi(x, t_1)$ and $\psi(x, T)$ are real up to a constant phase. Thus, the corresponding probability current densities can easily be seen to vanish.

(h) It is easy to see that at the time

$$t_2 = \frac{\pi}{2\omega}$$

we have

$$\psi(x, t_2) = -i\psi^*(-x, t_2)$$

Note that this is possible owing to the reality of the coefficients c_n. Thus, the probability density is even:

$$P(x, t_2) = P(-x, t_2)$$

Problem 4.2 A particle of mass m and electric charge q can move only in one dimension and is subject to a harmonic force and a homogeneous electrostatic field. The Hamiltonian operator for the system is

$$H = \frac{p^2}{2m} + \frac{m\omega^2}{2}x^2 - q\mathcal{E}x$$

(a) Solve the energy eigenvalue problem.

(b) If the system is initially in the ground state of the unperturbed harmonic oscillator, $|\psi(0)\rangle = |0\rangle$, what is the probability of finding it in the ground state of the full Hamiltonian?

(c) Assume again that the system is initially in the unperturbed harmonic oscillator ground state and calculate the probability of finding it in this state again at a later time.

(d) With the same initial condition calculate the probability of finding the particle at a later time in the first excited state of the unperturbed harmonic oscillator.

(e) Consider the *electric dipole moment* $d \equiv qx$ and calculate its vacuum expectation value in the evolved state $|\psi(t)\rangle$, assuming again that we start from the unperturbed vacuum state at $t = 0$.

Solution

(a) We can write the Hamiltonian as

$$H = \frac{p^2}{2m} + \frac{m\omega^2}{2}\left(x - \frac{q\mathcal{E}}{m\omega^2}\right)^2 - \frac{q^2\mathcal{E}^2}{2m\omega^2}$$

Note however that the exponential of the momentum operator p is a space-translation operator; we can prove in a straightforward fashion that

$$e^{ilp/\hbar} x \, e^{-ilp/\hbar} = x + l$$

Thus, we have

$$H = e^{-ilp/\hbar} \, H_0 \, e^{ilp/\hbar} - \frac{q^2 \mathcal{E}^2}{2m\omega^2}$$

where

$$H_0 = \frac{p^2}{2m} + \frac{m\omega^2}{2} x^2$$

is the standard harmonic-oscillator part of the Hamiltonian and

$$l \equiv q\mathcal{E}/m\omega^2$$

It is, therefore, clear that the energy eigenstates are

$$|\overline{n}\rangle = e^{-ilp/\hbar} |n\rangle$$

and the energy eigenvalues are

$$\overline{E}_n = \hbar\omega \left(n + \frac{1}{2} \right) - \frac{m\omega^2}{2} l^2$$

The corresponding eigenfunctions are just the standard harmonic oscillator eigen-functions translated by l:

$$\overline{\psi}_n(x) = \psi_n(x - l)$$

(b) The state of the system can be expanded in energy eigenstates as

$$|\psi(t)\rangle = \sum_{n=0}^{\infty} C_n e^{-i\overline{E}_n t/\hbar} |\overline{n}\rangle$$

The coefficients are

$$C_n = \langle \overline{n} | \psi(0) \rangle$$

and correspond to the probability amplitude for finding the system in each eigen-state. In our case, since the initial state is the unperturbed ground state,

$$C_n = \langle \overline{n} | 0 \rangle$$

The probability amplitude for finding the system in the true ground state $|\overline{0}\rangle$ is

$$\langle \overline{0} | \psi(t) \rangle = C_0 \, e^{-i\overline{E}_0 t/\hbar} = \langle \overline{0} | 0 \rangle \, e^{-i\overline{E}_0 t/\hbar}$$

The corresponding probability $\mathcal{P}_0$ is time independent:

$$\mathcal{P}_0 = \left| \langle 0 | \overline{0} \rangle \right|^2 = \left| \langle 0 | e^{-ilp/\hbar} | 0 \rangle \right|^2$$

Using the operator identity

$$e^{A+B} = e^A \, e^B \, e^{-[A,B]/2} = e^B \, e^A \, e^{-[B,A]/2}$$

we get, since $p = -i\sqrt{m\omega\hbar/2}\,(a - a^\dagger)$, where $a^\dagger$ and a are the creation and annihilation operators for the harmonic oscillator problem,

$$e^{-ilp/\hbar} = \exp\left(-l\sqrt{\frac{m\omega}{2\hbar}}\,a\right)\exp\left(l\sqrt{\frac{m\omega}{2\hbar}}\,a^\dagger\right)e^{m\omega l^2/4\hbar}$$

$$= \exp\left(l\sqrt{\frac{m\omega}{2\hbar}}\,a^\dagger\right)\exp\left(-l\sqrt{\frac{m\omega}{2\hbar}}\,a\right)e^{-m\omega l^2/4\hbar}$$

Therefore, we have

$$\langle 0|e^{-ilp/\hbar}|0\rangle = \left\langle 0\left|\exp\left(l\sqrt{\frac{m\omega}{2\hbar}}\,a^\dagger\right)\exp\left(-l\sqrt{\frac{m\omega}{2\hbar}}\,a\right)e^{-m\omega l^2/4\hbar}\right|0\right\rangle = e^{-m\omega l^2/4\hbar}$$

Thus, finally

$$\mathcal{P}_0 = \exp\left(-\frac{q^2\mathcal{E}^2}{2m\omega^3\hbar}\right)$$

(c) Let us consider again the expansion

$$|\psi(t)\rangle = \sum_{n=0}^{\infty} C_n\, e^{-i\overline{E}_n t/\hbar}\,|\overline{n}\rangle$$

The amplitude for finding the system in the unperturbed ground state at time t is

$$\langle 0|\psi(t)\rangle = \sum_{n=0}^{\infty} C_n\, e^{-i\overline{E}_n t/\hbar}\,\langle 0|\overline{n}\rangle = \sum_{n=0}^{\infty} |C_n|^2\, e^{-i\overline{E}_n t/\hbar}$$

The coefficients are given by

$$C_n^* = \langle 0|\overline{n}\rangle = \langle 0|e^{-ilp/\hbar}|n\rangle$$

$$= \left\langle 0\left|\exp\left(l\sqrt{\frac{m\omega}{2\hbar}}\,a^\dagger\right)\exp\left(-l\sqrt{\frac{m\omega}{2\hbar}}\,a\right)e^{-m\omega l^2/4\hbar}\right|n\right\rangle$$

$$= \left\langle 0\left|\exp\left(-l\sqrt{\frac{m\omega}{2\hbar}}\,a\right)\right|n\right\rangle e^{-m\omega l^2/4\hbar}$$

$$= \sum_{\ell=0}^{\infty} \frac{1}{\ell!}\left(-\sqrt{\frac{m\omega}{2\hbar}}\,l\right)^\ell \langle 0|a^\ell|n\rangle\, e^{-m\omega l^2/4\hbar}$$

$$= \sum_{\ell=0}^{\infty} \frac{1}{\sqrt{\ell!}}\left(-\sqrt{\frac{m\omega}{2\hbar}}\,l\right)^\ell \langle \ell|n\rangle\, e^{-m\omega l^2/4\hbar} = \frac{(-1)^n}{\sqrt{n!}}\left(\sqrt{\frac{m\omega}{2\hbar}}\,l\right)^n e^{-m\omega l^2/4\hbar}$$

Thus

$$C_n = \frac{(-1)^n}{\sqrt{n!}} \left(\sqrt{\frac{m\omega}{2\hbar}} l \right)^n e^{-m\omega l^2/4\hbar}$$

Substituting back into the amplitude, we get

$$\langle 0|\psi(t)\rangle = e^{-i\omega t/2} e^{-m\omega^2 l^2/2\hbar} e^{im\omega^2 l^2 t/2\hbar} \sum_{n=0} \frac{1}{n!} \left(\frac{m\omega l^2}{2\hbar} \right)^n e^{-in\omega t}$$

$$= e^{-i\omega t/2} e^{-m\omega l^2/2\hbar} e^{im\omega^2 l^2 t/2\hbar} \sum_{n=0} \frac{1}{n!} \left(\frac{m\omega l^2}{2\hbar} e^{-i\omega t} \right)^n$$

$$= e^{-i\omega t/2} e^{-m\omega l^2/2\hbar} e^{im\omega^2 l^2 t/2\hbar} \exp\left(\frac{m\omega l^2}{2\hbar} e^{-i\omega t} \right)$$

The corresponding probability is

$$\mathcal{P} = |\langle 0|\psi(t)\rangle|^2 = e^{-m\omega l^2/\hbar} \exp\left[\frac{m\omega l^2}{2\hbar}(e^{-i\omega t} + e^{i\omega t}) \right]$$

$$= \exp\left[\frac{m\omega}{\hbar} l^2 (\cos \omega t - 1) \right] = \exp\left(-2\frac{m\omega}{\hbar} l^2 \sin^2 \frac{\omega t}{2} \right)$$

or

$$\mathcal{P} = \exp\left(-2\frac{q^2 \mathcal{E}^2}{\hbar m \omega^3} \sin^2 \frac{\omega t}{2} \right)$$

(d) In this case the amplitude we want is

$$\langle 1|\psi(t)\rangle = \sum_{n=0}^{\infty} C_n e^{-i\bar{E}_n t/\hbar} \langle 1|\bar{n}\rangle$$

We have

$$\langle 1|\bar{n}\rangle = \langle 0|a\, e^{-ilp/\hbar}|n\rangle$$

It is not difficult to prove by induction that

$$[a,\, p^n] = in\sqrt{\frac{m\omega\hbar}{2}}\, p^{n-1}$$

and, consequently,

$$[a,\, e^{-ilp/\hbar}] = l\sqrt{\frac{m\omega}{2\hbar}}\, e^{-ilp/\hbar}$$

Using this in the matrix element at hand, we get

$$\langle 1|\bar{n}\rangle = \langle 0|e^{-ilp/\hbar}\, a|n\rangle + l\sqrt{\frac{m\omega}{2\hbar}} \langle 0|e^{-ilp/\hbar}|n\rangle$$

or

$$\langle 1|\bar{n}\rangle = \sqrt{n}\, C_{n-1} + l\sqrt{\frac{m\omega}{2\hbar}}\, C_n$$

Thus, the required amplitude is

$$\langle 1|\psi(t)\rangle = \sum_{n=0}^{\infty} C_n\, e^{-i\bar{E}_n t/\hbar} \left(\sqrt{n}\, C_{n-1} + l\sqrt{\frac{m\omega}{2\hbar}}\, C_n \right)$$

$$= e^{-i\omega t/2}\, e^{-im\omega^2 l^2 t/2\hbar} \sum_{n=0}^{\infty} e^{-in\omega t} \left\{ -\frac{\sqrt{n}}{\sqrt{n!}\sqrt{(n-1)!}} \left(l\sqrt{\frac{m\omega}{2\hbar}} \right)^{2n-1} \right.$$

$$\left. + l\sqrt{\frac{m\omega}{2\hbar}} \frac{1}{n!} \left(l\sqrt{\frac{m\omega}{2\hbar}} \right)^{2n} \right\} e^{-m\omega l^2/2\hbar}$$

$$= l\sqrt{\frac{m\omega}{2\hbar}} \left(-e^{-i\omega t} + 1 \right) \exp\left[-\frac{m\omega}{2\hbar} l^2 \left(1 - e^{-i\omega t} \right) \right] e^{-i\omega t/2}\, e^{-im\omega^2 l^2 t/2\hbar}$$

The corresponding probability is

$$\mathcal{P} = 2\left(\frac{q^2 \mathcal{E}^2}{\hbar m\omega^3} \right) \sin^2 \frac{\omega t}{2} \exp\left[-2\left(\frac{q^2 \mathcal{E}^2}{\hbar \omega^3 m} \right) \sin^2 \frac{\omega t}{2} \right]$$

(e) We have

$$d(t) = \langle \psi(t)|d|\psi(t)\rangle = q \sum_{n,n'=0}^{\infty} C_n^* C_{n'}\, e^{i\omega t(n-n')} \langle n|e^{ilp/\hbar} x e^{-ilp/\hbar}|n'\rangle$$

$$= q \sum_{n,n'=0}^{\infty} C_n^* C_{n'}\, e^{i\omega t(n-n')} \langle n|(x+l)|n'\rangle$$

$$= ql + q \sum_{n,n'=0}^{\infty} C_n^* C_{n'}\, e^{i\omega t(n-n')} \langle n|x|n'\rangle$$

We need the matrix element

$$\langle n|x|n'\rangle = \sqrt{\frac{\hbar}{2m\omega}} \langle n|(a+a^\dagger)|n'\rangle = \sqrt{\frac{\hbar}{2m\omega}} \left(\sqrt{n'}\delta_{n,n'-1} + \sqrt{n'+1}\delta_{n,n'+1} \right)$$

Substituting it into the expression for the dipole moment, we obtain

$$d(t) = ql + q\sqrt{\frac{\hbar}{2m\omega}} \sum_{n=0}^{\infty} \left(C_n^* C_{n+1}\sqrt{n+1}\, e^{-i\omega t} + C_n^* C_{n-1}\sqrt{n}\, e^{i\omega t} \right)$$

$$= ql + q\sqrt{\frac{\hbar}{2m\omega}}\, e^{-m\omega l^2/2\hbar} \sum_{n=0}^{\infty} \left[\frac{1}{n!} \left(-\sqrt{\frac{m\omega}{2\hbar}} l \right)^{2n+1} e^{-i\omega t} + \cdots \right]$$

$$= ql - ql\sqrt{\frac{\hbar}{2m\omega}} \sqrt{\frac{m\omega}{2\hbar}} (e^{-i\omega t} + e^{i\omega t})$$

Our final expression is

$$d(t) = ql - ql \cos \omega t = 2ql \sin^2 \frac{\omega t}{2}$$

Problem 4.3 Consider a simple harmonic oscillator with Hamiltonian

$$H = \frac{p^2}{2m} + \frac{m\omega^2}{2} x^2$$

(a) Show that

$$\left[H, \left[H, x^2 \right] \right] = (2\hbar\omega)^2 x^2 - \frac{4\hbar^2}{m} H$$

(b) Show that the matrix elements of the square of the position, $\langle n|x^2|n'\rangle$, vanish unless $n' = n \pm 2$ or $n' = n$.

(c) Compute the matrix element $\langle n|x|n'\rangle$ and verify the completeness of the energy eigenstates, checking explicitly that

$$\langle n|x^2|n'\rangle = \sum_{\ell=0}^{\infty} \langle n|x|\ell\rangle \langle \ell|x|n'\rangle$$

Solution

(a) It is straightforward to prove this operator relation.

(b) Substituting the expression for the position operator in terms of the creation and annihilation operators and using the properties of their action on the states, namely

$$a|n\rangle = \sqrt{n}|n-1\rangle, \qquad a^\dagger|n\rangle = \sqrt{n+1}\,|n+1\rangle$$

we arrive at

$$\langle n|x^2|n'\rangle = \frac{\hbar}{2m\omega} \left[(2n+1)\delta_{nn'} + \sqrt{(n+2)(n+1)}\,\delta_{n,\,n'-2} + \sqrt{n(n-1)}\,\delta_{n,\,n'+2} \right]$$

(c) Similarly, we obtain

$$\langle n|x|n'\rangle = \sqrt{\frac{\hbar}{2m\omega}} \left(\sqrt{n'}\,\delta_{n,\,n'-1} + \sqrt{n'+1}\,\delta_{n,\,n'+1} \right)$$

and by substitution we can check the given identity.

Problem 4.4 Consider a simple harmonic oscillator with Hamiltonian

$$H = \frac{p^2}{2m} + \frac{m\omega^2}{2} x^2$$

(a) Determine the expectation value $\langle x^2 \rangle_t$ by solving the corresponding time evolution equation and show that it is a periodic function of time with period $(2\omega)^{-1}$.

(b) Suppose that the initial wave function of the system is real and even, i.e.

$$\Psi(x, 0) \equiv \psi(x) = \psi(-x) = \psi^*(x)$$

Calculate the uncertainty at time t (the *dispersion*)

$$(\Delta x)_t^2 \equiv \langle (x - \langle x \rangle_t)^2 \rangle_t$$

in terms of $\langle x^2 \rangle_0$ and $\langle p^2 \rangle_0$.

(c) Consider the limit $t \ll \omega^{-1}$. If by T and V we symbolize the kinetic and potential energy, show that if $\langle V \rangle_0 > \langle T \rangle_0$ then $\langle x^2 \rangle_t < \langle x^2 \rangle_0$. In contrast, when $\langle V \rangle_0 < \langle T \rangle_0$ then $\langle x^2 \rangle_t > \langle x^2 \rangle_0$.

Solution

(a) From the solved expressions for the position and momentum Heisenberg operators,

$$x(t) = \cos \omega t \, x(0) + \sin \omega t \, \frac{p(0)}{m\omega}$$

$$p(t) = \cos \omega t \, p(0) - \sin \omega t \, m\omega x(0)$$

we obtain

$$\langle x^2 \rangle_t = \frac{1}{2} (1 + \cos 2\omega t) \langle x^2 \rangle_0 + \frac{1}{2(m\omega)^2} (1 - \cos 2\omega t) \langle p^2 \rangle_0$$

$$+ \frac{1}{2m\omega} \sin 2\omega t \langle xp + px \rangle_0$$

(b) For an even initial wave function we have

$$\langle \psi | x | \psi \rangle = \langle \psi | p | \psi \rangle = 0$$

If in addition the initial wave function is real then, using the fact that the momentum operator is imaginary, we get

$$\langle \psi | xp + px | \psi \rangle = -\langle \psi | xp + px | \psi \rangle^*$$
$$= -\langle \psi | (xp + px)^\dagger | \psi \rangle = -\langle \psi | xp + px | \psi \rangle = 0$$

Thus, we have

$$(\Delta x)_t^2 = \langle x^2 \rangle_0 \cos^2 \omega t + \frac{\langle p^2 \rangle_0}{(m\omega)^2} \sin^2 \omega t$$

(c) For small times, we have the approximate expression

$$(\Delta x)_t^2 \sim \langle x^2 \rangle_0 + (\omega t)^2 \left[\frac{\langle p^2 \rangle_0}{(m\omega)^2} - \langle x^2 \rangle_0 \right]$$

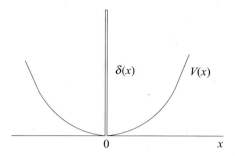

Fig. 20 Harmonic oscillator with attractive or repulsive delta function core.

from which we can immediately conclude that

$$\langle x^2\rangle_t < \langle x^2\rangle_0 \qquad \Longrightarrow \qquad \langle V\rangle_0 > \langle T\rangle_0$$

and

$$\langle x^2\rangle_t > \langle x^2\rangle_0 \qquad \Longrightarrow \qquad \langle V\rangle_0 < \langle T\rangle_0$$

Problem 4.5 Consider a harmonic oscillator potential with an extra delta function term at the origin (see Fig. 20):

$$V(x) = \frac{m\omega^2}{2}x^2 + \frac{\hbar^2 g}{2m}\delta(x)$$

(a) Notice that, owing to the parity invariance of the Hamiltonian, the energy eigenfunctions are even and odd functions. Notice also that the simple harmonic oscillator *odd-parity energy eigenstates* $\psi_{2v+1}(x)$ are still eigenstates of the system Hamiltonian, with eigenvalues $E_{2v+1} = \hbar\omega(2v + 1 + \frac{1}{2})$, for $v = 0, 1, \ldots$.

(b) Expand the even-parity energy eigenfunctions of the given system, $\psi_E(x)$, in terms of the even-parity harmonic oscillator eigenfunctions $\psi_{2v}(x)$:

$$\psi_E(x) = \sum_{v=0}^{\infty} C_v \psi_{2v}(x)$$

Substitute this expression into Schroedinger's equation and compute the coefficients C_v by multiplying by $\psi_{2v}(x)$ and integrating.

(c) Show that the energy eigenvalues that correspond to even eigenstates are solutions of the equation

$$\frac{2}{g} = -\sqrt{\frac{\hbar}{m\pi\omega}} \sum_{v=0}^{\infty} \frac{(2v)!}{2^{2v}(v!)^2} \left(2v + \frac{1}{2} - \frac{E}{\hbar\omega}\right)^{-1}$$

You can make use of the fact that, for the oscillator eigenfunctions, we have

$$\psi_{2\nu}(0) = \left(\frac{m\omega}{\pi\hbar}\right)^{1/4} \frac{\sqrt{(2\nu)!}}{2^\nu \nu!}$$

(d) Consider the following cases: (1) $g > 0$, $E > 0$; (2) $g < 0$, $E > 0$; (3) $g < 0$, $E < 0$ and show that the first and the second cases correspond to an infinity of energy eigenvalues. Can you place them relative to the set $E_{2\nu} = \hbar\omega(2\nu + \frac{1}{2})$? Show that in the third case, that of an attractive delta function core, there exists a single eigenvalue corresponding to the ground state of the system. Use the series summation[1]

$$\sum_{j=0}^{\infty} \frac{(2j)!}{4^j(j!)^2} \frac{1}{2j+1-x} = \frac{\sqrt{\pi}}{2} \frac{\Gamma(1/2 - x/2)}{\Gamma(1 - x/2)}$$

Solution

(a) The delta function term in Schroedinger's equation is proportional to $\psi(0)\delta(x)$, which vanishes for any odd function that satisfies the rest of the equation, such as the harmonic-oscillator odd eigenfunctions.

(b) The Schroedinger equation is

$$\sum_{\nu=0}^{\infty} C_\nu \left(-\frac{\hbar^2}{2m}\frac{d^2}{2m} + \frac{m\omega^2}{2}x^2 - E\right)\psi_{2\nu}(x) = -\frac{\hbar^2 g}{2m}\psi_E(0)\delta(x)$$

or

$$\sum_{\nu=0}^{\infty} C_\nu \left[\hbar\omega\left(2\nu + \frac{1}{2}\right) - E\right]\psi_{2\nu}(x) = -\frac{\hbar^2 g}{2m}\psi_E(0)\delta(x)$$

Multiplying by $\psi_{2\nu'}(x)$ and integrating gives, owing to the orthonormality of the harmonic oscillator eigenfunctions (remember that they are real),

$$\left[\hbar\omega\left(2\nu' + \frac{1}{2}\right) - E\right]C_{2\nu'} = -\frac{\hbar^2 g}{2m}\psi_E(0)\psi_{2\nu'}(0)$$

or

$$C_{2\nu} = -\frac{\hbar^2 g}{2m}\frac{\psi_E(0)\psi_{2\nu'}(0)}{\hbar\omega\left(2\nu + \frac{1}{2}\right) - E}$$

(c) Substituting into the expansion of $\psi_E(x)$, we obtain

$$\psi_E(x) = -\frac{\hbar^2 g}{2m}\psi_E(0)\sum_{\nu=0}^{\infty}\frac{\psi_{2\nu}(0)\psi_{2\nu}(x)}{\hbar\omega\left(2\nu + \frac{1}{2}\right) - E}$$

---

[1] You can use the mathematical fact that the gamma function $\Gamma(x)$ possesses poles at the points $x = 0$, $-1, -2, -3, \ldots$.

At the point $x = 0$, this expression is true provided that

$$\frac{1}{g} = -\frac{\hbar^2}{2m} \sum_{v=0}^{\infty} \frac{|\psi_{2v}(0)|^2}{\hbar\omega\left(2v + \frac{1}{2}\right) - E}$$

Using the given values of $\psi_{2v}(0)$, this is equivalent to

$$\frac{2}{g} = -\sqrt{\frac{\hbar}{m\omega\pi}} \sum_{v=0}^{\infty} \frac{(2v)!}{2^{2v}(v!)^2}\left(2v + \frac{1}{2} - \frac{E}{\hbar\omega}\right)^{-1}$$

(d) Using the given gamma function expression, we get

$$\frac{4}{g} = -\sqrt{\frac{\hbar}{m\omega}} \frac{\Gamma(1/4 - E/2\hbar\omega)}{\Gamma(3/4 - E/2\hbar\omega)}$$

The right-hand side has poles at the points

$$\frac{1}{4} - \frac{E}{2\hbar\omega} = -n \quad (n = 0, 1, 2, \ldots) \quad \Longrightarrow \quad E = \hbar\omega\left(2n + \frac{1}{2}\right)$$

and zeros at the points

$$\frac{3}{4} - \frac{E}{2\hbar\omega} = -n \quad (n = 0, 1, 2, \ldots) \quad \Longrightarrow \quad E = \hbar\omega\left(2n + 1 + \frac{1}{2}\right)$$

Since

$$\Gamma(z) = \frac{\Gamma(z+1)}{z}$$

we have for $z = -\epsilon$, with $\epsilon > 0$,

$$\Gamma(-\epsilon) = -\frac{1}{\epsilon}\Gamma(1) = -\frac{1}{\epsilon} < 0$$

For $E > 0$ the right-hand side can be plotted as shown in Fig. 21.

In the case $g > 0$, the left-hand side is a horizontal line in the upper half-plane cutting the right-hand side at an infinity of points $E_v > \hbar\omega(2v + \frac{1}{2})$. In the case

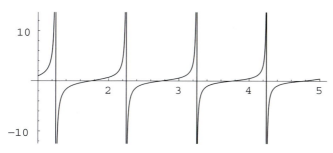

Fig. 21 Plot of $\Gamma(1/4 - x)/\Gamma(3/4 - x)$.

$g < 0$, the left-hand side is a horizontal line in the lower half-plane cutting the right-hand side at an infinity of points $E_\nu < \hbar\omega(2\nu + \frac{1}{2})$. Thus, the positive-energy eigenvalues are in one-to-one correspondence with those of the even-eigenfunction harmonic oscillator, lying higher or lower than them in the repulsive or attractive delta function case respectively.

In the attractive case ($g < 0$) there is also a single negative-energy eigenstate. For $E = -|E| < 0$, the right-hand side of

$$-\frac{4}{g}\sqrt{\frac{m\omega}{\hbar}} = \frac{\Gamma(1/4 + |E|/2\hbar\omega)}{\Gamma(3/4 + |E|/2\hbar\omega)}$$

is a monotonic function of $|E|$ that starts from the value

$$\frac{\Gamma(1/4)}{\Gamma(3/4)} \sim 2.96$$

at $E = 0$ and decreases to 1 at $|E| \to \infty$. The left-hand side is a horizontal line. There is a single solution, provided that the coupling is such that

$$\left[\frac{\Gamma(3/4)}{\Gamma(1/4)}\right]^2 < \frac{g^2\hbar}{16m\omega} < 1$$

Problem 4.6 Consider a simple harmonic oscillator in its ground state. An instantaneous force imparts momentum p_0 to the system. What is the probability that the system will stay in its ground state?

Solution

The new state of the system is

$$\psi_{p_0}(x) = e^{-ip_0 x/\hbar}\,\psi_0(x) = \left(\frac{m\omega}{\hbar\pi}\right)^{1/4} e^{-ip_0 x/\hbar}\, e^{-m\omega x^2/2\hbar}$$

In an expansion in the complete set of harmonic oscillator eigenfunctions,

$$\psi_{p_0}(x) = \sum_{n=0}^{\infty} C_n \psi_n(x)$$

the coefficients

$$C_n = \int_{-\infty}^{+\infty} dx\, \psi_n^*(x)\, \psi_{p_0}(x)$$

are the probability amplitudes for the system to be in the state ψ_n. Thus

$$\mathcal{P}_0 = \left|\int \psi_0(x)\psi_{p_0}(x)\right|^2 = \left|\int \psi_0^2(x)\, e^{-ip_0 x/\hbar}\right|^2$$

Calculating the Gaussian integral

$$\int_{-\infty}^{+\infty} dx \exp\left(-\frac{i}{\hbar} x\, p_0 - \frac{m\omega}{\hbar} x^2\right)$$

$$= \int_{-\infty}^{+\infty} dx \exp\left[-\frac{m\omega}{\hbar}\left(x - \frac{p_0}{2m\omega}\right)^2\right] \exp\left(-\frac{p_0^2}{4m\omega\hbar}\right)$$

we get

$$\mathcal{P}_0 = \exp\left(\frac{-p_0^2}{2m\omega\hbar}\right)$$

Another way to compute this probability is to start from the formula

$$\mathcal{P}_0 = \left|\langle 0|e^{-ip_0 x/\hbar}|0\rangle\right|^2$$

and use the operator identity

$$e^{A+B} = e^A\, e^B\, e^{-[A,\,B]/2}$$

which holds for operators that have a c-number commutator. Applying this identity for

$$x = \sqrt{\frac{\hbar}{2m\omega}}(a^\dagger + a)$$

we obtain

$$\mathcal{P}_0 = \left|\left\langle 0\left|\exp\left(-\frac{ip_0}{\sqrt{2m\omega\hbar}}a^\dagger\right)\exp\left(-\frac{ip_0}{\sqrt{2m\hbar\omega}}a\right)e^{-p_0^2/4m\hbar\omega}\right|0\right\rangle\right|^2$$

$$= \left|\langle 0|0\rangle\, e^{-p_0^2/4m\hbar\omega}\right|^2 = e^{-p_0^2/2m\hbar\omega}$$

since $\exp(\beta a)|0\rangle = |0\rangle$ and $\langle 0|\exp(\gamma a^\dagger) = \langle 0|$.

Problem 4.7 A simple harmonic oscillator is initially (at $t = 0$) in a state corresponding to the wave function

$$\psi(x, 0) = \left(\frac{\gamma}{\pi}\right)^{1/4} e^{-\gamma x^2/2}$$

where $\gamma \neq m\omega/\hbar$ is a positive parameter.

(a) Calculate the expectation values of the observables

$$x, \quad p, \quad xp + px, \quad x^2, \quad p^2$$

at $t = 0$.

(b) Calculate $\langle x^2 \rangle_t$ and $\langle p^2 \rangle_t$ for $t > 0$.

(c) Write down the uncertainty product $(\Delta x)_t (\Delta p)_t$. Verify that it is always greater than $\hbar/2$. Show that it is a periodic function of time.

Solution

(a) The expectation values of the position and momentum in the initial state vanish owing to parity, as can be checked explicitly. Similarly, the third quantity, being imaginary and Hermitian, leads to a vanishing expectation value in the initial (real) state. The remaining two expectation values are

$$\langle x^2 \rangle_0 = \sqrt{\frac{\gamma}{\pi}} \int_{-\infty}^{\infty} dx\, x^2 e^{-\gamma x^2} = -\sqrt{\frac{\gamma}{\pi}} \frac{\partial}{\partial \gamma} \int_{-\infty}^{\infty} dx\, e^{-\gamma x^2} = \frac{1}{2\gamma}$$

and

$$\langle p^2 \rangle_0 = -\hbar^2 \sqrt{\frac{\gamma}{\pi}} \int_{-\infty}^{\infty} dx\, e^{-\gamma x^2/2} \frac{\partial^2}{\partial x^2} e^{-\gamma x^2/2} = \frac{\hbar^2 \gamma}{2}$$

(b) From the solved Heisenberg operators

$$x(t) = \cos \omega t\, x(0) + \sin \omega t\, \frac{p(0)}{m\omega}$$

$$p(t) = \cos \omega t\, p(0) - \sin \omega t\, m\omega x(0)$$

we obtain

$$\langle x^2 \rangle_t = \tfrac{1}{2}(1 + \cos 2\omega t)\langle x^2 \rangle_0 + \frac{1}{2(m\omega)^2}(1 - \cos 2\omega t)\langle p^2 \rangle_0$$

or

$$\langle x^2 \rangle_t = \frac{1}{2\gamma} \cos^2 \omega t + \frac{\hbar^2 \gamma}{2(m\omega)^2} \sin^2 \omega t$$

Similarly, we get

$$\langle p^2 \rangle_t = \frac{\hbar^2 \gamma}{2} \cos^2 \omega t + \frac{(m\omega)^2}{2\gamma} \sin^2 \omega t$$

(c) The uncertainty product is

$$(\Delta x)_t^2 (\Delta p)_t^2 = \frac{\hbar^2}{4} + \frac{\sin^2 2\omega t}{16} \left(\frac{m\omega}{\gamma} - \frac{\hbar^2 \gamma}{m\omega} \right)^2$$

and its period is $T = 1/2\omega$. Note that for $\gamma = m\omega/\hbar$ we would be in a state of minimal uncertainty. This choice identifies our initial wave function with the ground state of the harmonic oscillator.

Problem 4.8 Consider a simple harmonic oscillator.

(a) Show that, for any function $f(x)$ of the position operator, we have

$$[a, \, f(x)] = \sqrt{\frac{\hbar}{2m\omega}} \, f'(x), \qquad \left[a^\dagger, \, f(x)\right] = -\sqrt{\frac{\hbar}{2m\omega}} \, f'(x)$$

(b) Choose $f(x) = e^{ikx}$ and show that

$$\langle n|e^{ikx}|0\rangle \propto \langle n-1|e^{ikx}|0\rangle \propto \cdots \propto \langle 0|e^{ikx}|0\rangle$$

Compute the proportionality coefficients as well as the matrix element $\langle 0|e^{ikx}|0\rangle$. Write down the final expression for the matrix elements $\langle n|e^{ikx}|0\rangle$ and show that they are properly normalized, satisfying

$$\sum_{n=0}^{\infty} |\langle n|e^{ikx}|0\rangle|^2 = 1$$

(c) Suppose now that initially (at $t = 0$) the system is in the state

$$|\psi(0)\rangle = e^{ikx}|0\rangle$$

Calculate the expectation values of the position and momentum observables in this state. What is the probability that an energy measurement at $t = 0$ will give the value $E_n = \hbar\omega(n + \frac{1}{2})$? Calculate the probability for such an outcome at time $t > 0$.

(d) Consider ℓ to be a given parameter with the dimensions of length. Work out the operator quantities

$$e^{ip\ell/\hbar}xe^{-ip\ell/\hbar}, \qquad [x, \, e^{ip\ell/\hbar}], \qquad [a, \, e^{ip\ell/\hbar}]$$

Show that

$$\langle n|e^{ip\ell/\hbar}|0\rangle \propto \langle n-1|e^{ip\ell/\hbar}|0\rangle \propto \cdots \propto \langle 0|e^{ip\ell/\hbar}|0\rangle$$

and compute the proportionality constants. Calculate the matrix element $\langle 0|e^{ip\ell/\hbar}|0\rangle$. Write down the final expression for the matrix elements $\langle n|e^{ip\ell/\hbar}|0\rangle$ and verify that they satisfy the normalization condition

$$\sum_{n=0}^{\infty} |\langle n|e^{ip\ell/\hbar}|0\rangle|^2 = 1$$

(e) Suppose that the system is initially (at $t = 0$) in the state

$$|\psi(0)\rangle = e^{i\ell p/\hbar}|0\rangle$$

If we make an energy measurement at time $t > 0$, what is the probability of finding the value $E_n = \hbar\omega(n + \frac{1}{2})$?

Solution

(b) Starting from the easily proved commutator relations in (a), we take $f(x) = e^{ikx}$ and consider the matrix element

$$\langle n-1|[a, e^{ikx}]|0\rangle = ik\sqrt{\frac{\hbar}{2m\omega}}\langle n-1|e^{ikx}|0\rangle$$

which gives

$$\sqrt{n}\langle n|e^{ikx}|0\rangle = ik\sqrt{\frac{\hbar}{2m\omega}}\langle n-1|e^{ikx}|0\rangle$$

or

$$\langle n|e^{ikx}|0\rangle = \frac{ik}{\sqrt{n}}\sqrt{\frac{\hbar}{2m\omega}}\langle n-1|e^{ikx}|0\rangle$$

Similarly, we get

$$\langle n-1|e^{ikx}|0\rangle = \frac{ik}{\sqrt{n-1}}\sqrt{\frac{\hbar}{2m\omega}}\langle n-2|e^{ikx}|0\rangle$$

and, by induction,

$$\langle n|e^{ikx}|0\rangle = \frac{(ik)^n}{\sqrt{n!}}\left(\sqrt{\frac{\hbar}{2m\omega}}\right)^n\langle 0|e^{ikx}|0\rangle$$

In order to compute the ground-state matrix element, we write it as

$$\langle 0|e^{ikx}|0\rangle = \left\langle 0\left|\exp\left[ik\sqrt{\frac{\hbar}{2m\omega}}(a+a^\dagger)\right]\right|0\right\rangle$$

and use the identity

$$e^{A+B} = e^A\, e^B\, e^{-[A,B]/2}$$

which is valid for any two operators with a c-number commutator. Therefore we obtain

$$\langle 0|e^{ikx}|0\rangle = \left\langle 0\left|\exp\left(ik\sqrt{\frac{\hbar}{2m\omega}}\,a^\dagger\right)\exp\left(ik\sqrt{\frac{\hbar}{2m\omega}}\,a\right)\right|0\right\rangle e^{-\hbar k^2/4m\omega}$$

$$= e^{-\hbar k^2/2m\omega}$$

Finally, we get

$$\langle n|e^{ikx}|0\rangle = \frac{(ik)^n}{\sqrt{n!}}\left(\sqrt{\frac{\hbar}{2m\omega}}\right)^n e^{-\hbar k^2/4m\omega}$$

which satisfies immediately the normalization property.

(c) The probability amplitude for finding the system in the state $|n\rangle$ at time $t = 0$ is

$$\langle n|\psi(0)\rangle = \langle n|e^{ikx}|0\rangle = \frac{(ik)^n}{\sqrt{n!}} \left(\sqrt{\frac{\hbar}{2m\omega}}\right)^n e^{-\hbar k^2/4m\omega}$$

The corresponding probability will be

$$P_n(0) = |\langle n|\psi(0)\rangle|^2 = \frac{k^{2n}}{n!} \left(\frac{\hbar}{2m\omega}\right)^n e^{-\hbar k^2/2m\omega}$$

At time $t > 0$ the state of the system can be expressed as

$$|\psi(t)\rangle = e^{-i\omega t/2} \sum_{j=0}^{\infty} C_j e^{-ij\omega t} |j\rangle$$

The coefficients of this expansion are

$$C_j = \langle j|\psi(0)\rangle = \frac{(ik)^j}{\sqrt{j!}} \left(\sqrt{\frac{\hbar}{2m\omega}}\right)^j e^{-\hbar k^2/4m\omega}$$

Thus, the evolved state will be

$$|\psi(t)\rangle = e^{-\hbar k^2/4m\omega} e^{-i\omega t/2} \sum_{j=0}^{\infty} \frac{(ik)^j}{\sqrt{j!}} \left(\sqrt{\frac{\hbar}{2m\omega}}\right)^j e^{-ij\omega t} |j\rangle$$

The probability amplitude for the system to be in the state $|n\rangle$ is

$$\langle n|\psi(t)\rangle = e^{-\hbar k^2/4m\omega} e^{-i\omega t/2} \frac{(ik)^n}{\sqrt{n!}} \left(\sqrt{\frac{\hbar}{2m\omega}}\right)^n e^{-in\omega t}$$

The corresponding probability is

$$P_n(t) = |\langle n|\psi(t)\rangle|^2 = \frac{k^{2n}}{n!} \left(\frac{\hbar}{2m\omega}\right)^n e^{-\hbar k^2/2m\omega} = P_n(0)$$

and it is time independent.

(d) It is easily seen that

$$e^{ip\ell/\hbar} x e^{-ip\ell/\hbar} = x + \ell$$

$$[x, e^{ip\ell/\hbar}] = -\ell e^{ip\ell/\hbar}, \qquad [a, e^{ip\ell/\hbar}] = -\ell \sqrt{\frac{m\omega}{2\hbar}} e^{ip\ell/\hbar}$$

Taking the matrix element $\langle n - 1| \cdots |0\rangle$ of the last expression, we obtain

$$\langle n|e^{ip\ell/\hbar}|0\rangle = -\frac{\ell}{\sqrt{n}} \sqrt{\frac{m\omega}{2\hbar}} \langle n - 1|e^{ip\ell/\hbar}|0\rangle$$

By induction we eventually get to

$$\langle n|e^{ip\ell/\hbar}|0\rangle = \frac{(-\ell)^n}{\sqrt{n!}}\left(\sqrt{\frac{m\omega}{2\hbar}}\right)^n \langle 0|e^{ip\ell/\hbar}|0\rangle$$

The ground-state matrix element is easily calculated as

$$\left\langle 0\left|\exp\left[\ell\sqrt{\frac{m\omega}{2\hbar}}(a-a^\dagger)\right]\right|0\right\rangle$$

$$= \left\langle 0\left|\exp\left(-\ell\sqrt{\frac{m\omega}{2\hbar}}a^\dagger\right)\exp\left(\ell\sqrt{\frac{m\omega}{2\hbar}}a\right)\right|0\right\rangle e^{-m\omega\ell^2/4\hbar} = e^{-m\omega\ell^2/4\hbar}$$

Thus

$$\langle n|e^{ip\ell/\hbar}|0\rangle = \frac{(-\ell)^n}{\sqrt{n!}}\left(\sqrt{\frac{m\omega}{2\hbar}}\right)^n e^{-m\omega\ell^2/4\hbar}$$

It is straightforward to verify that the normalization condition is true.

(e) The evolved state of the system is

$$|\psi(t)\rangle = e^{-i\omega t/2}\sum_{j=0}^{\infty}C_j e^{-ij\omega t}|j\rangle$$

with coefficients

$$C_j = \langle j|\psi(0)\rangle = \langle j|e^{i\ell p/\hbar}|0\rangle = \frac{(-\ell)^j}{\sqrt{j!}}\left(\sqrt{\frac{m\omega}{2\hbar}}\right)^j e^{-m\omega\ell^2/4\hbar}$$

Substituting, we get

$$|\psi(t)\rangle = e^{-m\omega\ell^2/4\hbar}\, e^{-i\omega t/2}\sum_{j=0}^{\infty}\frac{(-\ell)^j}{\sqrt{j!}}\left(\sqrt{\frac{m\omega}{2\hbar}}\right)^j e^{-ij\omega t}|j\rangle$$

The probability of finding the system in the state $|n\rangle$ is

$$P_n = \frac{\ell^{2n}}{n!}\left(\frac{m\omega}{2\hbar}\right)^n e^{-m\omega\ell^2/2\hbar}$$

and it is time independent.

Problem 4.9 For a simple harmonic oscillator, consider the set of *coherent states* defined as

$$|z\rangle \equiv e^{-|z|^2/2}\sum_{n=0}^{\infty}\frac{z^n}{\sqrt{n!}}|n\rangle$$

in terms of the complex number z.

(a) Show that they are normalized. Prove that they are eigenstates of the annihilation operator *a* with eigenvalue z.

(b) Calculate the expectation value $\mathcal{N} = \langle N \rangle$ and the uncertainty $\Delta\mathcal{N}$ in such a state. Show that in the limit $\mathcal{N} \to \infty$ of large occupation numbers the relative uncertainty $(\Delta\mathcal{N})/\mathcal{N}$ tends to zero.

(c) Suppose that the oscillator is initially in such a state at $t = 0$. Calculate the probability of finding the system in this state at a later time $t > 0$. Prove that the evolved state is still an eigenstate of the annihilation operator with a time-dependent eigenvalue. Calculate $\langle N \rangle$ and $\langle N^2 \rangle$ in this state and prove that they are time independent.

Solution

(a) It is straightforward to see that

$$\langle z | z \rangle = e^{-|z|^2} \sum_{n',n=0}^{\infty} \frac{(z^*)^{n'} z^n}{\sqrt{(n')! n!}} \langle n' | n \rangle = e^{-|z|^2} \sum_{n=0}^{\infty} \frac{|z|^{2n}}{n!} = 1$$

Acting on a coherent state with the annihilation operator gives

$$a|z\rangle = e^{-|z|^2/2} \sum_{n=0}^{\infty} \frac{z^n}{\sqrt{n!}} a|n\rangle = e^{-|z|^2/2} \sum_{n=1}^{\infty} \frac{\sqrt{n} z^n}{\sqrt{n!}} |n-1\rangle$$

$$= z e^{-|z|^2/2} \sum_{n=1}^{\infty} \frac{z^{n-1}}{\sqrt{(n-1)!}} |n-1\rangle = z|z\rangle$$

which proves that $|z\rangle$ is an eigenstate of *a* with eigenvalue z.

(b) Thanks to the property proved in (a), we have

$$\mathcal{N} \equiv \langle z | a^\dagger a | z \rangle = |z|^2$$

Similarly, we have

$$\langle z | N^2 | z \rangle = \langle z | a^\dagger a a^\dagger a | z \rangle = |z|^2 \langle z | a a^\dagger | z \rangle$$
$$= |z|^2 \langle z | (1 + a^\dagger a) | z \rangle = |z|^2 (1 + |z|^2)$$

Thus, the square of the uncertainty is

$$(\Delta\mathcal{N})^2 = \langle N^2 \rangle - \langle N \rangle^2 = |z|^2$$

The relative uncertainty is

$$\frac{\Delta\mathcal{N}}{\mathcal{N}} = \frac{1}{|z|} = \frac{1}{\sqrt{\mathcal{N}}}$$

In the limit $\mathcal{N} \to \infty$, it goes to zero.

(c) The time-evolved state of the system is

$$|\psi(t)\rangle = \sum_{n=0}^{\infty} C_n e^{-i E_n t/\hbar} |n\rangle$$

The coefficients are determined from the initial state through

$$C_n = \langle n|\psi(0)\rangle = \langle n|z\rangle = e^{-|z|^2/2} \sum_{n'=0}^{\infty} \frac{z^{n'}}{\sqrt{(n')!}} \langle n|n'\rangle = e^{-|z|^2/2} \frac{z^n}{\sqrt{n!}}$$

Thus, the evolved state is

$$|\psi(t)\rangle = e^{-i\omega t/2} e^{-|z|^2/2} \sum_{n=0}^{\infty} \frac{\left(z e^{-i\omega t}\right)^n}{\sqrt{n!}} |n\rangle$$

The probability amplitude for encountering the initial state $|z\rangle$ in the future is

$$\langle z|\psi(t)\rangle = e^{-i\omega t/2} e^{-|z|^2} \sum_{n=0}^{\infty} \frac{\left(|z|^2 e^{-i\omega t}\right)^n}{n!}$$

The corresponding probability is

$$\mathcal{P}_z(t) = e^{-4|z|^2 \sin^2(\omega t/2)}$$

Addressing the remaining issues, it is straightforward to show that

$$a|\psi(t)\rangle = z e^{-i\omega t}|\psi(t)\rangle$$

As a result of this relation, we obtain

$$\langle \psi(t)|a^\dagger a|\psi(t)\rangle = |z|^2 = \langle N\rangle_0, \qquad \langle \psi(t)|N^2|\psi(t)\rangle = |z|^2(1 + |z|^2) = \langle N^2\rangle_0$$

Problem 4.10 A simple harmonic oscillator is initially (at $t = 0$) in a state with wave function

$$\psi(x, 0) = N \sum_{n=0}^{\infty} c^n \psi_n(x)$$

where $\psi_n(x)$ are the harmonic oscillator energy eigenfunctions and c is a complex parameter.

(a) Calculate the normalization constant N.
(b) Find the wave function of the system at a later time $t > 0$.
(c) Calculate the probability of finding the system again in the initial state at a later time $t > 0$.
(d) Compute the expectation value of the energy.

Solution

(a) The normalization constant, up to a constant phase, is given by

$$N^{-2} = \sum_{n=0}^{\infty} |c|^{2n} = \frac{1}{1 - |c|^2} \qquad \Longrightarrow \qquad N = \sqrt{1 - |c|^2}$$

(b) The time-evolved wave function is

$$\psi(x, t) = N e^{-i\omega t/2} \sum_{n=0}^{\infty} c^n e^{-in\omega t} \, \psi_n(x)$$

(c) The probability amplitude for finding the system again in the initial state is

$$\langle \psi(0) | \psi(t) \rangle = |N|^2 e^{-i\omega t/2} \sum_{n=0}^{\infty} |c|^{2n} e^{-in\omega t} = e^{-i\omega t/2} \frac{1 - |c|^2}{1 - |c|^2 e^{-i\omega t}}$$

The corresponding probability is

$$\mathcal{P}(t) = |\langle \psi(0) | \psi(t) \rangle|^2 = \left[1 + \frac{4|c|^2}{(1 - |c|^2)^2} \sin^2 \frac{\omega t}{2} \right]^{-1}$$

(d) The expectation value of the energy is

$$\langle H \rangle = |N|^2 \sum_{n=0}^{\infty} E_n |c|^{2n} = \frac{\hbar\omega}{2} \left[1 + (1 - |c|^2) \sum_{n=0}^{\infty} 2n |c|^{2n} \right]$$

$$= \frac{\hbar\omega}{2} \left[1 + |c|(1 - |c|^2) \frac{\partial}{\partial |c|} \left(\sum_{n=0}^{\infty} |c|^{2n} \right) \right] = \frac{\hbar\omega}{2} \frac{1 + |c|^2}{1 - |c|^2}$$

Problem 4.11 A *polar* representation of the creation and annihilation operators for a simple harmonic oscillator can be introduced as

$$a \equiv \sqrt{N + 1} \, e^{i\phi}, \qquad a^{\dagger} \equiv e^{-i\phi} \sqrt{N + 1}$$

The operators N and ϕ are assumed to be Hermitian.

(a) Starting from the commutation relation $[a, a^{\dagger}] = 1$, show that

$$[e^{i\phi}, N] = e^{i\phi}, \qquad [e^{-i\phi}, N] = -e^{-i\phi}$$

Similarly, show that

$$[\cos \phi, N] = i \sin \phi, \qquad [\sin \phi, N] = -i \cos \phi$$

(b) Calculate the matrix elements

$$\langle n | e^{\pm i\phi} | n' \rangle, \qquad \langle n | \cos \phi | n' \rangle, \qquad \langle n | \sin \phi | n' \rangle$$

(c) Write down the Heisenberg uncertainty relation between the operators N and $\cos\phi$. Compute the quantities involved for the state

$$|\psi\rangle = (1 - |c|)^{1/2} \sum_{n=0}^{\infty} c^n |n\rangle$$

where c is a complex parameter. Show that the resulting inequality is always true.

(d) Consider the *coherent state*

$$|z\rangle = e^{-|z|^2/2} \sum_{n=0}^{\infty} \frac{z^n}{\sqrt{n!}} |n\rangle$$

and calculate the quantities $(\Delta N)^2$, $(\Delta \cos\phi)^2$ and $\langle \sin\phi \rangle$ in it. Show that the number–phase Heisenberg inequality in the limit of very large occupation numbers ($z \to \infty$) reduces to an equality.[2]

Solution

(a) Using the fundamental commutation relation we get

$$1 = [a, a^{\dagger}] = \sqrt{N+1} \, e^{i\phi} e^{-i\phi} \sqrt{N+1} - e^{-i\phi} \sqrt{N+1} \sqrt{N+1} \, e^{i\phi}$$
$$= N + 1 - e^{-i\phi} (N+1) e^{i\phi} = N - e^{-i\phi} N e^{i\phi} \qquad \Longrightarrow \qquad [e^{i\phi}, N] = e^{i\phi}$$

(b) Taking the matrix element of the commutation relation just proved, we get

$$\langle n | [e^{\pm i\phi}, N] | n' \rangle = \pm \langle n | e^{\pm i\phi} | n' \rangle$$

or

$$(n' - n \mp 1) \langle n | e^{\pm i\phi} | n' \rangle = 0$$

which shows that

$$\langle n | e^{\pm i\phi} | n' \rangle = C \delta_{n', n \pm 1}$$

The coefficient C is easily shown to be unity. Indeed,

$$C = \langle n | e^{i\phi} | n+1 \rangle = \frac{1}{\sqrt{n+1}} \langle n | e^{i\phi} a^{\dagger} | n \rangle = \frac{1}{\sqrt{n+1}} \langle n | \sqrt{N+1} | n \rangle = 1$$

Thus, we have

$$\langle n | e^{\pm i\phi} | n' \rangle = \delta_{n', n \pm 1}$$
$$\langle n | \cos\phi | n' \rangle = \tfrac{1}{2} \left(\delta_{n', n+1} + \delta_{n', n-1} \right), \qquad \langle n | \sin\phi | n' \rangle = \tfrac{1}{2} i \left(\delta_{n', n+1} - \delta_{n', n-1} \right)$$

---

[2] You may use the asymptotic formulae

$$\sum_{n=0}^{\infty} \frac{|z|^{2n}}{n! \sqrt{n+1}} \sim \frac{e^{|z|^2}}{|z|} \left(1 - \frac{1}{8|z|^2} + \cdots \right)$$

$$\sum_{n=0}^{\infty} \frac{|z|^{2n}}{n! \sqrt{(n+1)(n+2)}} \sim \frac{e^{|z|^2}}{|z|^2} \left(1 - \frac{1}{2|z|^2} + \cdots \right)$$

(c) The uncertainties are defined in the usual way

$$(\Delta \mathcal{N})^2 = \langle N^2 \rangle - \langle N \rangle^2$$
$$(\Delta \cos \phi)^2 = \langle \cos^2 \phi \rangle - \langle \cos \phi \rangle^2$$

Their product satisfies the Heisenberg inequality

$$(\Delta \mathcal{N})^2 (\Delta \cos \phi)^2 \geq \tfrac{1}{4} | \langle [\cos \phi, N] \rangle |^2 = \tfrac{1}{4} | \langle \sin \phi \rangle |^2$$

For the given state, we have on the one hand

$$\langle N \rangle = (1 - |c|^2) \sum_{n=0}^{\infty} |c|^{2n} n = |c|^2 (1 - |c|^2) \frac{\partial}{\partial |c|^2} \left(\sum_{n=0}^{\infty} |c|^{2n} \right)$$

$$= |c|^2 (1 - |c|^2) \frac{\partial}{\partial |c|^2} \left(\frac{1}{1 - |c|^2} \right) = \frac{|c|^2}{1 - |c|^2}$$

Similarly,

$$\langle N^2 \rangle = |c|^2 (1 - |c|^2) \frac{\partial}{\partial |c|^2} \left(\sum_{n=0}^{\infty} |c|^{2n} n \right)$$

$$= |c|^2 (1 - |c|^2) \frac{\partial}{\partial |c|^2} \left(\frac{\langle N \rangle}{1 - |c|^2} \right) = \frac{|c|^2 (1 + |c|^2)}{(1 - |c|^2)^2}$$

Thus, finally, we obtain

$$(\Delta \mathcal{N})^2 = \frac{|c|^2}{(1 - |c|^2)^2}$$

On the other hand, we have

$$\langle e^{i\phi} \rangle = (1 - |c|^2) \sum_{n,n'=0}^{\infty} (c^*)^n c^{n'} \langle n | e^{i\phi} | n' \rangle = (1 - |c|^2) \sum_{n=0}^{\infty} |c|^{2n} c = c$$

From this we get

$$\langle \cos \phi \rangle = \frac{1}{2}(c + c^*), \qquad \langle \sin \phi \rangle = \frac{1}{2i}(c - c^*)$$

We also have

$$\langle e^{2i\phi} \rangle = (1 - |c|^2) \sum_{n,n',n''}^{\infty} (c^*)^n c^{n'} \langle n | e^{i\phi} | n'' \rangle \langle n'' | e^{i\phi} | n' \rangle = (1 - |c|^2) \sum_{n=0}^{\infty} (c^*)^n c^{n+2} = c^2$$

Thus

$$\langle \cos^2 \phi \rangle = \frac{1}{2} + \frac{1}{4}[c^2 + (c^*)^2]$$

and

$$(\Delta \cos \phi)^2 = \frac{1}{2}(1 - |c|^2)$$

Substituting into the Heisenberg inequality, the latter then takes the form

$$\frac{1}{2(1 - |c|^2)} \geq \frac{|c - c^*|^2}{16}$$

Defining $c = |c|e^{i\beta}$, we arrive at

$$\sin^2 \beta \leq \frac{2}{1 - |c|^2}$$

which is always true.

(d) Since, as can be shown straightforwardly, a coherent state $|z\rangle$ is an eigenstate of the annihilation operator a with eigenvalue z, we have on the one hand

$$\langle z|N|z\rangle = |z|^2, \qquad \langle z|N^2|z\rangle = |z|^2(|z|^2 + 1)$$

and

$$(\Delta N)^2 = |z|^2$$

On the other hand, using the matrix element $\langle n|e^{i\phi}|n'\rangle = \delta_{n,n'+1}$, we get

$$\langle z|e^{i\phi}|z\rangle = z\, e^{-|z|^2} \sum_{n=0}^{\infty} \frac{|z|^{2n}}{n!\sqrt{n + 1}}$$

and

$$\langle z|e^{2i\phi}|z\rangle = z^2\, e^{-|z|^2} \sum_{n=0}^{\infty} \frac{|z|^{2n}}{n!\sqrt{(n + 1)(n + 2)}}$$

from which we obtain

$$\langle z|\cos\phi|z\rangle = \frac{1}{2}(z + z^*)\, e^{-|z|^2} \sum_{n=0}^{\infty} \frac{|z|^{2n}}{n!\sqrt{n + 1}}$$

and

$$\langle z|\cos^2\phi|z\rangle = \frac{1}{2} + \frac{1}{4}\big[z^2 + (z^*)^2\big] e^{-|z|^2} \sum_{n=0}^{\infty} \frac{|z|^{2n}}{n!\sqrt{(n + 1)(n + 2)}}$$

The resulting phase uncertainty is

$$(\Delta \cos\phi)^2 = \frac{1}{2} + \frac{1}{4}\big[z^2 + (z^*)^2\big] e^{-|z|^2} \sum_{n=0}^{\infty} \frac{|z|^{2n}}{n!\sqrt{(n + 1)(n + 2)}}$$
$$- \frac{1}{4}(z + z^*)^2 e^{-2|z|^2} \left(\sum_{n=0}^{\infty} \frac{|z|^{2n}}{n!\sqrt{n + 1}}\right)^2$$

In an analogous fashion we can also compute

$$|\langle z| \sin \phi |z\rangle|^2 = \tfrac{1}{4} e^{-2|z|^2} |z - z^*|^2 \left(\sum_{n=0}^{\infty} \frac{|z|^{2n}}{n!\sqrt{n+1}} \right)^2$$

Gathering all the above and substituting it into Heisenberg's inequality, we obtain

$$\frac{1}{2} + \frac{1}{4}[z^2 + (z^*)^2] e^{-|z|^2} \sum_{n=0}^{\infty} \frac{|z|^{2n}}{n!\sqrt{(n+1)(n+2)}}$$

$$-\frac{1}{4}(z + z^*)^2 e^{-2|z|^2} \left(\sum_{n=0}^{\infty} \frac{|z|^{2n}}{n!\sqrt{n+1}} \right)^2 \geq \frac{|z - z^*|^2}{16|z|^2} e^{-2|z|^2} \left(\sum_{n=0}^{\infty} \frac{|z|^{2n}}{n!\sqrt{n+1}} \right)^2$$

In the limit $z \to \infty$ we can replace the series appearing in the above relationship by their asymptotic approximations and get

$$\frac{1}{2} + \frac{1}{4}[z^2 + (z^*)^2] \frac{1}{|z|^2} \left(1 - \frac{1}{2|z|^2} + \cdots \right)$$

$$-\frac{1}{4}(z + z^*)^2 \frac{1}{|z|^2} \left(1 - \frac{1}{8|z|^2} + \cdots \right)^2 \geq \frac{|z - z^*|^2}{16|z|^4} \left(1 - \frac{1}{8|z|^2} + \cdots \right)^2$$

The $O(1)$ terms cancel out to give

$$\frac{1}{16|z|^4} \left[-2z^2 - 2(z^*)^2 + (z + z^*)^2 \right] \geq \frac{1}{16|z|^4}[2|z|^2 - z^2 - (z^*)^2] + O(z^{-4})$$

or

$$\frac{1}{|z|^4} \left[2|z|^2 - z^2 - (z^*)^2 \right] \geq \frac{1}{|z|^4} \left[2|z|^2 - z^2 - (z^*)^2 \right] + O(z^{-4})$$

Thus, for a coherent state, in the limit $z \to \infty$ Heisenberg's inequality for the phase and the occupation number reduces to an equality.

Problem 4.12 A harmonic oscillator is at a given moment ($t = 0$) in a state

$$|\psi_\ell(0)\rangle \equiv e^{i\ell p/\hbar}|0\rangle$$

where ℓ is a given length and $|0\rangle$ is the ground state.

(a) Starting from the translation property of the operator $e^{i\ell p/\hbar}$, that is,

$$e^{i\ell p/\hbar} \, x \, e^{-i\ell p/\hbar} = x + \ell$$

calculate the matrix elements $\langle n|e^{i\ell p/\hbar}|0\rangle$.

(b) Determine the wave function of the system at a later time $t > 0$ and show[3] that it corresponds to a Gaussian wave packet centred at a point that oscillates between ℓ and $-\ell$.

(c) Calculate the uncertainties $(\Delta x)^2$ and $(\Delta p)^2$ for the above wave packet and show that it has a minimal dispersion at all times.

Solution

(a) From the translational property of $e^{i\ell p/\hbar}$ it follows that

$$e^{i\ell p/\hbar} a e^{-i\ell p/\hbar} = a + \ell \sqrt{\frac{m\omega}{2\hbar}}$$

or

$$[e^{i\ell p/\hbar}, a] = \ell \sqrt{\frac{m\omega}{2\hbar}} \, e^{i\ell p/\hbar}$$

Taking the matrix element $\langle n - 1 | \cdots | 0 \rangle$ of the above commutator, we obtain

$$\langle n | e^{i\ell p/\hbar} | 0 \rangle = -\frac{\ell}{\sqrt{n}} \sqrt{\frac{m\omega}{2\hbar}} \langle n - 1 | e^{i\ell p/\hbar} | 0 \rangle$$

By repeated application of this procedure we get

$$\langle n | e^{i\ell p/\hbar} | 0 \rangle = \frac{(-\ell)^n}{\sqrt{n!}} \left(\frac{m\omega}{2\hbar} \right)^{n/2} \langle 0 | e^{i\ell p/\hbar} | 0 \rangle$$

The ground-state matrix element is

$$\langle 0 | e^{i\ell p/\hbar} | 0 \rangle = \int_{-\infty}^{\infty} dx \, \psi_0(x) \exp\left(\ell \frac{d}{dx} \right) \psi_0(x) = \int_{\infty}^{\infty} dx \, \psi_0(x) \psi_0(x + \ell)$$

$$= \left(\frac{m\omega}{\pi\hbar} \right)^{1/2} \int_{-\infty}^{\infty} dx \, \exp\left(-\frac{m\omega}{2\hbar} x^2 \right) \exp\left[-\frac{m\omega}{2\hbar}(x + \ell)^2 \right]$$

$$= e^{-m\omega\ell^2/4\hbar}$$

Thus, finally we have

$$\langle n | e^{i\ell p/\hbar} | 0 \rangle = e^{-m\omega\ell^2/4\hbar} \frac{(-\ell)^n}{\sqrt{n!}} \left(\frac{m\omega}{2\hbar} \right)^{n/2}$$

---

[3] You can use the fact that the harmonic oscillator energy eigenfunctions are expressed in terms of the Hermite polynomials as

$$\psi_n(x) = \left(\frac{m\omega}{\pi\hbar} \right)^{1/4} \frac{1}{\sqrt{2^n n!}} e^{-m\omega x^2/2\hbar} H_n\left(x \sqrt{\frac{m\omega}{\hbar}} \right)$$

and the Hermite polynomials are generated as follows:

$$\sum_{n=0}^{\infty} \frac{z^n}{n!} H_n(s) = e^{-z^2 + 2zs}$$

(b) The time-evolved wave function will be

$$\psi(x,t) = \sum_{n+0}^{\infty} C_n\, e^{-iE_n t/\hbar}\, \psi_n(x)$$

with

$$C_n = \langle n|\psi(0)\rangle = \langle n|e^{i\ell p/\hbar}|0\rangle = e^{-m\omega\ell^2/4\hbar}\,\frac{(-\ell)^n}{\sqrt{n!}}\left(\frac{m\omega}{2\hbar}\right)^{n/2}$$

Substituting, we obtain

$$\psi(x,t) = e^{-m\omega\ell^2/4\hbar}\, e^{-i\omega t/2}\sum_{n=0}^{\infty}\frac{(-\ell)^n}{\sqrt{n!}}\left(e^{-i\omega t}\sqrt{\frac{m\omega}{2\hbar}}\right)^n \psi_n(x)$$

$$= \left(\frac{m\omega}{\pi\hbar}\right)^{1/4} e^{-m\omega\ell^2/4\hbar}\, e^{-i\omega t/2}\, e^{-m\omega x^2/2\hbar}$$

$$\times \sum_{n=0}^{\infty}\frac{(-\ell)^n}{n!}\left(e^{-i\omega t}\sqrt{\frac{m\omega}{4\hbar}}\right)^n H_n\left(x\sqrt{\frac{m\omega}{\hbar}}\right)$$

Making use of the given mathematical property $\sum_{n=0}^{\infty}(z^n/n!)H_n(s) = e^{-z^2+2zs}$, we get

$$\psi(x,t) = \left(\frac{m\omega}{\pi\hbar}\right)^{1/4} e^{-i\omega t/2}\exp\left[-i\frac{m\omega\ell}{2\hbar}\sin\omega t(x+\ell\cos\omega t)\right]$$

$$\times \exp\left[-\frac{m\omega}{2\hbar}(x+\ell\cos\omega t)^2\right]$$

The associated probability density is

$$|\psi(x,t)|^2 = \left(\frac{m\omega}{\pi\hbar}\right)^{1/2}\exp\left[-\frac{m\omega}{\hbar}(x+\ell\cos\omega t)^2\right]$$

It is clearly a Gaussian function with an oscillating centre at $\ell\cos\omega t$.

(c) The expectation values of the position and its square at time t are

$$\langle x\rangle_t = \left(\frac{m\omega}{\pi\hbar}\right)^{1/2}\int_{-\infty}^{\infty}dx\,x\exp\left[-\frac{m\omega}{\hbar}(x+\ell\cos\omega t)^2\right]$$

$$= \left(\frac{m\omega}{\pi\hbar}\right)^{1/2}\int_{-\infty}^{\infty}dx\,x\,e^{-m\omega x^2/\hbar} - \ell\cos\omega t\left(\frac{m\omega}{\pi\hbar}\right)^{1/2}\int_{-\infty}^{\infty}dx\,e^{-m\omega x^2/\hbar}$$

$$= -\ell\cos\omega t$$

$$\langle x^2\rangle_t = \left(\frac{m\omega}{\pi\hbar}\right)^{1/2}\int_{-\infty}^{\infty}dx\,x^2\exp\left[-\frac{m\omega}{\hbar}(x+\ell\cos\omega t)^2\right]$$

$$= \left(\frac{m\omega}{\pi\hbar}\right)^{1/2}\int_{-\infty}^{\infty}dx\,\left(x^2+\ell^2\cos^2\omega t - 2\ell\cos\omega t\,x\right)e^{-m\omega x^2/\hbar}$$

$$= \ell^2\cos^2\omega t + \frac{\hbar}{2m\omega}$$

Thus, the spatial uncertainty comes out to be time independent and coincides with the ground-state value:

$$(\Delta x)^2_t = \frac{\hbar}{2m\omega}$$

The momentum and momentum-squared expectation values are

$$\langle p \rangle_t = -i\hbar \left(\frac{m\omega}{\hbar\pi}\right)^{1/2} \int_{-\infty}^{\infty} dx \, \exp\left[-\frac{m\omega}{\hbar}(x + \ell\cos\omega t)^2\right]$$

$$\times \left(\frac{m\omega}{2\hbar}\right)[-i\ell\sin\omega t - 2(x + \ell\cos\omega t)] = -\frac{m\omega}{2}\ell\sin\omega t$$

$$\langle p^2 \rangle_t = \frac{\hbar m\omega}{2} + \frac{m^2\omega^2}{4}\ell^2\sin^2\omega t$$

Finally, the momentum uncertainty comes out time independent and coincides with the ground-state value:

$$(\Delta p)^2_t = \frac{\hbar m\omega}{2}$$

The uncertainty product is therefore time independent and minimal:

$$(\Delta x)^2_t(\Delta p)^2_t = \frac{\hbar^2}{4}$$

Problem 4.13 The propagator $\mathcal{K}(x, x'; T) \equiv \langle x|e^{-iTH/\hbar}|x'\rangle$ of a free particle is proportional to the exponential of the *classical action* for the motion of the particle from point x to point x' in the time interval T, defined as $\mathcal{S} = \int_0^T dt \, \frac{1}{2}m\dot{x}^2(t)$, namely

$$\mathcal{K}(x, x'; T) \sim \exp\left(\frac{i\mathcal{S}}{\hbar}\right) = \exp\left[i\frac{m}{2\hbar T}(x - x')^2\right]$$

(a) In order to verify whether this property is true for a harmonic oscillator, first calculate the classical action

$$\mathcal{S} = \int_0^T dt \left[\frac{m}{2}\dot{x}^2(t) - \frac{m\omega^2}{2}x^2\right]$$

for the motion of a classical particle with conditions

$$x(0) = x, \qquad x(T) = x'$$

(b) Consider the expression

$$\mathcal{K}(x, x'; t) = f(t) \, \exp\left(\frac{i\mathcal{S}}{\hbar}\right)$$

with $f(t)$ an unknown function, and determine this function from the requirement that the propagator satisfies the Schroedinger equation for the harmonic oscillator. Determine the remaining multiplicative constant by demanding that the propagator expression reduces to the free propagator expression[4] in the limit of vanishing frequency.

(c) Check that the expression you have found for the harmonic oscillator propagator approaches the correct limit when $t \to 0 - i\epsilon$.

Solution

(a) The solution of the classical equation of motion $\ddot{x} = -\omega^2 x$, subject to the conditions $x(0) = x$, $x(T) = x'$, is easily found to be

$$x(t) = x \cos \omega t + \left(\frac{x'}{\sin \omega T} - x \cot \omega T \right) \sin \omega t$$

After a bit of algebra, the classical action is calculated to be

$$S = \frac{m\omega}{2} \left[(x^2 + x'^2) \cot \omega T - \frac{2xx'}{\sin \omega T} \right]$$

(b) Substituting the expression

$$K(x, x'; t) = f(t) \exp \left\{ i \frac{m\omega}{2\hbar} \left[(x^2 + x'^2) \cot \omega t - \frac{xx'}{\sin \omega t} \right] \right\}$$

into the harmonic oscillator Schroedinger equation

$$\left(i\hbar \frac{\partial}{\partial t} + \frac{\hbar^2}{2m} \frac{\partial^2}{\partial x^2} - \frac{m\omega^2}{2} x^2 \right) K(x, x'; t) = 0$$

we find that it is indeed satisfied provided that

$$\frac{df(t)}{dt} = -\frac{\omega}{2} f(t) \cot \omega t$$

This last equation has the solution

$$f(t) = \frac{C}{\sqrt{\sin \omega t}}$$

and the propagator is

$$K(x, x'; t) = \frac{C}{\sqrt{\sin \omega t}} \exp \left\{ i \frac{m\omega}{2\hbar} \left[(x^2 + x'^2) \cot \omega t - \frac{2xx'}{\sin \omega t} \right] \right\}$$

---

[4] The complete expression for the free propagator is

$$K_0(x, x'; t) = \sqrt{\frac{m}{2\pi i\hbar t}} \exp \left[i \frac{m}{2\hbar t} (x - x')^2 \right]$$

The constant C is determined from the $\omega \to 0$ limit to be $C = \sqrt{\omega m / 2\pi i \hbar}$. The complete expression for the harmonic oscillator propagator is thus

$$\mathcal{K}(x, x'; t) = \sqrt{\frac{m\omega}{2i\pi \sin \omega t}} \exp\left\{ i \frac{m\omega}{2\hbar} \left[(x^2 + x'^2) \cot \omega t - \frac{2xx'}{\sin \omega t} \right] \right\}$$

(c) As a final check, let us consider the zero-time limit, in which the propagator should reduce to a delta function:

$$\mathcal{K}(x, x'; 0) = \langle x | x' \rangle = \delta(x - x')$$

In order to take this limit, we regularize our expression, giving a small imaginary part to the time via $t \to t - i\epsilon$. Our expression becomes

$$\mathcal{K}(x, x'; 0) = \lim_{\epsilon \sim 0} \left\{ \sqrt{\frac{m}{2\pi \hbar \epsilon}} \exp\left[-\frac{m}{2\hbar \epsilon} (x - x')^2 \right] \right\} = \delta(x - x')$$

Problem 4.14 Consider a one-dimensional harmonic oscillator in the presence of a time-dependent force, i.e., a *driven* harmonic oscillator:

$$H = \frac{p^2}{2m} + \frac{m\omega^2}{2} x^2 - x\, F(t)$$

The propagator $\mathcal{K}_F(x', x; t_f, t_i)$ for the motion from a point x at an initial time t_i to a point x' at a final time t_f can be obtained in terms of the classical action. A straightforward but tedious calculation gives[5]

$$\mathcal{K}_F(x', x; t_f, t_i) = \mathcal{K}(x', x; T)\, e^{iX/\hbar}$$

where

$$X \equiv \frac{J_3}{4m\omega} - \frac{J_1 J_2 - 2x' m\omega J_1 - 2x m\omega J_2}{2m\omega \sin \omega T}$$

Here J_1, J_2 and J_3 are the following integrals:

$$J_1 \equiv \int_{t_i}^{t_f} dt\, F(t) \sin \omega(t - t_i), \qquad J_2 \equiv \int_{t_i}^{t_f} dt\, F(t) \sin \omega(t_f - t)$$

and

$$J_3 \equiv \int_{t_i}^{t_f} dt\, F(t) \int_{t_i}^{t_f} dt'\, F(t') \sin \omega |t - t'|$$

---

[5] $\mathcal{K}$ stands for the unperturbed harmonic oscillator propagator,

$$\mathcal{K}(x', x; T) = \sqrt{\frac{m\omega}{2\pi i \hbar \sin \omega T}} \exp\left\{ i \frac{m\omega}{2\hbar} \left[(x^2 + x'^2) \cot \omega T - \frac{2xx'}{\sin \omega T} \right] \right\}$$

with $T = t_f - t_i$.

(a) Verify explicitly that the above expression for $\mathcal{K}_F$ satisfies the Schroedinger equation.
(b) Consider the case of an instantaneous force

$$F(t) = p_0 \, \delta(t - t_0)$$

that acts at a moment $t_i < t_0 < t_f$ and imparts momentum p_0 to the system. Write down the expression of the propagator in this particular case.
(c) Assume that initially (at $t = t_i$) the system is in its ground state and at time $t_0 > 0$ it is subject to the above instantaneous force. Calculate the probability of finding the system again in the ground state at some later time $t_f > t_0$. You could use the matrix elements

$$\langle n|e^{ikx}|0\rangle = \frac{1}{\sqrt{n!}} \exp\left[-\frac{\hbar}{2m\omega}k^2\right] \left(ik\sqrt{\frac{\hbar}{2m\omega}}\right)^n$$

Solution

(a) The Schroedinger equation reads

$$\left[i\hbar\frac{\partial}{\partial T} + \frac{\hbar^2}{2m}\frac{\partial^2}{\partial x'^2} - \frac{m\omega^2}{2}x'^2 + F(T)x'\right]\mathcal{K}\,e^{iX/\hbar} = 0$$

Using the fact that $\mathcal{K}(x', 0; T, 0)$ satisfies the Schroedinger equation for $F = 0$, we arrive at

$$-\frac{\partial X}{\partial T} - \omega\cot\omega T\, x'\frac{\partial X}{\partial x'} - \frac{1}{2m}\left(\frac{\partial X}{\partial x'}\right)^2 + F(T)x' = 0$$

Noting first the properties

$$\frac{dJ_3}{dT} = 2F(T)J_2, \qquad \frac{dJ_1}{dT} = F(T)\sin\omega T$$

and

$$\frac{dJ_2}{dT} = \omega\cot\omega T\, J_2 + \frac{\omega}{\sin\omega T}J_1$$

we have

$$\frac{\partial X}{\partial T} = x'F(T) - \frac{1}{2m}\left(\frac{J_1}{\sin\omega T}\right)^2 - \omega\frac{\cot\omega T}{\sin\omega T}J_1 x'$$

which, when substituted back into the second equation together with

$$\frac{\partial X}{\partial x'} = \frac{J_1}{\sin\omega T}$$

shows it to be always true.

(b) For the force $F(t) = p_0 \, \delta(t - t_0)$ with $t_i < t_0 < t_f$, we have

$$J_1 = p_0 \sin\omega(t_0 - t_i), \qquad J_2 = p_0 \sin\omega(t_f - t_0), \qquad J_3 = 0$$

We have for the exponent X

$$X = -\gamma \frac{p_0^2}{2m\omega} + \alpha p_0 x' + \beta p_0 x \qquad \text{with} \qquad \gamma \equiv \frac{\sin \omega(t_f - t_0) \sin \omega(t_0 - t_i)}{\sin \omega(t_f - t_i)}$$

and

$$\alpha \equiv \frac{\sin \omega(t_0 - t_i)}{\sin \omega(t_f - t_i)}, \qquad \beta \equiv \frac{\sin \omega(t_f - t_0)}{\sin \omega(t_f - t_i)}$$

Thus, the propagator is

$$\mathcal{K}_{p_0}(x', x; t_f, t_i) = \mathcal{K}(x', x; t_f - t_i) \exp\left\{ \frac{i}{\hbar}\left[-\gamma \frac{p_0^2}{2m\omega} + \alpha p_0 x' + \beta p_0 x \right] \right\}$$

(c) The time-evolved wave function will be, setting $T \equiv t_f - t_i$,

$$\psi(x', t_f)$$

$$= \int_{-\infty}^{\infty} dx\, \mathcal{K}(x', x; t_f, t_i) \exp\left\{ \frac{i}{\hbar}\left[-\gamma \frac{p_0^2}{2m\omega} + \alpha p_0 x' + \beta p_0 x \right] \right\} \psi_0(x)$$

$$= e^{i\omega T/2} \exp\left(-i\frac{\gamma}{2m\hbar\omega} p_0^2 \right) \sum_{n=0}^{\infty} e^{-in\omega T}\, \psi_n(x')\, e^{i\alpha p_0 x'/\hbar}\, \langle n | e^{i\beta p_0 x/\hbar} | 0 \rangle$$

$$= e^{-i\omega T/2} \exp\left(-\frac{\beta^2 + i\gamma}{2\hbar m\omega} p_0^2 \right) \sum_{n=0}^{\infty} e^{-in\omega T} \frac{1}{\sqrt{n!}} \psi_n(x')\, e^{i\alpha p_0 x'/\hbar} \left(\frac{i\beta p_0}{\sqrt{2\hbar m\omega}} \right)^n$$

The amplitude for finding the system in the ground state again is

$$\langle 0 | \psi(t_f) \rangle$$

$$= e^{-i\omega T/2} \exp\left(-\frac{\beta^2 + i\gamma}{2\hbar m\omega} p_0^2 \right) \sum_{n=0}^{\infty} e^{-in\omega T} \frac{1}{\sqrt{n!}} \langle 0 | e^{i\alpha p_0 x/\hbar} | n \rangle \left(\frac{i\beta p_0}{\sqrt{2\hbar m\omega}} \right)^n$$

$$= e^{-i\omega T/2} \exp\left(-\frac{\alpha^2 + \beta^2 + i\gamma}{2\hbar m\omega} p_0^2 \right) \sum_{n=0}^{\infty} \frac{1}{n!} \left(-\frac{\alpha\beta p_0^2 e^{-i\omega T}}{2\hbar m\omega} \right)^n$$

$$= e^{-i\omega T/2} \exp\left[-\frac{p_0^2}{2\hbar m\omega}(\alpha^2 + \beta^2 + \alpha\beta e^{-i\omega T} + i\gamma) \right]$$

The corresponding probability will be

$$\mathcal{P} = \exp\left[-\frac{p_0^2}{\hbar m\omega}(\alpha^2 + \beta^2 + \alpha\beta \cos\omega T) \right]$$

Substituting for α and β in the exponent, we have

$$\alpha^2 + \beta^2 + \alpha\beta \cos\omega T = \frac{\sin^2 \omega(t_0 - t_i)}{\sin^2 \omega T} + \frac{\sin^2 \omega(t_f - t_0)}{\sin^2 \omega T}$$

$$+ \sin \omega(t_0 - t_i) \sin \omega(t_f - t_0) \frac{\cot \omega T}{\sin \omega T}$$

and it never vanishes. Thus, the system never returns to the ground state.

Problem 4.15 A simple harmonic oscillator is in a state

$$|\psi\rangle = \frac{1}{\sqrt{1+\lambda^2}} \left(|1\rangle + e^{i\nu}\lambda|2\rangle \right)$$

(a) Calculate the uncertainties $(\Delta x)^2$ and $(\Delta p)^2$ in this state.
(b) Write down the uncertainty product

$$(\Delta x)^2(\Delta p)^2 = \frac{\hbar^2}{4} X(\lambda^2, \nu)$$

and determine the values ν_0 and λ_0 for which it becomes minimum. Obtain this value.
(c) Consider the uncertainty product for the value $\tilde{\nu}_0$ corresponding to $\partial X/\partial \nu = 0$ but $\partial^2 X/\partial \nu^2 < 0$ and then determine its minimum with respect to λ^2. Compare its final value with the value obtained at the absolute minimum. What do you conclude about the dependence on the phase ν? How do both values compare with the uncertainty-product value at the absolute maximum?

Solution
(a) The expectation value of the position in the above state is

$$\langle x\rangle = \frac{1}{1+\lambda^2} \left(\langle 1|x|1\rangle + \lambda^2\langle 2|x|2\rangle + \lambda\, e^{i\nu}\langle 1|x|2\rangle + \lambda\, e^{-i\nu}\langle 2|x|1\rangle \right)$$

Using the fact that

$$x = \sqrt{\frac{\hbar}{2m\omega}}(a+a^\dagger) \qquad \Longrightarrow \qquad \langle 1|x|1\rangle = \langle 2|x|2\rangle = 0, \quad \langle 1|x|2\rangle = \sqrt{\frac{\hbar}{m\omega}}$$

we end up with

$$\langle x\rangle = \frac{2\lambda\cos\nu}{1+\lambda^2}\sqrt{\frac{\hbar}{m\omega}}$$

The expectation value of the square of the position operator is

$$\langle x^2\rangle = \frac{1}{1+\lambda^2} \left(\langle 1|x^2|1\rangle + \lambda^2\langle 2|x^2|2\rangle + \lambda\, e^{i\nu}\,\langle 1|x^2|2\rangle + \lambda\, e^{-i\nu}\,\langle 2|x^2|1\rangle \right)$$

Using the fact that

$$x^2 = \frac{\hbar}{2m\omega}\left[a^2 + (a^\dagger)^2 + 2N + 1 \right]$$

$$\Longrightarrow \qquad \langle 1|x^2|1\rangle = \frac{3\hbar}{2m\omega}, \qquad \langle 2|x^2|2\rangle = \frac{5\hbar}{2m\omega}, \qquad \langle 1|x^2|2\rangle = 0$$

we obtain

$$\langle x^2\rangle = \frac{\hbar}{2m\omega}\frac{3+5\lambda^2}{1+\lambda^2}$$

The uncertainty is then

$$(\Delta x)^2 = \frac{\hbar}{2m\omega} F(\lambda^2, \nu)$$

with

$$F(\lambda^2, \nu) = \frac{1}{(1 + \lambda^2)^2} \left(3 + 5\lambda^4 + 8\lambda^2 \sin^2 \nu\right)$$

Similarly, for the momentum operator we obtain

$$p = -i\sqrt{\frac{m\omega\hbar}{2}}(a - a^\dagger) \qquad \Longrightarrow \qquad \langle 1|p|2\rangle = -i\sqrt{m\omega\hbar}$$

$$p^2 = \frac{m\omega\hbar}{2}\left[-a^2 - (a^\dagger)^2 + 2N + 1\right] \qquad \Longrightarrow \qquad \langle n|p^2|n\rangle = \frac{(2n + 1)m\omega\hbar}{2}$$

Thus

$$(\Delta p)^2 = \frac{m\omega\hbar}{2} G(\lambda^2, \nu)$$

with

$$G(\lambda^2, \nu) = \frac{1}{(1 + \lambda^2)^2} \left(3 + 5\lambda^4 + 8\lambda^2 \cos^2 \nu\right)$$

(b) The uncertainty product is

$$(\Delta x)^2 (\Delta p)^2 = \frac{\hbar^2}{4} X(\lambda^2, \nu)$$

with

$$X(\lambda^2, \nu) = FG = \frac{(3 + 5\lambda^4 + 8\lambda^2 \sin^2 \nu)(3 + 5\lambda^4 + 8\lambda^2 \cos^2 \nu)}{(1 + \lambda^2)^4}$$

Considering the partial derivative with respect to the angle ν we get

$$\frac{\partial X}{\partial \nu} = \frac{32\lambda^4}{(1 + \lambda^2)^4} \sin 4\nu$$

This vanishes for the values

$$\nu = 0, \quad \frac{\pi}{4}, \quad \frac{\pi}{2}$$

The second derivative is

$$\frac{\partial^2 X}{\partial \nu^2} = \frac{128\lambda^4}{(1 + \lambda^2)^4} \cos 4\nu$$

and we can see that $\nu = \pi/4$ corresponds to a maximum in the uncertainty product while $\nu_0 = 0$, $\pi/2$ correspond to a minimum. For these values we have

$$X(\lambda^2, \nu_0) = \frac{(3 + 5\lambda^4)(3 + 5\lambda^4 + 8\lambda^2)}{(1 + \lambda^2)^4}$$

Considering the derivative with respect to λ^2, we get

$$\frac{dX}{d\lambda^2} = \frac{60(\lambda^4 - \frac{1}{5})}{(1 + \lambda^2)^4}$$

which vanishes for the value

$$\lambda_0^2 = \frac{1}{\sqrt{5}}$$

This corresponds to a minimum, since

$$\frac{d^2X}{d(\lambda^2)^2} = \frac{120}{(1 + \lambda^2)^5}\left(-\lambda^4 + \lambda^2 + \frac{2}{5}\right)$$

is positive at λ_0^2.

The uncertainty product at $\nu_0 = 0$, $\pi/2$ and $\lambda_0^2 = 1/\sqrt{5}$ has the value

$$(\Delta x)(\Delta p) = \frac{\hbar}{2}\sqrt{X_0} = \frac{\hbar}{2}\left(10\frac{5 + 2\sqrt{5}}{7 + 3\sqrt{5}}\right)^{1/2} \sim \frac{\hbar}{2}2.63$$

(c) At the maximum $\nu = \pi/4$, we have

$$X = \left(\frac{3 + 5\lambda^4 + 4\lambda^2}{(1 + \lambda^2)^2}\right)^2 = F^2(\lambda^2)$$

From

$$F'(\lambda^2) = \frac{6(\lambda^2 - \frac{1}{3})}{(1 + \lambda^2)^3} = 0, \qquad F''(\lambda^2) = \frac{12(1 - \lambda^2)}{(1 + \lambda^2)^4} > 0$$

we get the value

$$\tilde{\lambda}_0^2 = \frac{1}{3}$$

for which

$$F(\tilde{\lambda}_0^2) = \frac{11}{4} \sim 2.75$$

We observe that this, although larger, does not differ much from the value 2.63 obtained at the absolute minimum. The dependence on the phase ν is weak. In contrast, at the absolute maximum $\lambda^2 \to \infty$, the corresponding value is 5.

5

Angular momentum

Problem 5.1 Consider an electron bound in a hydrogen atom under the influence of a homogeneous magnetic field $\mathbf{B} = \hat{\mathbf{z}} B$. Ignore the electron spin. The Hamiltonian of the system is

$$H = H_0 - \omega L_z$$

with $\omega \equiv |e|B/2m_e c$. The eigenstates $|n \ \ell \ m\rangle$ and eigenvalues $E_n^{(0)}$ of the unperturbed hydrogen atom Hamiltonian H_0 are to be considered as known. Assume that initially (at $t = 0$) the system is in the state

$$|\psi(0)\rangle = \tfrac{1}{\sqrt{2}} (|2 \ 1 \ -1\rangle - |2 \ 1 \ 1\rangle)$$

(a) For each of the following states, calculate the probability of finding the system, at some later time $t > 0$, in that state:

$$|2p_x\rangle = \tfrac{1}{\sqrt{2}} (|2 \ 1 \ -1\rangle - |2 \ 1 \ 1\rangle)$$
$$|2p_y\rangle = \tfrac{1}{\sqrt{2}} (|2 \ 1 \ -1\rangle + |2 \ 1 \ 1\rangle)$$
$$|2p_z\rangle = |2 \ 1 \ 0\rangle$$

When does each probability become equal to 1?

(b) Consider a state $|\hat{\mathbf{n}}\rangle$ defined by

$$(\hat{\mathbf{n}} \cdot \mathbf{L})|\hat{\mathbf{n}}\rangle = \hbar|\hat{\mathbf{n}}\rangle, \qquad \mathbf{L}^2|\hat{\mathbf{n}}\rangle = 2\hbar^2|\hat{\mathbf{n}}\rangle$$

Here $|\hat{\mathbf{n}}\rangle$ is an angular momentum eigenstate with angular-momentum-magnitude quantum number $\ell = 1$ and angular momentum eigenvalue along the direction $\hat{\mathbf{n}}$ equal to $+\hbar$. Calculate the probability of finding the system in this state and show that it is a periodic function of time. What is the period? What are the maximum and minimum values of this probability? For simplicity consider the direction $\hat{\mathbf{n}}$ to be in the xy-plane.

(c) Calculate the expectation value of the magnetic dipole moment associated with the orbital angular momentum at time t.

Solution

(a) The eigenstates $|n \ \ell \ m\rangle$ of H_0 are also eigenstates of the complete Hamiltonian with modified eigenvalues

$$E_{nm} = E_n^{(0)} - m\hbar\omega$$

The evolved state of the system is

$$|\psi(t)\rangle = \frac{1}{\sqrt{2}} e^{-iE_2^{(0)}t/\hbar} \left(e^{i\omega t} |2 \ 1 \ -1\rangle - e^{-i\omega t} |2 \ 1 \ 1\rangle \right)$$

The probability of finding the system in the state $|2p_x\rangle$ is

$$\mathcal{P}_x = |\langle 2p_x | \psi(t)\rangle|^2 = |e^{-iE_n^{(0)}t/\hbar} \cos\omega t|^2 = \cos^2\omega t$$

Similarly,

$$\mathcal{P}_y = |\langle 2p_y | \psi(t)\rangle|^2 = |ie^{-iE_n^{(0)}t/\hbar} \sin\omega t|^2 = \sin^2\omega t$$
$$\mathcal{P}_z = |\langle 2p_z | \psi(t)\rangle|^2 = 0$$

The probability $\mathcal{P}_x$ becomes 1 at $t_n = n\pi/\omega$, i.e.

$$t = \frac{\pi}{\omega}, \quad \frac{2\pi}{\omega}, \quad \frac{3\pi}{\omega}, \quad \cdots$$

while $\mathcal{P}_y = 1$ at $t_n = (2n+1)\pi/2\omega$, i.e.

$$t = \frac{\pi}{2\omega}, \quad \frac{3\pi}{2\omega}, \quad \frac{5\pi}{2\omega}, \quad \cdots$$

(b) The state $|\hat{\mathbf{n}}\rangle$ can be expanded in terms of the states $|2 \ 1 \ m\rangle$ with $m = \pm 1, 0$ as follows:

$$|\hat{\mathbf{n}}\rangle = C_1|2 \ 1 \ 1\rangle + C_{-1}|2 \ 1 \ -1\rangle + C_0|2 \ 1 \ 0\rangle$$

The coefficients C_m are given by

$$C_m = \langle 2 \ 1 \ m|\hat{\mathbf{n}}\rangle$$

They can be obtained from the eigenvalue condition:

$$\langle 2 \ 1 \ m|(\hat{\mathbf{n}} \cdot \mathbf{L})|\hat{\mathbf{n}}\rangle = \hbar\langle 2 \ 1 \ m|\hat{\mathbf{n}}\rangle = \hbar C_m$$

or

$$C_m = \frac{1}{2\hbar}(\hat{\mathbf{n}}_x - i\hat{\mathbf{n}}_y)\langle 2 \ 1 \ m|L_+|\hat{\mathbf{n}}\rangle + \frac{1}{2\hbar}(\hat{\mathbf{n}}_x + i\hat{\mathbf{n}}_y)\langle 2 \ 1 \ m|L_-|\hat{\mathbf{n}}\rangle$$

Since we have

$$\langle 2\ 1\ 1|L_+ = \hbar\sqrt{2}\langle 2\ 1\ 0|, \qquad\qquad \langle 2\ 1\ 1|L_- = 0$$
$$\langle 2\ 1\ 0|L_+ = \hbar\sqrt{2}\langle 2\ 1\ -1|, \qquad\quad \langle 2\ 1\ 0|L_- = \hbar\sqrt{2}\langle 2\ 1\ 1|$$
$$\langle 2\ 1\ -1|L_+ = 0, \qquad\qquad\qquad \langle 2\ 1\ -1|L_- = \hbar\sqrt{2}\langle 2\ 1\ 0|$$

we get

$$C_1 = \tfrac{1}{\sqrt{2}}(\hat{\mathbf{n}}_x - i\hat{\mathbf{n}}_y)\langle 2\ 1\ 0|\hat{\mathbf{n}}\rangle$$
$$C_0 = \tfrac{1}{\sqrt{2}}\langle 2\ 1\ -1|\hat{\mathbf{n}}\rangle + \tfrac{1}{\sqrt{2}}(\hat{\mathbf{n}}_x + i\hat{\mathbf{n}}_y)\langle 2\ 1\ 1|\hat{\mathbf{n}}\rangle$$
$$C_{-1} = \tfrac{\hbar}{\sqrt{2}}(\hat{\mathbf{n}}_x + i\hat{\mathbf{n}}_y)\langle 2\ 1\ 0|\hat{\mathbf{n}}\rangle$$

and so

$$C_1 = \tfrac{1}{\sqrt{2}}(\hat{\mathbf{n}}_x - i\hat{\mathbf{n}}_y)C_0$$
$$C_{-1} = \tfrac{1}{\sqrt{2}}(\hat{\mathbf{n}}_x + i\hat{\mathbf{n}}_y)C_0$$

The third relation is satisfied trivially ($\hat{\mathbf{n}}_x^2 + \hat{\mathbf{n}}_y^2 = 1$). C_0 can be obtained from the normalization requirement

$$1 = |C_0|^2 + |C_1|^2 + |C_{-1}|^2 = |C_0|^2\left(1 + \tfrac{1}{2} + \tfrac{1}{2}\right) \implies C_0 = \tfrac{1}{\sqrt{2}}$$

It is convenient to define

$$\hat{\mathbf{n}} = \cos\phi\,\hat{\mathbf{x}} + \sin\phi\,\hat{\mathbf{y}}$$

Then we can write

$$|\hat{\mathbf{n}}\rangle = \tfrac{1}{\sqrt{2}}\left(|2\ 1\ 0\rangle + \tfrac{1}{\sqrt{2}}e^{-i\phi}|2\ 1\ 1\rangle + \tfrac{1}{\sqrt{2}}e^{i\phi}|2\ 1\ -1\rangle\right)$$

The amplitude for finding the system in this state is

$$\langle\hat{\mathbf{n}}|\psi(t)\rangle = \tfrac{1}{2\sqrt{2}}e^{-iE_2^{(0)}t/\hbar}(e^{i\omega t+i\phi} - e^{-i\omega t-i\phi}) = \tfrac{1}{\sqrt{2}}ie^{-iE_2^{(0)}t/\hbar}\sin(\omega t+\phi)$$

The corresponding probability is

$$\mathcal{P}_n = \tfrac{1}{2}\sin^2(\omega t+\phi)$$

The probability of finding the system in the state $|\hat{\mathbf{n}}\rangle$ is thus seen to be a periodic function of time with period $2\pi/\omega$. Its minimum value is zero, while its maximum value is $1/2$. These values are attained at times

$$t_n^{(\text{min})} = -\phi + n\frac{\pi}{\omega}, \qquad t_n^{(\text{max})} = -\phi + (2n+1)\frac{\pi}{2\omega}$$

(c) It is simpler to solve first Heisenberg's equations of motion for the Heisenberg operator $\mathbf{L}(t)$. They are

$$\frac{d\mathbf{L}}{dt} = \frac{i}{\hbar}[H, \mathbf{L}] = -\frac{i\omega}{\hbar}[L_z, \mathbf{L}]$$

or

$$\frac{dL_x}{dt} = \omega L_y, \qquad \frac{dL_y}{dt} = -\omega L_x$$

while L_z is a constant of the motion. The solution of the above system is

$$L_x(t) = L_x(0)\cos \omega t + L_y(0)\sin \omega t$$
$$L_y(t) = L_y(0)\cos \omega t - L_x(0)\sin \omega t$$
$$L_z(t) = L_z(0)$$

The expectation values of these quantities are

$$\langle L_x \rangle_t = \tfrac{1}{2}\langle\psi(0)|\left[e^{-i\omega t}L_+(0) + e^{i\omega t}L_-(0)\right]|\psi(0)\rangle$$
$$= \tfrac{1}{2}\hbar\langle\psi(0)|\left[e^{-i\omega t}|0\rangle - e^{i\omega t}|0\rangle\right] = 0$$
$$\langle L_y \rangle_t = -\tfrac{1}{2}i\langle\psi(0)|\left[e^{-i\omega t}L_+(0) - e^{i\omega t}L_-(0)\right]|\psi(0)\rangle$$
$$= -\tfrac{1}{2}i\hbar\langle\psi(0)|\left[e^{-i\omega t}|0\rangle + e^{-i\omega t}|0\rangle\right] = 0$$
$$\langle L_z \rangle_t = \tfrac{1}{2}\left(\langle-1| - \langle1|\right)L_z(0)\left(|-1\rangle - |1\rangle\right)$$
$$= \tfrac{1}{2}\hbar\left(\langle-1| - \langle1|\right)\left(-|-1\rangle - |1\rangle\right) = 0$$

Thus finally we get

$$\langle\mathbf{L}\rangle_t = 0$$

which also implies the vanishing of the magnetic dipole moment associated with the orbital angular momentum:

$$\langle\boldsymbol{\mu}\rangle_t = \frac{e}{2m_e c}\langle\mathbf{L}\rangle_t = 0$$

Problem 5.2 The most general spin state of an electron is

$$|\chi\rangle = a|\uparrow\rangle + b|\downarrow\rangle$$

where $|\uparrow\rangle$, $|\downarrow\rangle$ are the eigenstates of S_z corresponding respectively to eigenvalues $\pm\hbar/2$. Determine the direction $\hat{\mathbf{n}}$ such that the above state is an eigenstate of $\hat{\mathbf{n}} \cdot \mathbf{S}$ with eigenvalue $-\hbar/2$. Calculate the expectation value of $\mathbf{S}$ in that state.

Solution

Using the previous problem, we can conclude that

$$a = \sin(\theta/2), \qquad b = -\cos(\theta/2)\,e^{i\phi}$$

The spin expectation value in such a state is

$$\langle\chi|\mathbf{S}|\chi\rangle = \frac{\hbar}{2}\left[-\sin\theta\,(\cos\phi\,\hat{\mathbf{x}} + \sin\phi\,\hat{\mathbf{y}}) - \cos\theta\,\hat{\mathbf{z}}\right]$$

Note that, as expected,

$$\langle\chi|(\hat{\mathbf{n}}\cdot\mathbf{S})|\chi\rangle = \frac{\hbar}{2}\left[-\sin^2\theta\left(\cos^2\phi + \sin^2\phi\right) - \cos^2\theta\right] = -\frac{\hbar}{2}$$

Problem 5.3 An electron is described by a Hamiltonian that does not depend on spin. The electron's spin wave function is an eigenstate of S_z with eigenvalue $+\hbar/2$. The operator $\hat{\mathbf{n}}\cdot\mathbf{S}$ represents the spin projection along a direction $\hat{\mathbf{n}}$. We can express this direction as $\hat{\mathbf{n}} = \sin\theta(\cos\phi\,\hat{\mathbf{x}} + \sin\phi\,\hat{\mathbf{y}}) + \cos\theta\,\hat{\mathbf{z}}$.

(a) Solve the eigenvalue problem of $\hat{\mathbf{n}}\cdot\mathbf{S}$. What is the probability of finding the electron in each $\hat{\mathbf{n}}\cdot\mathbf{S}$ eigenstate?

(b) Assume now that the system is subject to a homogeneous magnetic field $\mathbf{B} = \hat{\mathbf{n}}B$. The Hamiltonian is $H = H_0 + \omega\hat{\mathbf{n}}\cdot\mathbf{S}$. The original spatial state of the electron continues as an eigenstate of the modified system. Calculate the spin state of the system at later times $t > 0$. What is the probability of finding the system again in the initial state? What is the probability of finding it with inverted spin?

Solution

In terms of the Pauli representation we write

$$\hat{\mathbf{n}}\cdot\mathbf{S} = \frac{\hbar}{2}\left[\sin\theta\cos\phi\begin{pmatrix} 0 & 1 \\ 1 & 0 \end{pmatrix} + \sin\theta\sin\phi\begin{pmatrix} 0 & -i \\ i & 0 \end{pmatrix} + \cos\theta\begin{pmatrix} 1 & 0 \\ 0 & -1 \end{pmatrix}\right]$$

$$\implies \qquad \hat{\mathbf{n}}\cdot\mathbf{S} = \frac{\hbar}{2}\begin{pmatrix} \cos\theta & \sin\theta\,e^{-i\phi} \\ \sin\theta\,e^{i\phi} & -\cos\theta \end{pmatrix}$$

The eigenvalues of this matrix are $\pm\hbar/2$, with corresponding eigenvectors

$$\begin{pmatrix} \cos(\theta/2) \\ \sin(\theta/2)\,e^{i\phi} \end{pmatrix}, \qquad \begin{pmatrix} \sin(\theta/2) \\ -\cos(\theta/2)\,e^{i\phi} \end{pmatrix}$$

The probabilities of finding the electron in the above states are respectively $\cos^2(\theta/2)$ and $\sin^2(\theta/2)$.

(b) The evolved state will be

$$|\psi(t)\rangle = e^{-iE^{(0)}t/\hbar}\,e^{-i\omega t(\hat{\mathbf{n}}\cdot\mathbf{S})/\hbar}\,|\psi(0)\rangle$$

Ignoring the spatial part, we have

$$\Psi(t) = e^{-iE^{(0)}t/\hbar}\,e^{-i\omega t(\hat{\mathbf{n}}\cdot\boldsymbol{\sigma})/2}\begin{pmatrix} 1 \\ 0 \end{pmatrix}$$

or

$$\Psi(t) = e^{-iE^{(0)}t/\hbar}\,[\cos(\omega t/2) - i\hat{\mathbf{n}}\cdot\boldsymbol{\sigma}\,\sin(\omega t/2)]\begin{pmatrix}1\\0\end{pmatrix}$$

so that

$$\Psi(t) = e^{-iE^{(0)}t/\hbar}\begin{pmatrix}\cos(\omega t/2) - i\sin(\omega t/2)\cos\theta\\-i\sin(\omega t/2)\sin\theta\,e^{i\phi}\end{pmatrix}$$

The probability of finding the system again with spin up is

$$\mathcal{P}_\uparrow = \big|\cos(\omega t/2) - i\sin(\omega t/2)\cos\theta\big|^2 = 1 - \sin^2(\omega t/2)\sin^2\theta$$

Note that at times $t = 2\pi/\omega,\ 4\pi/\omega,\ \dots$ this probability becomes unity. The probability of finding the system with spin down must be

$$\mathcal{P}_\downarrow = \sin^2(\omega t/2)\sin^2\theta$$

Problem 5.4 Consider a state with orbital angular momentum (quantum number) $\ell = 1$ (for example, an $n = 2$, $\ell = 1$ state of the hydrogen atom),

$$|\psi\rangle = C_0|1\ 0\rangle + C_{-1}|1\ -1\rangle + C_1|1\ 1\rangle$$

(a) Find a direction $\hat{\mathbf{n}}$ such that this state is an eigenstate of the operator $\hat{\mathbf{n}}\cdot\mathbf{L}$. Express the coefficients C in terms of the angles θ, ϕ that define the direction $\hat{\mathbf{n}}$.

(b) Write down expressions for the eigenvectors of L_y,

$$|1, L_y = 0\rangle,\quad |1, L_y = \hbar\rangle,\quad |1, L_y = -\hbar\rangle$$

Do the same for the eigenstates $|1, L_x\rangle$.

Solution

(a) The given operator can be written as follows:

$$\hat{\mathbf{n}}\cdot\mathbf{L} = \hat{n}_z L_z + \tfrac{1}{2}(\hat{n}_x - i\hat{n}_y)L_+ + \tfrac{1}{2}(\hat{n}_x + i\hat{n}_y)L_-$$

or

$$\hat{\mathbf{n}}\cdot\mathbf{L} = \cos\theta\,L_z + \tfrac{1}{2}\sin\theta\,e^{-i\phi}\,L_+ + \tfrac{1}{2}\sin\theta\,e^{i\phi}\,L_-$$

It is easy to see that

$$\hat{\mathbf{n}}\cdot\mathbf{L}|1\ 0\rangle = \tfrac{1}{\sqrt{2}}\hbar\sin\theta\,e^{-i\phi}|1\ 1\rangle + \tfrac{1}{\sqrt{2}}\hbar\sin\theta\,e^{i\phi}|1\ -1\rangle$$

$$\hat{\mathbf{n}}\cdot\mathbf{L}|1\ 1\rangle = \hbar\cos\theta|1\ 1\rangle + \tfrac{1}{\sqrt{2}}\hbar\sin\theta\,e^{i\phi}|1\ 0\rangle$$

$$\hat{\mathbf{n}}\cdot\mathbf{L}|1\ -1\rangle = -\hbar\cos\theta|1\ -1\rangle + \tfrac{1}{\sqrt{2}}\hbar\sin\theta\,e^{-i\phi}|1\ 0\rangle$$

Thus

$$\hat{\mathbf{n}} \cdot \mathbf{L}|\psi\rangle = C_0 \left(\tfrac{1}{\sqrt{2}}\hbar \sin\theta\, e^{-i\phi}|1\ 1\rangle + \tfrac{1}{\sqrt{2}}\hbar \sin\theta\, e^{i\phi}|1\ -1\rangle\right)$$
$$+ C_1 \left(\hbar \cos\theta|1\ 1\rangle + \tfrac{1}{\sqrt{2}}\hbar \sin\theta\, e^{i\phi}|1\ 0\rangle\right)$$
$$+ C_{-1} \left(-\hbar \cos\theta|1\ -1\rangle + \tfrac{1}{\sqrt{2}}\hbar \sin\theta\, e^{-i\phi}|1\ 0\rangle\right)$$
$$= \tfrac{1}{\sqrt{2}}\hbar \left(C_1 e^{i\phi} + C_{-1}e^{-i\phi}\right)\sin\theta\,|1\ 0\rangle$$
$$+ \left(\tfrac{1}{\sqrt{2}}\hbar C_0 \sin\theta\, e^{-i\phi} + \hbar C_1 \cos\theta\right)|1\ 1\rangle$$
$$+ \left(\tfrac{1}{\sqrt{2}}\hbar C_0 \sin\theta\, e^{i\phi} - \hbar C_{-1} \cos\theta\right)|1\ -1\rangle$$

The eigenvalue condition translates to

$$\tfrac{1}{\sqrt{2}}\hbar \left(C_1 e^{i\phi} + C_{-1}e^{-i\phi}\right)\sin\theta = \hbar\alpha C_0$$
$$\tfrac{1}{\sqrt{2}}\hbar C_0 \sin\theta\, e^{-i\phi} + \hbar C_1 \cos\theta = \hbar\alpha C_1$$
$$\tfrac{1}{\sqrt{2}}\hbar C_0 \sin\theta\, e^{i\phi} - \hbar C_{-1} \cos\theta = \hbar\alpha C_{-1}$$

where $\hbar\alpha = 0, \pm\hbar$ are the eigenvalues. Solving the above for the three eigenvalues we obtain

$$|\psi\rangle_0 = \cos\theta\,|1\ 0\rangle + \tfrac{1}{\sqrt{2}}\sin\theta\left(-e^{-i\phi}|1\ 1\rangle + e^{i\phi}|1\ -1\rangle\right)$$
$$|\psi\rangle_1 = \tfrac{1}{\sqrt{2}}\sin\theta\,|1\ 0\rangle + \cos^2(\theta/2)\,e^{-i\phi}|1\ 1\rangle + \sin^2(\theta/2)\,e^{i\phi}|1\ -1\rangle$$
$$|\psi\rangle_{-1} = \tfrac{1}{\sqrt{2}}\sin\theta\,|1\ 0\rangle - \sin^2(\theta/2)\,e^{-i\phi}|1\ 1\rangle - \cos^2(\theta/2)\,e^{i\phi}|1\ -1\rangle$$

(b) Taking $\theta = \pi/2$ and $\phi = \pi/2$ we get

$$|1, L_y = 0\rangle = \tfrac{1}{\sqrt{2}}i\left(|1\ 1\rangle + |1\ -1\rangle\right)$$
$$|1, L_y = \pm\hbar\rangle = \tfrac{1}{\sqrt{2}}|1\ 0\rangle + \left(\mp\tfrac{i}{2}\right)\left(|1\ 1\rangle - |1\ -1\rangle\right)$$

Similarly, for $\theta = \pi/2$ and $\phi = 0$, we get

$$|1, L_x = 0\rangle = \tfrac{1}{\sqrt{2}}\left(-|1\ 1\rangle + |1\ -1\rangle\right)$$
$$|1, L_x = \pm\hbar\rangle = \tfrac{1}{\sqrt{2}}|0\rangle + \left(\pm\tfrac{1}{2}\right)\left(|1\ 1\rangle + |1\ -1\rangle\right)$$

Problem 5.5 Consider an operator describing a rotation around the y-axis by $\pi/2$ and apply it to an eigenstate of $\mathbf{L}^2$ and L_x with $\ell = 1$:

$$e^{-i\pi L_y/2\hbar}\,|1, L_x\rangle$$

Show that the resulting state is an eigenstate of L_z. Prove the rotation operator relation

$$e^{i\pi L_y/2\hbar} \, L_z \, e^{-i\pi L_y/2\hbar} = -L_x$$

Show also that

$$e^{-i\pi L_y/2\hbar} \, e^{-i\pi L_x/2\hbar} \, e^{i\pi L_y/2\hbar} \, e^{-i\pi L_z/2\hbar} = 1$$

Solution
Begin with the state

$$e^{-i\pi L_y/2\hbar}|1, L_x = \hbar\rangle$$
$$= e^{-i\pi L_y/2\hbar}\left[\tfrac{1}{2}\left(|1, L_y = \hbar\rangle + |1, L_y = -\hbar\rangle\right) - \tfrac{1}{\sqrt{2}}i|1, L_y = 0\rangle\right]$$
$$= -\tfrac{1}{2}i\left(|1, L_y = \hbar\rangle - |1, L_y = -\hbar\rangle\right) - \tfrac{1}{\sqrt{2}}i|1, L_y = 0\rangle = |1 -1\rangle$$

In an analogous way we can show that any L_x eigenstate rotated by $\pi/2$ around $\hat{\mathbf{y}}$ is an eigenstate of L_z. For the above state we can write

$$e^{-i\pi L_y/2\hbar}|1, L_x = \hbar\rangle = |1 -1\rangle = -\frac{1}{\hbar}L_z|1 -1\rangle = -\frac{1}{\hbar}L_z \, e^{-i\pi L_y/2\hbar}|1, L_x = \hbar\rangle$$

giving

$$e^{-i\pi L_y/2\hbar} L_x |1, L_x = \hbar\rangle = -L_z \, e^{-i\pi L_y/2\hbar}|1, L_x = \hbar\rangle$$

or

$$e^{-i\pi L_y/2\hbar} L_x \, e^{i\pi L_y/2\hbar} = -L_z$$

since the same holds also for $|1, L_x = 0\rangle$ and $|1, L_x = -\hbar\rangle$.

It is clear that

$$e^{-i\pi L_y/2\hbar} L_x^n \, e^{i\pi L_y/2\hbar} = (-1)^n L_z^n$$

Thus,

$$e^{-i\pi L_y/2\hbar} \, e^{-i\pi L_x/2\hbar} \, e^{i\pi L_y/2\hbar} = e^{i\pi L_z/2\hbar}$$

Problem 5.6 Consider the state $|j_1 \, j_2 \, j \, m\rangle$, which is a common eigenstate of the angular momentum operators $\mathbf{J}_1^2$, $\mathbf{J}_2^2$, $\mathbf{J}^2$ and J_z, where $\mathbf{J} = \mathbf{J}_1 + \mathbf{J}_2$. Show that this state is also an eigenstate of the inner product operator $\mathbf{J}_1 \cdot \mathbf{J}_2$ and find its eigenvalues. Do the same for the operators $\mathbf{J} \cdot \mathbf{J}_1$ and $\mathbf{J} \cdot \mathbf{J}_2$.

Solution
Since

$$\mathbf{J}_1 \cdot \mathbf{J}_2 = \tfrac{1}{2}\left(\mathbf{J}^2 - \mathbf{J}_1^2 - \mathbf{J}_2^2\right)$$

it is clear that

$$\mathbf{J}_1 \cdot \mathbf{J}_2 |j_1 \ j_2 \ j \ m\rangle = \tfrac{1}{2}\hbar \left[j(j+1) - j_1(j_1+1) - j_2(j_2+1) \right] |j_1 \ j_2 \ j \ m\rangle$$

The eigenvalues are independent of the eigenvalues m of J_z, since

$$[(\mathbf{J}_1 \cdot \mathbf{J}_2), \ J_\pm] = \tfrac{1}{2}[\mathbf{J}^2, \ J_\pm] - \tfrac{1}{2}[\mathbf{J}_1^2, \ J_{1\pm}] - \tfrac{1}{2}[\mathbf{J}_2^2, \ J_{2\pm}] = 0$$

Similarly, for $\mathbf{J} \cdot \mathbf{J}_1$ we have

$$\mathbf{J} \cdot \mathbf{J}_1 = \mathbf{J}_1^2 + \mathbf{J}_2 \cdot \mathbf{J}_1$$

Thus,

$$\mathbf{J} \cdot \mathbf{J}_1 |j_1 \ j_2 \ j \ m\rangle = \hbar^2 \left[j(j+1) - j_2(j_2+1) - \tfrac{1}{2}j_1(j_1+1) \right] |j_1 \ j_2 \ j \ m\rangle$$

Analogously for $\mathbf{J} \cdot \mathbf{J}_2$.

Problem 5.7 Consider an operator $\mathbf{V}$ that satisfies the commutation relation

$$[\mathbf{L}_i, \ V_j] = i\hbar\epsilon_{ijk}V_k$$

This is by definition a *vector operator* (for example, $\mathbf{V} = \mathbf{r}, \ \mathbf{p}, \ \mathbf{L}$).

(a) Prove that the operator $e^{-i\phi L_x/\hbar}$ is a *rotation operator* corresponding to a rotation around the x-axis by an angle ϕ, by showing that

$$e^{-i\phi L_x/\hbar} \ V_i \ e^{i\phi L_x/\hbar} = \mathcal{R}_{ij}(\phi) \ V_j$$

where $\mathcal{R}(\phi)$ is the corresponding rotation matrix.

(b) Prove that

$$e^{i\pi L_x/\hbar} |\ell \ m\rangle = |\ell \ -m\rangle$$

(c) Show that a rotation by π around the z-axis can also be achieved by first rotating around the x-axis by $\pi/2$, then rotating around the y-axis by π and, finally, rotating back by $-\pi/2$ around the x-axis. In terms of rotation operators, this is expressed as

$$e^{i\pi L_x/2\hbar} \ e^{-i\pi L_y/\hbar} \ e^{-i\pi L_x/2\hbar} = e^{-i\pi L_z/\hbar}$$

(d) Now consider an electron. How is its state modified if we rotate it by π around the z-axis, then by π around the y-axis and, finally, by π around the x-axis?

Solution
(a) Consider the operator

$$X_i = e^{-i\phi L_x/\hbar} \ V_i \ e^{i\phi L_x/\hbar}$$

as a function of ϕ and differentiate it with respect to ϕ. We get

$$\frac{dX_i}{d\phi} = -\frac{i}{\hbar}e^{-i\phi L_x/\hbar}\,[L_x,\,V_i]\,e^{i\phi L_x/\hbar} = \epsilon_{xij}X_j$$

From this we obtain

$$X_x(\phi) = X_x(0) = V_x$$
$$X_y(\phi) = X_y(0)\cos\phi + X_z(0)\sin\phi = V_y\cos\phi + V_z\sin\phi$$
$$X_z(\phi) = X_z(0)\cos\phi - X_y(0)\sin\phi = V_z\cos\phi - V_y\sin\phi$$

or

$$e^{-i\phi L_x/\hbar}\,V_i\,e^{i\phi L_x/\hbar} = \begin{pmatrix} 1 & 0 & 0 \\ 0 & \cos\phi & \sin\phi \\ 0 & -\sin\phi & \cos\phi \end{pmatrix}\begin{pmatrix} V_x \\ V_y \\ V_z \end{pmatrix} = \mathcal{R}_{ij}V_j$$

Clearly, the matrix $\mathcal{R}$ is a rotation matrix corresponding to a rotation around the x-axis by an angle ϕ.

(b) Putting $\phi = \pi$ in the above expression, we get

$$e^{-i\pi L_x/\hbar}\,L_z\,e^{i\pi L_x/\hbar} = -L_z$$

Acting on the rotated state with L_z, we get

$$L_z\,e^{i\pi L_x/\hbar}\,|\ell\;m\rangle = -e^{-i\pi L_x/\hbar}\,L_z\,|\ell\;m\rangle = -\hbar m\,e^{i\pi L_x/\hbar}\,|\ell\;m\rangle$$

Thus

$$e^{i\pi L_x/\hbar}\,|\ell\;m\rangle \propto |\ell\;-m\rangle$$

Since the rotation operator is unitary and it is acting on a normalized state, the proportionality coefficient is just a phase, which we take to be unity. Thus,

$$e^{i\pi L_x/\hbar}\,|\ell\;m\rangle = |\ell\;-m\rangle$$

(c) Putting $\phi = \pi/2$ in the rotation matrix, we get

$$e^{i\pi L_x/2\hbar}\begin{pmatrix} L_x \\ L_y \\ L_z \end{pmatrix}e^{-i\pi L_x/2\hbar} = \begin{pmatrix} L_x \\ L_z \\ -L_y \end{pmatrix}$$

Thus, we obtain

$$e^{i\pi L_x/2\hbar}\left(L_y\right)^n e^{-i\pi L_x/2\hbar} = (L_z)^n$$

and finally

$$e^{i\pi L_x/2\hbar}\,e^{-i\pi L_y/\hbar}\,e^{-i\pi L_x/2\hbar} = e^{-i\pi L_z/\hbar}$$

(d) In the case of an electron the rotation operators involve the *total angular momentum* $\mathbf{J} = \mathbf{L} + \mathbf{S}$. Thus, the action of the rotations on the electron state can be written as

$$
e^{-i\pi J_x/\hbar}\, e^{-i\pi J_y/\hbar}\, e^{-i\pi J_z/\hbar}\, |\Psi\rangle
$$

$$
= e^{-i\pi\sigma_1/2}\, e^{-i\pi\sigma_2/2}\, e^{-i\pi\sigma_3/2}\, e^{-i\pi L_x/\hbar}\, e^{-i\pi L_y/\hbar}\, e^{-i\pi L_z/\hbar}\, |\Psi\rangle
$$

$$
= i\sigma_1\sigma_2\sigma_3\, e^{-i\pi L_x/2\hbar} \left(e^{-i\pi L_x/2\hbar}\, e^{-i\pi L_y/\hbar} \right) e^{-i\pi L_z/\hbar}|\Psi\rangle
$$

$$
= -e^{-i\pi L_x/2\hbar} \left(e^{-i\pi L_z/\hbar} e^{-i\pi L_x/2\hbar} \right) e^{-i\pi L_z/\hbar}|\Psi\rangle
$$

$$
= -e^{-i\pi L_x/2\hbar} \left(e^{-i\pi L_z/\hbar} e^{-i\pi L_x/2\hbar} \right) e^{i\pi L_z/\hbar}|\psi\rangle
$$

$$
= -e^{-i\pi L_x/2\hbar} \left(e^{-i\pi L_z/\hbar} e^{-i\pi L_x/2\hbar} e^{i\pi L_z/\hbar} \right) |\Psi\rangle
$$

$$
= -e^{-i\pi L_x/2\hbar} e^{i\pi L_x/2\hbar}|\Psi\rangle = -|\Psi\rangle
$$

Problem 5.8 Consider again a vector operator $\mathbf{V}$.[1]

(a) Prove the property

$$
\left[\mathbf{J}^2,\ \mathbf{J}\times\mathbf{V}\right] = 2i\hbar\left(\mathbf{J}^2\mathbf{V} - (\mathbf{J}\cdot\mathbf{V})\mathbf{J}\right)
$$

(b) Demonstrate that

$$
\langle j\, m'|\mathbf{V}|j\, m\rangle = \frac{1}{\hbar^2\, j(j+1)}\langle j\, m'|(\mathbf{J}\cdot\mathbf{V})\mathbf{J}|j\, m\rangle
$$

$$
= \frac{1}{\hbar^2\, j(j+1)}\langle j\, m'|\mathbf{J}|j\, m\rangle\langle j\, m|(\mathbf{J}\cdot\mathbf{V})|j\, m\rangle
$$

(c) Assume now that the states $|j\, m\rangle$ correspond to two angular momentum operators $\mathbf{J}_1$, $\mathbf{J}_2$, being eigenstates of J_1^2 and J_2^2 in addition to $\mathbf{J}^2$, J_z, namely

$$
|j\, m\rangle \Longrightarrow |j_1\, j_2\, j\, m\rangle
$$

Calculate the matrix elements

$$
\langle j_1\, j_2\, j\, m|\mathbf{J}_1|j_1\, j_2\, j\, m\rangle, \qquad \langle j_1\, j_2\, j\, m|\mathbf{J}_2|j_1\, j_2\, j\, m\rangle
$$

Solution

(b) Let us introduce a complete set of states $|j'\, m''\rangle$. Then we get

$$
\langle j\, m'|\,(\mathbf{J}\cdot\mathbf{V})\,\mathbf{J}|j\, m\rangle = \left\langle j\, m'\left|\,(\mathbf{J}\cdot\mathbf{V})\sum_{j'\, m''}\right|j'\, m''\right\rangle\langle j'\, m''|\mathbf{J}|j\, m\rangle
$$

Since $[\mathbf{J}^2, \mathbf{J}] = 0$, we have

$$
0 = \langle j'\, m''|[\mathbf{J}^2, \mathbf{J}]|j\, m\rangle = \hbar^2\left[j'(j'+1) - j(j+1)\right]\langle j'\, m''|\mathbf{J}|j\, m\rangle
$$

---

[1] By definition a *vector operator* satisfies the commutation relation $[J_i,\ V_j] = i\hbar\epsilon_{ijk}V_k$, where $\mathbf{J}$ is the total angular momentum.

and, thus, $j' = j$. Note however that we can easily show that

$$[\mathbf{J}, (\mathbf{J} \cdot \mathbf{V})] = 0$$

which implies

$$[J_z, (\mathbf{J} \cdot \mathbf{V})] = [J_\pm, (\mathbf{J} \cdot \mathbf{V})] = 0$$

Therefore, we have

$$\langle j\, m' | (\mathbf{J} \cdot \mathbf{V}) | j\, m'' \rangle \neq 0$$

only for $m' = m''$.

(c) We can write

$$\mathbf{J} = \tfrac{1}{2}(\hat{\mathbf{x}} - i\hat{\mathbf{y}})J_+ + \tfrac{1}{2}(\hat{\mathbf{x}} + i\hat{\mathbf{y}})J_- + \hat{\mathbf{z}}J_z$$

and calculate the matrix element

$$\begin{aligned}
\langle j\, m' | \mathbf{J} | j\, m \rangle &= \tfrac{1}{2}\hbar(\hat{\mathbf{x}} - i\hat{\mathbf{y}})\sqrt{j(j+1) - m(m+1)}\, \delta_{m',m+1} \\
&\quad + \tfrac{1}{2}\hbar(\hat{\mathbf{x}} + i\hat{\mathbf{y}})\sqrt{j(j+1) - m(m-1)}\, \delta_{m',m-1} + \hat{\mathbf{z}}\hbar m\, \delta_{m',m}
\end{aligned}$$

From the relation previously proved we get

$$\langle j\, m' | \mathbf{J}_a | j\, m \rangle = \frac{1}{\hbar^2 j(j+1)} \langle j\, m' | \mathbf{J} \cdot \mathbf{J}_a | j\, m \rangle \langle j\, m' | \mathbf{J} | j\, m \rangle$$

We also have

$$\mathbf{J} \cdot \mathbf{J}_1 = J_1^2 + \mathbf{J}_2 \cdot \mathbf{J}_1 = J_1^2 + \tfrac{1}{2}\left(J^2 - J_1^2 - J_2^2\right) = \tfrac{1}{2}\left(J^2 + J_1^2 - J_2^2\right)$$

and, analogously,

$$\mathbf{J} \cdot \mathbf{J}_2 = \tfrac{1}{2}\left(J^2 + J_2^2 - J_1^2\right)$$

Therefore, the matrix elements in question are

$$\langle j\, m | \mathbf{J}_1 | j\, m \rangle = \hat{\mathbf{z}}\frac{\hbar m}{2}\left[1 + \frac{j_1(j_1+1)}{j(j+1)} - \frac{j_2(j_2+1)}{j(j+1)}\right]$$

$$\langle j\, m | \mathbf{J}_2 | j\, m \rangle = \hat{\mathbf{z}}\frac{\hbar m}{2}\left[1 - \frac{j_1(j_1+1)}{j(j+1)} + \frac{j_2(j_2+1)}{j(j+1)}\right]$$

Problem 5.9 Consider an electron. We know its orbital angular momentum ℓ and the z-component m of its total angular momentum j. What are the possible values of j? Calculate the expectation value of the magnetic dipole moment of the electron in the state $|\ell\, \tfrac{1}{2}; j\, m\rangle$.

Solution

The possible values of the total angular momentum are

$$j = \ell + \tfrac{1}{2}, \qquad j = \ell - \tfrac{1}{2}$$

From problem 5.8, making the correspondences

$$\mathbf{J}_1 \rightarrow \mathbf{L}, \qquad \mathbf{J}_2 \rightarrow \mathbf{S}$$

we have the following diagonal matrix elements:

$$\langle \ell + \tfrac{1}{2}, \, m; \, \ell \, \tfrac{1}{2} \, |\mathbf{S}| \, \ell + \tfrac{1}{2}, \, m; \, \ell \, \tfrac{1}{2} \rangle$$

$$= \hat{\mathbf{z}} \frac{\hbar m}{2} \left[1 - \frac{\ell(\ell+1)}{\left(\ell + \tfrac{1}{2}\right)\left(\ell + \tfrac{3}{2}\right)} + \frac{3}{4\left(\ell + \tfrac{1}{2}\right)\left(\ell + \tfrac{3}{2}\right)} \right] = \hat{\mathbf{z}} \frac{\hbar m}{2\ell + 1}$$

$$\langle \ell - \tfrac{1}{2}, \, m; \, \ell \, \tfrac{1}{2} \, |\mathbf{S}| \, \ell - \tfrac{1}{2}, \, m; \, \ell \, \tfrac{1}{2} \rangle$$

$$= \hat{\mathbf{z}} \frac{\hbar m}{2} \left[1 - \frac{\ell(\ell+1)}{\left(\ell - \tfrac{1}{2}\right)\left(\ell + \tfrac{1}{2}\right)} + \frac{3}{4\left(\ell - \tfrac{1}{2}\right)\left(\ell + \tfrac{1}{2}\right)} \right] = -\hat{\mathbf{z}} \frac{\hbar m}{2\ell + 1}$$

$$\langle \ell + \tfrac{1}{2}, \, m; \, \ell \, \tfrac{1}{2} \, |\mathbf{L}| \, \ell + \tfrac{1}{2}, \, m; \, \ell \, \tfrac{1}{2} \rangle$$

$$= \hat{\mathbf{z}} \frac{\hbar m}{2} \left[1 + \frac{\ell(\ell+1)}{\left(\ell + \tfrac{1}{2}\right)\left(\ell + \tfrac{3}{2}\right)} - \frac{3}{4\left(\ell + \tfrac{1}{2}\right)\left(\ell + \tfrac{3}{2}\right)} \right] = \hat{\mathbf{z}} \hbar m \frac{2\ell}{2\ell + 1}$$

$$\langle \ell - \tfrac{1}{2}, \, m; \, \ell \, \tfrac{1}{2} \, |\mathbf{L}| \, \ell - \tfrac{1}{2}, \, m; \, \ell \, \tfrac{1}{2} \rangle$$

$$= \hat{\mathbf{z}} \frac{\hbar m}{2} \left[1 + \frac{\ell(\ell+1)}{\left(\ell - \tfrac{1}{2}\right)\left(\ell + \tfrac{1}{2}\right)} - \frac{3}{4\left(\ell - \tfrac{1}{2}\right)\left(\ell + \tfrac{1}{2}\right)} \right] = \hat{\mathbf{z}} \hbar m \frac{\ell + 1}{\ell + \tfrac{1}{2}}$$

The magnetic dipole moment operator of the electron is

$$\boldsymbol{\mu} \equiv \frac{e}{2m_e c} \, (\mathbf{L} + 2\mathbf{S})$$

Thus, we get

$$\langle \ell + \tfrac{1}{2}, \, m; \, \ell \, \tfrac{1}{2} \, |(\mathbf{L} + 2\mathbf{S})| \, \ell + \tfrac{1}{2}, \, m; \, \ell \, \tfrac{1}{2} \rangle = \hat{\mathbf{z}} \hbar m \frac{\ell + 1}{\ell + \tfrac{1}{2}}$$

$$\langle \ell - \tfrac{1}{2}, \, m; \, \ell \, \tfrac{1}{2} \, |(\mathbf{L} + 2\mathbf{S})| \, \ell - \tfrac{1}{2}, \, m; \, \ell \, \tfrac{1}{2} \rangle = \hat{\mathbf{z}} \hbar m \frac{\ell}{\ell + \tfrac{1}{2}}$$

and, finally,

$$\langle \boldsymbol{\mu} \rangle = \hat{\mathbf{z}} \frac{e\hbar}{2m_e c} \frac{2m}{2\ell + 1} \begin{cases} (\ell + 1), & j = \ell + \tfrac{1}{2} \\ \ell, & j = \ell - \tfrac{1}{2} \end{cases}$$

Problem 5.10 A hydrogen atom is under the influence of a homogeneous magnetic field $\mathbf{B} = \hat{\mathbf{z}}B$. Assume that the atom is initially (at $t = 0$) in a state $|n; \ell \, \frac{1}{2}; j \, m\rangle$ with $j = \ell + \frac{1}{2}$. Calculate the probability of finding the atom at a later time $t > 0$ in the state $|n'; \ell' \, \frac{1}{2}; j' \, m'\rangle$, with $j' = \ell' + \frac{1}{2}$ or $j' = \ell' - \frac{1}{2}$.

Solution

The Hamiltonian operator of the system is

$$H = H_0 + \omega(J_z + S_z)$$

where H_0 is the unperturbed hydrogen-atom Hamiltonian. Note that $[H_0, J_z] = [H_0, S_z] = 0$.

The evolved state of the system will be

$$|\psi(t)\rangle = e^{-iE_n t/\hbar} \, e^{-i\omega m t} \, e^{-i\omega t S_z/\hbar} |n; \ell \, \tfrac{1}{2}; j = \ell + \tfrac{1}{2}, m\rangle$$

The probability that we require is[2]

$$\mathcal{P}(t) = \left| \langle n'; \ell' \, \tfrac{1}{2}; j' \, m' | e^{-i\omega t S_z/\hbar} |n; \ell \, \tfrac{1}{2}; j = \ell + \tfrac{1}{2}, m\rangle \right|^2$$

with $j' = \ell' \pm \frac{1}{2}$. This gives

$$\mathcal{P}(t) = \delta_{n'n}\delta_{m'm}\delta_{\ell'\ell} \left| \cos \frac{\omega t}{2} - \frac{2i}{\hbar} \sin \frac{\omega t}{2} \langle S_z \rangle \right|^2$$

$$= \delta_{n'n}\delta_{m'm}\delta_{\ell'\ell} \left[\cos^2 \frac{\omega t}{2} + \frac{4}{\hbar^2} \sin^2 \frac{\omega t}{2} |\langle S_z \rangle|^2 \right]$$

using the fact that

$$\langle S_z \rangle = \langle \ell \, \tfrac{1}{2}; j' \, m | S_z | \ell + \tfrac{1}{2}, \, m; \ell \, \tfrac{1}{2} \rangle$$

is real. Proceeding with the determination of this matrix element, we note that the case $j' = \ell + \frac{1}{2}$ was calculated in problem 5.9. The second case, $j' = \ell - \frac{1}{2}$, can be obtained by starting from

$$\frac{\hbar^2}{4} = \langle \ell + \tfrac{1}{2}, \, m; \ell \, \tfrac{1}{2} | S_z^2 | \ell + \tfrac{1}{2}, \, m; \ell \, \tfrac{1}{2} \rangle$$

$$= \sum_{j'm'} \langle \ell + \tfrac{1}{2}, \, m; \ell \, \tfrac{1}{2} | S_z | j' \, m'; \ell \, \tfrac{1}{2} \rangle \langle j' \, m'; \ell \, \tfrac{1}{2} | S_z | \ell + \tfrac{1}{2}, \, m; \ell \, \tfrac{1}{2} \rangle$$

$$= \langle \ell + \tfrac{1}{2}, \ldots | S_z | \ell + \tfrac{1}{2}, \ldots \rangle \langle \ell + \tfrac{1}{2}, \ldots | S_z | \ell + \tfrac{1}{2}, \ldots \rangle$$

$$+ \langle \ell + \tfrac{1}{2}, \ldots | S_z | \ell - \tfrac{1}{2}, \ldots \rangle \langle \ell - \tfrac{1}{2}, \ldots | S_z | \ell + \tfrac{1}{2}, \ldots \rangle$$

Thus

$$\left| \langle \ell + \tfrac{1}{2}, \ldots | S_z | \ell + \tfrac{1}{2}, \ldots \rangle \right|^2 + \left| \langle \ell - \tfrac{1}{2}, \ldots | S_z | \ell + \tfrac{1}{2}, \ldots \rangle \right|^2 = \frac{\hbar^2}{4}$$

---

[2] Applying the formula (b) of problem 5.8 for $\mathbf{V} \to J_z$, we obtain that the matrix elements $\langle j \, m' | \mathbf{L} | j \, m \rangle$, $\langle j \, m' | \mathbf{S} | j \, m \rangle$ vanish unless $m' = m$.

Substituting the known first term, we get

$$\left| \langle \ell - \tfrac{1}{2}, \dots | S_z | \ell + \tfrac{1}{2}, \dots \rangle \right|^2 = \frac{\hbar^2}{4} \left[1 - \left(\frac{2m}{2\ell + 1} \right)^2 \right]$$

Thus, finally, we obtain

$$\mathcal{P}_+(t) = \delta_{nn'} \delta_{\ell\ell'} \delta_{mm'} \left[\cos^2 \frac{\omega t}{2} + \sin^2 \frac{\omega t}{2} \left(\frac{2m}{2\ell + 1} \right)^2 \right]$$

and

$$\mathcal{P}_-(t) = \delta_{nn'} \delta_{\ell\ell'} \delta_{mm'} \sin^2 \frac{\omega t}{2} \left[1 - \left(\frac{2m}{2\ell + 1} \right)^2 \right]$$

Problem 5.11 Consider a spinless particle of mass μ and charge q under the simultaneous influence of a uniform magnetic field and a uniform electric field. The interaction Hamiltonian consists of the two terms

$$H_{\mathrm{M}} = -\frac{q}{2\mu c} \, (\mathbf{B} \cdot \mathbf{L}), \quad H_{\mathrm{E}} = -q \, (\boldsymbol{\mathcal{E}} \cdot \mathbf{r})$$

Show that

$$\left| \langle \ell' \, m' | H_{\mathrm{M}} + H_{\mathrm{E}} | \ell \, m \rangle \right|^2 = \left| \langle \ell' \, m' | H_{\mathrm{M}} | \ell \, m \rangle \right|^2 + \left| \langle \ell' \, m' | H_{\mathrm{E}} | \ell \, m \rangle \right|^2$$

and that, always, one of the matrix elements $\langle \ell' \, m' | H_{\mathrm{M}} | \ell \, m \rangle$ and $\langle \ell' \, m' | H_{\mathrm{E}} | \ell \, m \rangle$ vanishes.

Solution

We can always take the plane defined by the vectors $\mathbf{E}$ and $\mathbf{B}$ to be the $\hat{x}$ plane; then the electric and magnetic field will not have components along this direction. A rotation by π around the x-axis will change the sign of $\mathbf{B} \cdot \mathbf{L}$. The state $e^{-i\pi L_x/\hbar} \, | \ell \, m \rangle$ can easily be shown to be an eigenstate of L_z with eigenvalue $-\hbar m$, namely

$$L_z \left(e^{-i\pi L_x/\hbar} \, | \ell \, m \rangle \right) = -e^{-i\pi L_x/\hbar} \, L_z \, | \ell \, m \rangle = -\hbar m \left(e^{-i\pi L_x/\hbar} \, | \ell \, m \rangle \right)$$

Furthermore,

$$e^{-i\pi L_x/\hbar} \, | \ell \, m \rangle = | \ell \, {-m} \rangle$$

Thus, we have

$$\langle \ell' \, m' | H_{\mathrm{M}} | \ell \, m \rangle = -\frac{q}{2\mu} \mathbf{B} \cdot \langle \ell' \, m' | \mathbf{L} | \ell \, m \rangle$$

$$= \frac{q}{2\mu} \mathbf{B} \cdot \langle \ell' \, m' | e^{i\pi L_x/\hbar} \mathbf{L} e^{-i\pi L_x/\hbar} | \ell \, m \rangle$$

$$= \frac{q}{2\mu} \mathbf{B} \cdot \langle \ell' \, {-m'} | \mathbf{L} | \ell \, {-m} \rangle$$

The states $|\ell -m\rangle$ can be obtained via

$$|\ell -m\rangle \propto L_-^m|\ell\, 0\rangle$$

and we have, noting that $L_j^* = -L_j$,

$$|\ell\, m\rangle \propto L_+^m|\ell\, 0\rangle \propto \left(L_-^m\right)^*|\ell\, 0\rangle \propto (|\ell -m\rangle)^*$$

Considering the square of the quantities in the last line we can see that the proportionality constant is just a phase. Thus, we have

$$\langle \ell'\, m'|H_M|\ell\, m\rangle = -\frac{q}{2\mu}\mathbf{B}\cdot\langle \ell'\, -m'|\mathbf{L}^*|\ell\, -m\rangle$$

$$= -\frac{q}{2\mu}\mathbf{B}\cdot\left(\langle \ell'\, m'|\mathbf{L}|\ell\, m\rangle\right)^*$$

$$= \langle \ell'\, m'|H_M|\ell\, m\rangle^*$$

In an analogous fashion we have

$$\langle \ell'\, m'|H_E|\ell\, m\rangle = -q\mathbf{E}\cdot\langle \ell'\, m'|\mathbf{r}|\ell\, m\rangle$$

$$= q\mathbf{E}\cdot\langle \ell'\, m'|e^{i\pi L_x/\hbar}\mathbf{r}e^{-i\pi L_x/\hbar}|\ell\, m\rangle$$

$$= q\mathbf{E}\cdot\langle \ell'\, -m'|\mathbf{r}|\ell\, -m\rangle$$

$$= q\mathbf{E}\cdot\langle \ell'\, m'|\mathbf{r}|\ell\, m\rangle^*$$

$$= -\langle \ell'\, m'|H_E|\ell\, m\rangle^*$$

Consequently,

$$\left|\langle \ell'\, m'|H_M + H_E|\ell\, m\rangle\right|^2 = \left|\langle \ell'\, m'|H_M|\ell\, m\rangle\right|^2 + \left|\langle \ell'\, m'|H_E|\ell\, m\rangle\right|^2$$
$$+ \langle \ell'\, m'|H_M|\ell\, m\rangle^*\, \langle \ell'\, m'|H_E|\ell\, m\rangle$$
$$+ \langle \ell'\, m'|H_M|\ell\, m\rangle\, \langle \ell'\, m'|H_E|\ell\, m\rangle^*$$

Since, however,

$$\langle \ell'\, m'|H_M|\ell\, m\rangle^*\, \langle \ell'\, m'|H_E|\ell\, m\rangle + \langle \ell'\, m'|H_M|\ell\, m\rangle\, \langle \ell'\, m'|H_E|\ell\, m\rangle^*$$
$$= \langle \ell'\, m'|H_M|\ell\, m\rangle\, \langle \ell'\, m'|H_E|\ell\, m\rangle - \langle \ell'\, m'|H_M|\ell\, m\rangle\, \langle \ell'\, m'|H_E|\ell\, m\rangle = 0$$

we obtain as required

$$\left|\langle \ell'\, m'|H_M + H_E|\ell\, m\rangle\right|^2 = \left|\langle \ell'\, m'|H_M|\ell\, m\rangle\right|^2 + \left|\langle \ell'\, m'|H_E|\ell\, m\rangle\right|^2$$

The states $|\ell\ m\rangle$ are parity eigenstates.[3] Thus, on the one hand,

$$\langle\ell'\ m'|H_\mathrm{M}|\ell\ m\rangle = \varpi'\varpi\,\langle\ell'\ m'|\mathcal{P}H_\mathrm{M}\mathcal{P}|\ell\ m\rangle = \varpi'\varpi\,\langle\ell'\ m'|H_\mathrm{M}|\ell\ m\rangle$$

which implies that $\varpi\varpi' = 1$. On the other hand,

$$\langle\ell'\ m'|H_\mathrm{E}|\ell\ m\rangle = \varpi'\varpi\,\langle\ell'\ m'|\mathcal{P}H_\mathrm{E}\mathcal{P}|\ell\ m\rangle = -\varpi\varpi'\langle\ell'\ m'|H_\mathrm{E}|\ell\ m\rangle$$

which implies that $\varpi\varpi' = -1$. The two conclusions are mutually exclusive and thus, always, one of the two matrix elements must vanish.

Problem 5.12 Consider an electron under the simultaneous influence of a uniform magnetic field and a uniform electric field. Calculate the matrix elements

$$\left\langle\ell\ m;\tfrac{1}{2}\ m_s|H_\mathrm{M}|\ell\ m;\tfrac{1}{2}\ m_s'\right\rangle, \qquad \left\langle\ell\ m;\tfrac{1}{2}\ m_s|H_\mathrm{E}|\ell\ m;\tfrac{1}{2}\ m_s'\right\rangle$$

where H_M, H_E are the interaction-Hamiltonian terms:

$$H_\mathrm{M} = -\frac{e}{2m_e c}\mathbf{B}\cdot(\mathbf{L} + 2\mathbf{S}), \qquad H_\mathrm{E} = -e\mathbf{E}\cdot\mathbf{r}$$

Solution

The electric matrix element vanishes because of parity:

$$\begin{aligned}
\left\langle\ell\ m;\tfrac{1}{2}\ m_s|\mathbf{r}|\ell\ m;\tfrac{1}{2}\ m_s'\right\rangle &= -\left\langle\ell\ m;\tfrac{1}{2}\ m_s|\mathcal{P}\mathbf{r}\mathcal{P}|\ell\ m;\tfrac{1}{2}\ m_s'\right\rangle \\
&= -(-1)^{2\ell}\left\langle\ell\ m;\tfrac{1}{2}\ m_s|\mathbf{r}|\ell\ m;\tfrac{1}{2}\ m_s'\right\rangle \\
&= -\left\langle\ell\ m;\tfrac{1}{2}\ m_s|\mathbf{r}|\ell\ m;\tfrac{1}{2}\ m_s'\right\rangle = 0
\end{aligned}$$

The magnetic matrix element consists of two parts. The first is[4]

$$\begin{aligned}
-\frac{e}{2m_e c}\mathbf{B}\cdot\left\langle\ell\ m;\tfrac{1}{2}\ m_s|\mathbf{L}|\ell\ m;\tfrac{1}{2}\ m_s'\right\rangle &= -\frac{e}{2m_e c}\delta_{m_s m_s'}\mathbf{B}\cdot\left\langle\ell\ m;\tfrac{1}{2}\ m_s|\mathbf{L}|\ell\ m;\tfrac{1}{2}\ m_s'\right\rangle \\
&= -\frac{m\hbar e}{2m_e c}\delta_{m_s m_s'}B_z
\end{aligned}$$

The second contribution,

$$-\frac{e}{m_e c}\mathbf{B}\cdot\left\langle\ell\ m;\tfrac{1}{2}\ m_s|\mathbf{S}|\ell\ m;\tfrac{1}{2}\ m_s'\right\rangle$$

can be written as

$$-\frac{e\hbar}{2m_e c}\begin{pmatrix} B_z & B_x - iB_y \\ B_x + iB_y & -B_z \end{pmatrix}_{m_s m_s'}$$

---

[3] The parity eigenvalue of $|\ell\ m\rangle$ is $\varpi = (-1)^\ell$.
[4] $\mathbf{L} = \hat{\mathbf{x}}L_x + \hat{\mathbf{y}}L_y + \hat{\mathbf{z}}L_z = \tfrac{1}{2}(\hat{\mathbf{x}} - i\hat{\mathbf{y}})L_+ + \tfrac{1}{2}(\hat{\mathbf{x}} + i\hat{\mathbf{y}})L_- + \hat{\mathbf{z}}L_z$

Thus, finally, we have

$$\langle \ell\, m; \tfrac{1}{2}\uparrow\, |H_{\mathrm{M}}|\ell\, m; \tfrac{1}{2}\uparrow\rangle = -\frac{(m+1)\hbar e}{2m_{\mathrm{e}}}B_z$$

$$\langle \ell\, m; \tfrac{1}{2}\downarrow\, |H_{\mathrm{M}}|\ell\, m; \tfrac{1}{2}\downarrow\rangle = -\frac{(m-1)\hbar e}{2m_{\mathrm{e}}}B_z$$

$$\langle \ell\, m; \tfrac{1}{2}\uparrow\, |H_{\mathrm{M}}|\ell\, m; \tfrac{1}{2}\downarrow\rangle = \langle \ell\, m; \tfrac{1}{2}\downarrow\, |H_{\mathrm{M}}|\ell\, m; \tfrac{1}{2}\uparrow\rangle^*$$
$$= -\frac{\hbar e}{2m_{\mathrm{e}}}(B_x - i B_y)$$

Problem 5.13 Consider a particle with spin quantum number $s = 1$. Ignore all spatial degrees of freedom and assume that the particle is subject to an external magnetic field $\mathbf{B} = \hat{\mathbf{x}}B$. The Hamiltonian operator of the system is $H = g\mathbf{B}\cdot\mathbf{S}$.

(a) Obtain explicitly the spin matrices in the basis of the $\mathbf{S}^2$, S_z eigenstates, $|s, m_s\rangle$.
(b) If the particle is initially (at $t = 0$) in the state $|1\,1\rangle$, find the evolved state of the particle at times $t > 0$.
(c) What is the probability of finding the particle in the state $|1\,-1\rangle$?

Solution
(a) In order to obtain the spin matrices for $s = 1$, we consider the relations

$$S_+|1\,0\rangle = \hbar\sqrt{2}|1\,1\rangle, \qquad S_+|1\,-1\rangle = \hbar\sqrt{2}|1\,0\rangle$$
$$S_-|1\,1\rangle = \hbar\sqrt{2}|1\,0\rangle, \qquad S_-|1\,0\rangle = \hbar\sqrt{2}|1\,-1\rangle$$

which lead to

$$S_+ = \hbar\sqrt{2}\begin{pmatrix} 0 & 1 & 0 \\ 0 & 0 & 1 \\ 0 & 0 & 0 \end{pmatrix}, \qquad S_- = (S_+)^\dagger = \hbar\sqrt{2}\begin{pmatrix} 0 & 0 & 0 \\ 1 & 0 & 0 \\ 0 & 1 & 0 \end{pmatrix}$$

From these we obtain

$$S_x = \frac{1}{2}(S_+ + S_-) = \frac{\hbar}{\sqrt{2}}\begin{pmatrix} 0 & 1 & 0 \\ 1 & 0 & 1 \\ 0 & 1 & 0 \end{pmatrix}$$

Similarly, we get

$$S_y = \frac{1}{2i}(S_+ - S_-) = \frac{\hbar}{\sqrt{2}}\begin{pmatrix} 0 & -i & 0 \\ i & 0 & -i \\ 0 & i & 0 \end{pmatrix}$$

Their commutator is

$$[\mathcal{S}_x, \mathcal{S}_y] = i\hbar^2 \begin{pmatrix} 1 & 0 & 0 \\ 0 & 0 & 0 \\ 0 & 0 & -1 \end{pmatrix}$$

implying that

$$\mathcal{S}_z = \hbar \begin{pmatrix} 1 & 0 & 0 \\ 0 & 0 & 0 \\ 0 & 0 & -1 \end{pmatrix}$$

(b) From the matrix $\mathcal{S}_x$ we can derive that its eigenvectors[5] correspond to the states

$$|S_x = \hbar\rangle = \tfrac{1}{2}\left(|1\ 1\rangle + \sqrt{2}|1\ 0\rangle + |1\ -1\rangle\right)$$

$$|S_x = 0\rangle = \tfrac{1}{\sqrt{2}}\left(|1\ 1\rangle - |1\ -1\rangle\right)$$

$$|S_x = -\hbar\rangle = \tfrac{1}{2}\left(|1\ 1\rangle - \sqrt{2}|1\ 0\rangle + |1\ -1\rangle\right)$$

The inverse relations, in an obvious notation, are

$$|1\ 1\rangle = \tfrac{1}{2}\left(|\hbar\rangle + |-\hbar\rangle + \sqrt{2}|0\rangle\right)$$

$$|1\ 0\rangle = \tfrac{1}{\sqrt{2}}\left(|\hbar\rangle - |-\hbar\rangle\right)$$

$$|1\ -1\rangle = \tfrac{1}{2}\left(|\hbar\rangle + |-\hbar\rangle - \sqrt{2}|0\rangle\right)$$

The evolved state of the particle will be

$$|\psi(t)\rangle = e^{-igBtS_x/\hbar}\,|1\ 1\rangle = \tfrac{1}{2}\left(e^{-igBt}|\hbar\rangle + e^{igBt}|-\hbar\rangle + \sqrt{2}|0\rangle\right)$$

(c) Transforming back to the $\mathcal{S}_z$ eigenstates, we get

$$|\psi(t)\rangle = \cos^2(gBt/2)\,|1\ 1\rangle - \sin^2(gBt/2)\,|1\ -1\rangle$$
$$- i\sqrt{2}\sin(gBt/2)\cos(gBt/2)\,|1\ 0\rangle$$

The probability of finding the particle in the $\mathcal{S}_z$ eigenstate $|1\ -1\rangle$ is

$$\mathcal{P}_\downarrow = \sin^4(gBt/2)$$

---

[5] The eigenvalues are, of course, $\hbar, 0, -\hbar$.

Problem 5.14 Consider a pair of particles with opposite electric charges that have a magnetic-dipole-moment interaction

$$H_I = A \left(\boldsymbol{\mu}_1 \cdot \boldsymbol{\mu}_2 \right) = -\frac{e^2 g_1 g_2}{2m_1 m_2} A \left(\mathbf{S}^{(1)} \cdot \mathbf{S}^{(2)} \right)$$

The system is subject to an external uniform magnetic field **B**, which introduces the interaction

$$-\mathbf{B} \cdot (\boldsymbol{\mu}_1 + \boldsymbol{\mu}_2) = -\frac{e}{2m_1 m_2} \mathbf{B} \cdot \left(m_2 g_1 \mathbf{S}^{(1)} - m_1 g_2 \mathbf{S}^{(2)} \right)$$

Ignore all degrees of freedom other than those due to spin.

(a) Determine the energy eigenvalues and eigenstates.[6] Express the results in terms of the parameters $a = e^2 g_1 g_2 A / 4 m_1 m_2$ and $b_i = e B g_i / 4 m_i$.
(b) The system is initially (at $t = 0$) in the state $| \uparrow \rangle^{(1)} | \downarrow \rangle^{(2)}$. Calculate the probability of finding the system in the state $| \downarrow \rangle^{(1)} | \uparrow \rangle^{(2)}$ at a later time $t > 0$. What is the maximum value of this probability and at what time is it attained?
(c) Find the expectation values of the individual spins and of the total spin at any time.

Solution

(a) The Hamiltonian of the system is

$$H = -a \mathbf{S}^{(1)} \cdot \mathbf{S}^{(2)} - b_1 S_z^{(1)} + b_2 S_z^{(2)}$$
$$= -\frac{a}{2} \left(S_+^{(1)} S_-^{(2)} + S_-^{(1)} S_+^{(2)} \right) - a S_z^{(1)} S_z^{(2)} - b_1 S_z^{(1)} + b_2 S_z^{(2)}$$

Acting on the products of the one-particle spin states, we get

$$H | \uparrow \rangle^{(1)} | \uparrow \rangle^{(2)} = \left(-\frac{a\hbar^2}{4} - \frac{b_1 \hbar}{2} + \frac{b_2 \hbar}{2} \right) | \uparrow \rangle^{(1)} | \uparrow \rangle^{(2)}$$

$$H | \downarrow \rangle^{(1)} | \downarrow \rangle^{(2)} = \left(-\frac{a\hbar^2}{4} + \frac{b_1 \hbar}{2} - \frac{b_2 \hbar}{2} \right) | \downarrow \rangle^{(1)} | \downarrow \rangle^{(2)}$$

$$H | \uparrow \rangle^{(1)} | \downarrow \rangle^{(2)} = -\frac{a\hbar^2}{2} | \downarrow \rangle^{(1)} | \uparrow \rangle^{(2)} + \left(\frac{a\hbar^2}{4} - \frac{b_1 \hbar}{2} - \frac{b_2 \hbar}{2} \right) | \uparrow \rangle^{(1)} | \downarrow \rangle^{(2)}$$

$$H | \downarrow \rangle^{(1)} | \uparrow \rangle^{(2)} = -\frac{a\hbar^2}{2} | \uparrow \rangle^{(1)} | \downarrow \rangle^{(2)} + \left(\frac{a\hbar^2}{4} + \frac{b_1 \hbar}{2} + \frac{b_2 \hbar}{2} \right) | \downarrow \rangle^{(1)} | \uparrow \rangle^{(2)}$$

---

[6] Expressed in terms of the spin eigenstates $| \uparrow \rangle^{(1)} | \uparrow \rangle^{(2)}, \ldots, | \downarrow \rangle^{(1)} | \downarrow \rangle^{(2)}$.

Thus, we obtain the following eigenstates and corresponding eigenvalues:

$$|\Psi_{++}\rangle \equiv |\uparrow\rangle^{(1)} |\uparrow\rangle^{(2)}$$

$$E_{++} = \frac{\hbar}{4}(2b_2 - 2b_1 - a\hbar)$$

$$|\Psi_{--}\rangle \equiv |\downarrow\rangle^{(1)} |\downarrow\rangle^{(2)}$$

$$E_{--} = -\frac{\hbar}{4}(2b_2 - 2b_1 + a\hbar)$$

$$|\Psi_{+-}\rangle = \cos\gamma \, |\uparrow\rangle^{(1)} |\downarrow\rangle^{(2)} + \sin\gamma \, |\downarrow\rangle^{(1)} |\uparrow\rangle^{(2)}$$

$$E_{+-} = \frac{a\hbar^2}{4}(1 - 2\cot 2\gamma - 2\tan\gamma)$$

$$|\Psi_{-+}\rangle = \sin\gamma \, |\uparrow\rangle^{(1)} |\downarrow\rangle^{(2)} - \cos\gamma \, |\downarrow\rangle^{(1)} |\uparrow\rangle^{(2)}$$

$$E_{-+} = \frac{a\hbar^2}{4}(1 - 2\cot 2\gamma + 2\cot\gamma)$$

where

$$\cot 2\gamma \equiv \frac{b_1 + b_2}{a\hbar}$$

(b) The evolved state of the system is

$$|\Psi(t)\rangle = e^{-iHt/\hbar} |\uparrow\rangle^{(1)} |\downarrow\rangle^{(2)} = e^{-iHt/\hbar} \left(\cos\gamma |\Psi_{+-}\rangle + \sin\gamma |\Psi_{-+}\rangle\right)$$

$$= e^{-iE_{+-}t/\hbar} \cos\gamma \, |\Psi_{+-}\rangle + e^{-iE_{-+}t/\hbar} \sin\gamma \, |\Psi_{-+}\rangle$$

$$= \left(\cos^2\gamma \, e^{-iE_{+-}t/\hbar} + \sin^2\gamma \, e^{-iE_{-+}t/\hbar}\right) |\uparrow\rangle^{(1)} |\downarrow\rangle^{(2)}$$

$$+ \cos\gamma \, \sin\gamma \left(e^{-iE_{+-}t/\hbar} - e^{-iE_{-+}t/\hbar}\right) |\downarrow\rangle^{(1)} |\uparrow\rangle^{(2)}$$

The probability for the *flipped* state can be read off as

$$\mathcal{P}_{\downarrow\uparrow} = \sin^2 2\gamma \sin^2 \left[\frac{(E_{+-} - E_{-+})t}{2\hbar}\right] = \sin^2 2\gamma \sin^2 \left(\frac{a\hbar t}{2\sin 2\gamma}\right)$$

This probability achieves its maximum value $\sin^2 2\gamma$ at times

$$t_n = \frac{(2n+1)\pi}{\hbar a} \sin 2\gamma \qquad (n = 0, 1, \ldots)$$

Total flip would be possible only for $\gamma = \pi/4$.

(c) The evolved state is

$$|\Psi(t)\rangle = C(t)|\uparrow\rangle^{(1)} |\downarrow\rangle^{(2)} + D(t)|\downarrow\rangle^{(1)} |\uparrow\rangle^{(2)}$$

where $C(t)$ and $D(t)$ can be read off from the expression for $|\Psi(t)\rangle$ in part (b). The expectation value of the total spin is

$$\langle\Psi(t)|\mathbf{S}|\Psi(t)\rangle = |C(t)|^2 \left(\langle\uparrow|\mathbf{S}_1|\uparrow\rangle\langle\downarrow|\downarrow\rangle + \langle\downarrow|\mathbf{S}_2|\downarrow\rangle\langle\uparrow|\uparrow\rangle\right)$$
$$+ |D(t)|^2 \left(\langle\downarrow|\mathbf{S}_1|\downarrow\rangle\langle\uparrow|\uparrow\rangle + \langle\uparrow|\mathbf{S}_2|\uparrow\rangle\langle\downarrow|\downarrow\rangle\right)$$
$$+ C^*(t)D(t)\left(\langle\uparrow|\mathbf{S}_1|\downarrow\rangle\langle\downarrow|\uparrow\rangle + \langle\downarrow|\mathbf{S}_2|\uparrow\rangle\langle\uparrow|\downarrow\rangle\right) + \text{h.c.}$$
$$= |C(t)|^2 \left(\frac{\hbar}{2}\hat{\mathbf{z}} - \frac{\hbar}{2}\hat{\mathbf{z}}\right) + |D(t)|^2 \left(\frac{\hbar}{2}\hat{\mathbf{z}} - \frac{\hbar}{2}\hat{\mathbf{z}}\right)$$
$$+ C^*(t)D(t)\,(0+0) + C(t)D^*(t)\,(0+0) = 0$$

Thus, we get

$$\langle\mathbf{S}\rangle_t = 0$$

The individual spin expectation values are

$$\langle\mathbf{S}_1\rangle_t = \frac{\hbar}{2}\left(|C(t)|^2 - |D(t)|^2\right)\hat{\mathbf{z}} = \frac{\hbar}{2}\left(1 - 2|D(t)|^2\right)\hat{\mathbf{z}} = \frac{\hbar}{2}\left(1 - 2\mathcal{P}_{\downarrow\uparrow}\right)\hat{\mathbf{z}}$$

and we have

$$\langle\mathbf{S}_2\rangle_t = -\langle\mathbf{S}_1\rangle_t$$

Problem 5.15 A beam of neutrons with energy E_0 and spin along the positive z-axis enters a region where there is a uniform magnetic field $\mathbf{B}$ (see Fig. 22). The Hamiltonian interaction term with the magnetic field is

$$H = -\mathbf{B}\cdot\boldsymbol{\mu}_n = 2\omega\hat{\mathbf{n}}\cdot\mathbf{S}$$

where $\hat{\mathbf{n}}$ is the direction of the magnetic field and $\omega = B\mu_n/\hbar$.

Ignore the spatial degrees of freedom and find the state of the system at any time $t > 0$. Compute the expectation value of the spin $\mathbf{S}$.

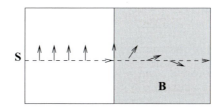

Fig. 22 A neutron spin in a uniform magnetic field.

Solution

If the width of the region in which the change in magnetic field value from zero to B occurs is small, namely,

$$\Delta x \ll \frac{v_0}{\omega} = \frac{2^{1/2}\hbar\sqrt{E_0}}{B\mu_n m_n^{1/2}}$$

the transition can be considered as being instantaneous. Thus, we can assume that at the moment of entrance ($t = 0$) the spin wave function does not change and from then on evolves according to the Hamiltonian H.

Let us characterize the direction $\hat{\mathbf{n}}$ by the angles θ and ϕ. Then, we can write

$$\hat{\mathbf{n}} = \hat{\mathbf{x}}\cos\phi\sin\theta + \hat{\mathbf{y}}\sin\phi\sin\theta + \hat{\mathbf{z}}\cos\theta$$

The matrix $\hat{\mathbf{n}} \cdot \mathbf{S}$ is given by

$$\hat{\mathbf{n}} \cdot \mathbf{S} = \frac{\hbar}{2}\begin{pmatrix} \cos\theta & \sin\theta\, e^{-i\phi} \\ \sin\theta\, e^{i\phi} & -\cos\theta \end{pmatrix}$$

The corresponding evolution operator will be

$$\exp\left(-\frac{i}{\hbar}2\omega\hat{\mathbf{n}} \cdot \mathbf{S}t\right) = \exp\left[-i\omega t\begin{pmatrix} \cos\theta & \sin\theta\, e^{-i\phi} \\ \sin\theta\, e^{i\phi} & -\cos\theta \end{pmatrix}\right]$$

$$= \cos\omega t - i\begin{pmatrix} \cos\theta & \sin\theta\, e^{-i\phi} \\ \sin\theta\, e^{i\phi} & -\cos\theta \end{pmatrix}\sin\omega t$$

$$= \begin{pmatrix} \cos\omega t - i\sin\omega t\cos\theta & -i\sin\omega t\sin\theta\, e^{-i\phi} \\ -i\sin\omega t\sin\theta\, e^{i\phi} & \cos\omega t + i\sin\omega t\cos\theta \end{pmatrix}$$

The evolved state will be

$$\Psi(t) = \begin{pmatrix} \cos\omega t - i\sin\omega t\cos\theta & -i\sin\omega t\sin\theta\, e^{-i\phi} \\ -i\sin\omega t\sin\theta\, e^{i\phi} & \cos\omega t + i\sin\omega t\cos\theta \end{pmatrix}\begin{pmatrix} 1 \\ 0 \end{pmatrix}$$

or

$$\Psi(t) = \begin{pmatrix} \cos\omega t - i\sin\omega t\cos\theta \\ -i\sin\omega t\sin\theta\, e^{i\phi} \end{pmatrix}$$

The expectation value of the spin in this state is

$$\langle \mathbf{S} \rangle_t = \frac{\hbar}{2}\left[\hat{\mathbf{x}}\left(-\sin\theta\sin\phi\,\sin 2\omega t + \sin 2\theta\cos\phi\,\sin^2\omega t\right)\right.$$
$$+ \hat{\mathbf{y}}\left(\sin\theta\cos\phi\,\sin 2\omega t + \sin 2\theta\sin\phi\,\sin^2\omega t\right)$$
$$\left.+ \hat{\mathbf{z}}\left(\cos^2\omega t + \sin^2\omega t\,\cos 2\theta\right)\right]$$

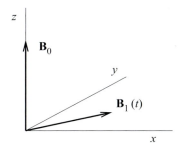

Fig. 23 Combination of static and oscillating magnetic fields.

As a particular case, let us take $\phi = 0$, $\theta = \pi/2$, namely $\mathbf{B} = \hat{\mathbf{x}}B$. Then we get

$$\langle \mathbf{S} \rangle_t = \frac{\hbar}{2} \left(\sin 2\omega t \, \hat{\mathbf{y}} + \cos 2\omega t \, \hat{\mathbf{z}} \right)$$

which corresponds to a rotation in the zy-plane.

Problem 5.16 An electron is subject to a static uniform magnetic field $\mathbf{B}_0 = B_0 \hat{\mathbf{z}}$ (see Fig. 23) and occupies the spin eigenstate $| \uparrow \rangle$. At a given moment ($t = 0$) an additional time-dependent, spatially uniform, magnetic field $\mathbf{B}_1(t) = B_1 (\cos \omega t \, \hat{\mathbf{x}} + \sin \omega t \, \hat{\mathbf{y}})$ is turned on. Calculate the probability of finding the electron with its spin along the negative z-axis at time $t > 0$. Ignore spatial degrees of freedom.

Solution

The Schroedinger equation for the system is

$$i\hbar \frac{d}{dt} |\psi(t)\rangle = -\frac{e}{m_e} \left(B_0 S_z + B_1 \cos \omega t \, S_x + B_1 \sin \omega t \, S_y \right) |\psi(t)\rangle$$

or

$$i\hbar \dot{\psi}(t) = -\frac{e\hbar}{2m_e} \begin{pmatrix} B_0 & B_1 e^{-i\omega t} \\ B_1 e^{i\omega t} & -B_0 \end{pmatrix} \psi(t)$$

Setting

$$\psi(t) = \begin{pmatrix} a(t) \\ b(t) \end{pmatrix}$$

we obtain

$$\dot{a} = i \frac{e}{2m_e} \left(B_0 a + B_1 e^{-i\omega t} b \right)$$

$$\dot{b} = i \frac{e}{2m_e} \left(B_1 e^{i\omega t} a - B_0 b \right)$$

Introducing

$$\omega_0 = \frac{|e|B_0}{2m_e}, \qquad \omega_1 = \frac{|e|B_1}{2m_e}$$

we obtain the above equations in the form

$$\dot{a} = -i\omega_0\, a - i\omega_1 e^{-i\omega t} b$$
$$\dot{b} = i\omega_0\, b - i\omega_1 e^{i\omega t} a$$

At this point, let us substitute the trial solutions

$$a(t) = e^{-i\omega t/2 + i\Omega t}\, A, \qquad b(t) = e^{i\omega t/2 + i\Omega t}\, B$$

We immediately obtain

$$\Omega = \pm\frac{1}{2}\sqrt{(\omega - 2\omega_0)^2 + 4\omega_1^2} \equiv \pm\frac{1}{2}\gamma$$

and

$$B = -\frac{\pm\gamma - (\omega - 2\omega_0)}{2\omega_1} A$$

Finally, we obtain

$$a(t) = e^{-i\omega t/2}\left(A_+ e^{i\gamma t/2} + A_- e^{-i\gamma t/2}\right)$$
$$b(t) = e^{i\omega t/2}\left[-\frac{\gamma - (\omega - 2\omega_0)}{2\omega_1} A_+ e^{i\gamma t/2} + \frac{\gamma + (\omega - 2\omega_0)}{2\omega_1} A_- e^{-i\gamma t/2}\right]$$

Applying the initial condition $a(0) = 1$, $b(0) = 0$, we arrive at

$$A_\pm = \frac{\gamma \pm (\omega - 2\omega_0)}{2\gamma}$$

and

$$\psi(t) = \begin{pmatrix} e^{-i\omega t/2}\left(\cos\dfrac{\gamma t}{2} + i\dfrac{\omega - 2\omega_0}{\gamma}\sin\dfrac{\gamma t}{2}\right) \\ -2i\, e^{i\omega t/2}\dfrac{\omega_1}{\gamma}\sin\dfrac{\gamma t}{2} \end{pmatrix}$$

The probability of finding the spin of the electron pointing along $-z$ is

$$\mathcal{P}_\downarrow = \frac{4\omega_1^2}{(\omega - 2\omega_0)^2 + 4\omega_1^2}\sin^2\frac{\gamma t}{2}$$

Note that for $\omega = 2\omega_0$ this probability exhibits the resonance phenomenon. For this choice the probability of spin flip is

$$\mathcal{P}_\downarrow\big|_{\omega=2\omega_0} = \sin^2 \omega_1 t$$

and becomes unity at $t = (2n + 1)\pi/2\omega_1$.

Problem 5.17 A particle with total angular momentum $j = \frac{3}{2}$ is in an eigenstate $|j\ m\rangle$ of $\mathbf{J}^2$, J_z. Determine the probability of finding the particle in an eigenstate $|j\ m_n\rangle$ of the operator $\hat{\mathbf{n}} \cdot \mathbf{J}$ corresponding to the angular momentum component along an arbitrary direction $\hat{\mathbf{n}} = \sin\theta\cos\phi\,\hat{\mathbf{x}} + \sin\theta\sin\phi\,\hat{\mathbf{y}} + \cos\theta\,\hat{\mathbf{z}}$.

Solution

We can write

$$\hat{\mathbf{n}} \cdot \mathbf{J} = \tfrac{1}{2}\sin\theta\,e^{-i\phi}J_+ + \tfrac{1}{2}\sin\theta\,e^{i\phi}J_- + \cos\theta\,J_z$$

As implied in the question, this operator has common eigenstates with $\mathbf{J}^2$. Indeed, we have

$$\mathbf{J}^2|j\ m_n\rangle = \hbar^2 j(j+1)|j\ m_n\rangle$$
$$\hat{\mathbf{n}} \cdot \mathbf{J}|j\ m_n\rangle = \hbar m_n|j\ m_n\rangle$$

Each of these states can be expanded as follows, setting $j = \frac{3}{2}$:

$$\left|\tfrac{3}{2}\ m_n\right\rangle = \sum_{m=-\frac{3}{2}}^{m=\frac{3}{2}} C_m|3/2\ m\rangle$$

Acting on it with $\hat{\mathbf{n}} \cdot \mathbf{J}$ gives

$$\sum_{m=-\frac{3}{2}}^{m=\frac{3}{2}} C_m \left(\tfrac{1}{2}\sin\theta\,e^{-i\phi}C_m^{(+)}|\tfrac{3}{2},\ m+1\rangle + \tfrac{1}{2}\sin\theta\,e^{i\phi}C_m^{(-)}|\tfrac{3}{2},\ m-1\rangle\right.$$

$$\left. + \cos\theta\,\hbar m|\tfrac{3}{2}\ m\rangle\right) = \hbar m_n \sum_{m=-\frac{3}{2}}^{m=\frac{3}{2}} C_m|\tfrac{3}{2}\ m\rangle$$

with

$$C_m^{(\pm)} = \hbar\sqrt{15/4 - m(m\pm 1)}$$

Thus

$$C_{3/2}^{(+)} = C_{-3/2}^{(-)} = 0$$
$$C_{1/2}^{(+)} = C_{-3/2}^{(+)} = C_{3/2}^{(-)} = C_{-1/2}^{(-)} = \hbar\sqrt{3}$$
$$C_{-1/2}^{(+)} = C_{1/2}^{(-)} = 2\hbar$$

Equating the coefficients on both sides of the expansion we obtain

$$C_{3/2}\left(\tfrac{3}{2}\cos\theta - m_n\right) + C_{1/2}\left(\tfrac{\sqrt{3}}{2}\sin\theta\,e^{-i\phi}\right) = 0$$

$$C_{3/2}\left(\tfrac{\sqrt{3}}{2}\sin\theta\,e^{i\phi}\right) + C_{1/2}\left(\tfrac{1}{2}\cos\theta - m_n\right) + C_{-1/2}\left(\sin\theta\,e^{-i\phi}\right) = 0$$

$$C_{1/2}\left(\sin\theta\,e^{i\phi}\right) + C_{-1/2}\left(-\tfrac{1}{2}\cos\theta - m_n\right) + C_{-3/2}\left(\tfrac{\sqrt{3}}{2}\sin\theta\,e^{-i\phi}\right) = 0$$

$$C_{-1/2}\left(\tfrac{\sqrt{3}}{2}\sin\theta\,e^{i\phi}\right) + C_{-3/2}\left(-\tfrac{3}{2}\cos\theta - m_n\right) = 0$$

A non-trivial solution for these coefficients requires a vanishing determinant, which, after some algebra, gives as expected

$$m_n^4 - \tfrac{5}{2}m_n^2 + \tfrac{9}{16} = 0 \qquad \Longrightarrow \qquad m_n = \pm\tfrac{3}{2}, \pm\tfrac{1}{2}$$

As an example, let us work out the $m_n = \tfrac{3}{2}$ case. We get

$$\frac{C_{-1/2}}{C_{-3/2}} = \frac{1}{\sqrt{3}} \cot\frac{\theta}{2}\, e^{-i\phi}, \qquad \frac{C_{1/2}}{C_{-3/2}} = \frac{2}{\sqrt{3}} \cot\theta \cot\frac{\theta}{2}\, e^{-2i\phi}$$

$$\frac{C_{3/2}}{C_{-3/2}} = \frac{2}{3}\cot\theta \cot^2\frac{\theta}{2}\, e^{-3i\phi}$$

$C_{-3/2}$ is determined from the normalization condition as

$$\left(\frac{1}{C_{-3/2}}\right)^2 = \left(1 + \frac{1}{3}\cot^2\frac{\theta}{2}\right)\left(1 + \frac{4}{3}\cot^2\theta \cot^2\frac{\theta}{2}\right)$$

The probability of finding the particle in the state corresponding to the eigenvalue $m_n = \tfrac{3}{2}$ is

$$\mathcal{P}\left(m, m_n = \tfrac{3}{2}\right) = \left|\left\langle \tfrac{3}{2} m \,\middle|\, \tfrac{3}{2},\, m_n = \tfrac{3}{2}\right\rangle\right|^2 = |C_m|^2$$

Problem 5.18 A particle with $s = 1$ is in the S_z eigenstate with eigenvalue $+\hbar$. Consider the spin operator $S_{z'}$ in a direction $\hat{z}'$ that makes an angle θ with $\hat{z}$.

(a) Find the eigenstates of $S_{z'}$ and express the state of the system in terms of them.
(b) Calculate the uncertainty in $S_{z'}$ for the given state of the system.

Solution
(a) Acting with

$$S_{z'} = \sin\theta\,\cos\phi\, S_x + \sin\theta\,\sin\phi\, S_y + \cos\theta\, S_z$$
$$= \tfrac{1}{2}\sin\theta\left(e^{-i\phi} S_+ + e^{i\phi} S_-\right) + \cos\theta\, S_z$$

on the ansatz

$$|1\ m_{z'}\rangle = C_{-1}|1-1\rangle + C_0|1\ 0\rangle + C_1|1\ 1\rangle$$

we obtain

$$m_{z'}|1\ m_{z'}\rangle = \sin\theta\left(\tfrac{1}{\sqrt{2}}C_{-1}e^{-i\phi} + \tfrac{1}{\sqrt{2}}C_1 e^{i\phi}\right)|1\ 0\rangle$$

$$+ \left(-C_{-1}\cos\theta + \tfrac{1}{\sqrt{2}}C_0\sin\theta\, e^{i\phi}\right)|1-1\rangle$$

$$+ \left(\tfrac{1}{\sqrt{2}}C_0\sin\theta\, e^{-i\phi} + \cos\theta\, C_1\right)|1\ 1\rangle$$

This corresponds to the system of equations

$$\frac{1}{\sqrt{2}} \sin\theta\, e^{i\phi} C_1 - m_{z'} C_0 + \frac{1}{\sqrt{2}} \sin\theta\, e^{-i\phi} C_{-1} = 0$$

$$\frac{1}{\sqrt{2}} \sin\theta\, e^{i\phi} C_0 - (\cos\theta + m_{z'}) C_{-1} = 0$$

$$(m_{z'} - \cos\theta) C_1 - \frac{1}{\sqrt{2}} \sin\theta\, e^{-i\phi} C_0 = 0$$

The vanishing of the determinant of this system gives, as expected,

$$m_{z'} = -1,\ 0,\ 1$$

Solving for the coefficients, we obtain

$$C_{-1} = \frac{C_0}{\sqrt{2}} \left(\frac{\sin\theta}{m_{z'} + \cos\theta} \right) e^{i\phi}, \qquad C_1 = \frac{C_0}{\sqrt{2}} \left(\frac{\sin\theta}{m_{z'} - \cos\theta} \right) e^{-i\phi}$$

The coefficient C_0 is determined from normalization. In the case $m_{z'} = \pm 1$ it equals $\sin\theta/\sqrt{2}$, while in the case $m_{z'} = 0$, it equals $\cos\theta$.

The eigenstates $|1\ m_{z'}\rangle$ are given by

$$|1,\ m_{z'} = 1\rangle = \frac{\sin\theta}{\sqrt{2}} |1\ 0\rangle + \sin^2\frac{\theta}{2}\, e^{i\phi} |1\ -1\rangle + \cos^2\frac{\theta}{2}\, e^{-i\phi} |1\ 1\rangle$$

$$|1,\ m_{z'} = 0\rangle = \cos\theta\, |1\ 0\rangle + \frac{\sin\theta}{\sqrt{2}} \left(e^{i\phi} |1\ -1\rangle - e^{-i\phi} |1\ 1\rangle \right)$$

$$|1,\ m_{z'} = -1\rangle = \frac{\sin\theta}{\sqrt{2}} |1\ 0\rangle - \cos^2\frac{\theta}{2}\, e^{i\phi} |1\ -1\rangle - \sin^2\frac{\theta}{2}\, e^{-i\phi} |1\ 1\rangle$$

Since $\langle 1|0'\rangle = -\sin\theta\, e^{-i\phi}/\sqrt{2}$, $\langle 1|-1'\rangle = -\sin^2(\theta/2)\, e^{-i\phi}$ and $\langle 1|1'\rangle = \cos^2\theta/2\, e^{-i\phi}$, the state of the particle can be written as

$$|1\ 1\rangle = \frac{1}{\sqrt{2}} e^{-i\phi} \left(-\sin\theta\, |1,\ m_{z'} = 0\rangle - \sqrt{2} \sin^2\frac{\theta}{2} |1,\ m_{z'} = -1\rangle \right.$$

$$\left. + \sqrt{2} \cos^2\frac{\theta}{2} |1,\ m_{z'} = 1\rangle \right)$$

(b) It is straightforward to obtain

$$\langle 1\ 1|S_{z'}|1\ 1\rangle = \hbar \left(\cos^4\frac{\theta}{2} - \sin^4\frac{\theta}{2} \right)$$

and

$$\langle 1\ 1|S_{z'}^2|1\ 1\rangle = \hbar^2 \left(\cos^4\frac{\theta}{2} + \sin^4\frac{\theta}{2} \right)$$

and, finally,

$$(\Delta S_{z'})^2 = \frac{\hbar^2}{2}\sin^2\theta$$

Problem 5.19 An electron under the influence of a uniform magnetic field $\mathbf{B} = \hat{y}B$ has its spin initially (at $t = 0$) pointing in the positive x-direction. That is, it is in an eigenstate of S_x with eigenvalue $+\hbar/2$. Calculate the probability of finding the electron with its spin pointing in the positive z-direction. Ignore all Hamiltonian terms apart from the interaction of the magnetic dipole moment due to spin and the magnetic field, $H = -\mu \cdot \mathbf{B} = \omega S_y$.

Solution

Employing the Pauli representation, it is easy to see that

$$|S_x = \hbar/2\rangle = \tfrac{1}{\sqrt{2}}(|\uparrow\rangle + |\downarrow\rangle)$$
$$|S_y = \pm\hbar/2\rangle = \tfrac{1}{\sqrt{2}}(|\uparrow\rangle \pm i|\downarrow\rangle)$$

which can be written as

$$|\uparrow\rangle = \tfrac{1}{\sqrt{2}}\left(|S_y = \hbar/2\rangle + |S_y = -\hbar/2\rangle\right)$$
$$|\downarrow\rangle = -\tfrac{1}{\sqrt{2}}i\left(|S_y = \hbar/2\rangle - |S_y = -\hbar/2\rangle\right)$$

Furthermore,

$$|\psi(0)\rangle = |S_x = \hbar/2\rangle = \tfrac{1}{\sqrt{2}}\left(e^{-i\pi/4}|S_y = \hbar/2\rangle + e^{i\pi/4}|S_y = -\hbar/2\rangle\right)$$

The evolved state will be

$$|\psi(t)\rangle = e^{-i\omega t S_y/\hbar}|\psi(0)\rangle$$
$$= \tfrac{1}{\sqrt{2}}\left(e^{-i\pi/4 - i\omega t/2}|S_y = +\hbar/2\rangle + e^{i\pi/4 + i\omega t/2}|S_y = -\hbar/2\rangle\right)$$
$$= \cos\left(\frac{\pi}{4} + \frac{\omega t}{2}\right)|\uparrow\rangle + \sin\left(\frac{\pi}{4} + \frac{\omega t}{2}\right)|\downarrow\rangle$$

The probability of finding the electron in the state $|\uparrow\rangle$ is

$$\mathcal{P}_\uparrow(t) = \cos^2\left(\frac{\pi}{4} + \frac{\omega t}{2}\right)$$

This probability becomes unity at times

$$t = \frac{3\pi}{2\omega}, \quad \ldots, \quad \frac{\pi}{2\omega}(4n - 1)$$

Problem 5.20 A system of two particles with spins $s_1 = \frac{3}{2}$ and $s_2 = \frac{1}{2}$ is described by the approximate Hamiltonian $H = \alpha \, \mathbf{S}_1 \cdot \mathbf{S}_2$, with α a given constant. The system is initially (at $t = 0$) in the following eigenstate of $\mathbf{S}_1^2$, $\mathbf{S}_2^2$, S_{1z}, S_{2z}:

$$\left| \tfrac{3}{2} \, \tfrac{1}{2}; \tfrac{1}{2} \, \tfrac{1}{2} \right\rangle$$

Find the state of the system at times $t > 0$. What is the probability of finding the system in the state $\left| \tfrac{3}{2} \, \tfrac{3}{2}; \tfrac{1}{2} \, -\tfrac{1}{2} \right\rangle$?

Solution The Hamiltonian is

$$H = \tfrac{1}{2}\alpha \left(\mathbf{S}^2 - \mathbf{S}_1^2 - \mathbf{S}_2^2 \right) = \tfrac{1}{2}\alpha\hbar^2 \left[S(S+1) - \tfrac{9}{2} \right]$$

The eigenstates of S^2, S_z will also be stationary states; the allowed values of the total spin quantum number s are 1, 2. These states can be expressed in terms of the S_1^2, S_2^2, S_{1z}, S_{2z} eigenstates through the Clebsch–Gordan coefficients. In particular, we have

$$\left| 1 \ 1 \right\rangle = a \left| \tfrac{3}{2} \, \tfrac{3}{2}; \tfrac{1}{2} \, -\tfrac{1}{2} \right\rangle + b \left| \tfrac{3}{2} \, \tfrac{1}{2}; \tfrac{1}{2} \, \tfrac{1}{2} \right\rangle$$

$$\left| 2 \ 1 \right\rangle = c \left| \tfrac{3}{2} \, \tfrac{1}{2}; \tfrac{1}{2} \, \tfrac{1}{2} \right\rangle + d \left| \tfrac{3}{2} \, \tfrac{3}{2}; \tfrac{1}{2} \, -\tfrac{1}{2} \right\rangle$$

The coefficients a, b, c, d are easily determined from

$$S_+ |1 \ 1\rangle = 0 = a \, S_2^{(+)} \left| \tfrac{3}{2} \, \tfrac{3}{2}; \tfrac{1}{2} \, -\tfrac{1}{2} \right\rangle + b \, S_1^{(+)} \left| \tfrac{3}{2} \, \tfrac{1}{2}; \tfrac{1}{2} \, \tfrac{1}{2} \right\rangle$$

$$= a\hbar \left| \tfrac{3}{2} \, \tfrac{3}{2}; \tfrac{1}{2} \, \tfrac{1}{2} \right\rangle + b\hbar\sqrt{3} \left| \tfrac{3}{2} \, \tfrac{3}{2}; \tfrac{1}{2} \, \tfrac{1}{2} \right\rangle$$

which gives

$$a = -\tfrac{\sqrt{3}}{2}, \qquad b = \tfrac{1}{2}$$

Similarly, we have

$$S_+ |2 \ 1\rangle = 2\hbar |2 \ 2\rangle = 2\hbar \left| \tfrac{3}{2} \, \tfrac{3}{2}; \tfrac{1}{2} \, \tfrac{1}{2} \right\rangle$$

$$= c\hbar\sqrt{3} \left| \tfrac{3}{2} \, \tfrac{3}{2}; \tfrac{1}{2} \, \tfrac{1}{2} \right\rangle + d\hbar \left| \tfrac{3}{2} \, \tfrac{3}{2}; \tfrac{1}{2} \, \tfrac{1}{2} \right\rangle$$

which gives

$$d = \tfrac{1}{2}, \qquad c = \tfrac{\sqrt{3}}{2}$$

Thus, we have

$$\left| 1 \ 1 \right\rangle = -\tfrac{\sqrt{3}}{2} \left| \tfrac{3}{2} \, \tfrac{3}{2}; \tfrac{1}{2} \, -\tfrac{1}{2} \right\rangle + \tfrac{1}{2} \left| \tfrac{3}{2} \, \tfrac{1}{2}; \tfrac{1}{2} \, \tfrac{1}{2} \right\rangle$$

$$\left| 2 \ 1 \right\rangle = \tfrac{\sqrt{3}}{2} \left| \tfrac{3}{2} \, \tfrac{1}{2}; \tfrac{1}{2} \, \tfrac{1}{2} \right\rangle + \tfrac{1}{2} \left| \tfrac{3}{2} \, \tfrac{3}{2}; \tfrac{1}{2} \, -\tfrac{1}{2} \right\rangle$$

The inverse relations are

$$\left| \tfrac{3}{2} \, \tfrac{1}{2}; \tfrac{1}{2} \, \tfrac{1}{2} \right\rangle = \tfrac{\sqrt{3}}{2} |2 \ 1\rangle + \tfrac{1}{2} |1 \ 1\rangle$$

$$\left| \tfrac{3}{2} \, \tfrac{3}{2}; \tfrac{1}{2} \, -\tfrac{1}{2} \right\rangle = \tfrac{1}{2} |2 \ 1\rangle - \tfrac{\sqrt{3}}{2} |1 \ 1\rangle$$

The evolved state of the system will be

$$|\psi(t)\rangle = \frac{1}{2}\left(\sqrt{3}\,e^{-iE_2t/\hbar}|2\ 1\rangle + e^{-iE_1t/\hbar}|1\ 1\rangle\right)$$

with $E_1 = -5\alpha\hbar/4$ and $E_2 = 3\alpha\hbar/4$ as the two energy eigenvalues. The probability of finding the system in the state $|\frac{3}{2}\ \frac{3}{2};\ \frac{1}{2}\ -\frac{1}{2}\rangle$ is

$$\mathcal{P} = \left|\langle\tfrac{3}{2}\ \tfrac{3}{2};\ \tfrac{1}{2}\ -\tfrac{1}{2}|\psi(t)\rangle\right|^2 = \tfrac{3}{4}\sin^2\frac{(E_2 - E_1)t}{\hbar}$$

$$= \tfrac{3}{4}\sin^2 2\alpha t$$

Problem 5.21 Consider a particle with spin $s = 1$.

(a) Derive the spin matrices in the basis $|1\ m\rangle$ of S^2, S_z eigenstates.
(b) Find the eigenstates of the spin component operator $\hat{\mathbf{n}}\cdot\mathbf{S}$ along the arbitrary direction
$\hat{\mathbf{n}} = \sin\theta\,(\hat{\mathbf{x}}\cos\phi + \hat{\mathbf{y}}\sin\phi) + \cos\theta\,\hat{\mathbf{z}}$.

Solution

(a) The relations discussed in problem 5.13,

$$S_+|1\ 0\rangle = \hbar\sqrt{2}|1\ 1\rangle, \qquad S_+|1\ -1\rangle = \hbar\sqrt{2}|1\ 0\rangle$$
$$S_-|1\ 1\rangle = \hbar\sqrt{2}|1\ 0\rangle, \qquad S_-|1\ 0\rangle = \hbar\sqrt{2}|1\ -1\rangle$$

imply that

$$S_+ = \hbar\sqrt{2}\begin{pmatrix} 0 & 1 & 0 \\ 0 & 0 & 1 \\ 0 & 0 & 0 \end{pmatrix}, \qquad S_- = \hbar\sqrt{2}\begin{pmatrix} 0 & 0 & 0 \\ 1 & 0 & 0 \\ 0 & 1 & 0 \end{pmatrix}$$

Thus, we get

$$S_x = \frac{\hbar}{\sqrt{2}}\begin{pmatrix} 0 & 1 & 0 \\ 1 & 0 & 1 \\ 0 & 1 & 0 \end{pmatrix}, \qquad S_y = \frac{\hbar}{\sqrt{2}}\begin{pmatrix} 0 & -i & 0 \\ i & 0 & -i \\ 0 & i & 0 \end{pmatrix}$$

$$S_z = \hbar\begin{pmatrix} 1 & 0 & 0 \\ 0 & 0 & 0 \\ 0 & 0 & -1 \end{pmatrix}$$

(b) Using the above matrices, we get

$$\hat{\mathbf{n}}\cdot\mathbf{S} = \hbar\begin{pmatrix} \cos\theta & \frac{1}{\sqrt{2}}\sin\theta\,e^{-i\phi} & 0 \\ \frac{1}{\sqrt{2}}\sin\theta\,e^{i\phi} & 0 & \frac{1}{\sqrt{2}}\sin\theta\,e^{-i\phi} \\ 0 & \frac{1}{\sqrt{2}}\sin\theta\,e^{i\phi} & -\cos\theta \end{pmatrix}$$

Solving the eigenvalue problem of this matrix, we get, as expected,

$$m_n = -1, \; 0, \; 1$$

The corresponding eigenvectors are

$$\chi_0 = \begin{pmatrix} \frac{1}{\sqrt{2}} \sin\theta \, e^{-i\phi} \\ -\cos\theta \\ -\frac{1}{\sqrt{2}} \sin\theta \, e^{i\phi} \end{pmatrix}$$

$$\chi_{-1} = \begin{pmatrix} \sin^2(\theta/2) \, e^{-i\phi} \\ -\frac{1}{\sqrt{2}} \sin\theta \\ \cos^2(\theta/2) \, e^{i\phi} \end{pmatrix}, \qquad \chi_1 = \begin{pmatrix} \cos^2(\theta/2) \, e^{-i\phi} \\ \frac{1}{\sqrt{2}} \sin\theta \\ \sin^2(\theta/2) \, e^{i\phi} \end{pmatrix}$$

Problem 5.22 A beam of particles is subject to a simultaneous measurement of the angular momentum variables $\mathbf{L}^2$, L_z. The measurement gives pairs of values $\ell = m = 0$ and $\ell = 1$, $m = -1$ with probabilities $3/4$ and $1/4$ respectively.

(a) Reconstruct the state of the beam immediately before the measurement.
(b) The particles in the beam with $\ell = 1$, $m = -1$ are separated out and subjected to a measurement of L_x. What are the possible outcomes and their probabilities?
(c) Construct the spatial wave functions of the states that could arise from the second measurement.

Solution

(a) The state of the beam is, in terms of the eigenstates of L_z,

$$|\psi\rangle = \tfrac{\sqrt{3}}{2} |0 \; 0\rangle + \tfrac{1}{2} e^{i\alpha} |1 \; -1\rangle$$

where α is an arbitrary phase.

(b) The possible outcomes will correspond to the common eigenstates of $\mathbf{L}^2$, L_x,

$$|1, \; m_x = 1\rangle, \qquad |1, \; m_x = 0\rangle, \qquad |1, \; m_x = -1\rangle$$

Each of these states can be expanded in terms of L_z eigenstates:

$$|1, \; m_x\rangle = C_1(m_x)|1 \; 1\rangle + C_0(m_x)|1 \; 0\rangle + C_{-1}(m_x)|1 \; -1\rangle$$

Acting on this state with

$$L_x = \tfrac{1}{2}(L_+ + L_-)$$

we should get $m_x\hbar$. Doing this, we obtain the following relations between the coefficients:

$$C_0 = m_x\sqrt{2}C_1 = m_x\sqrt{2}C_{-1}, \qquad C_1 + C_{-1} = m_x\sqrt{2}C_0$$

Therefore, we are led to

$$|1, m_x = 1\rangle = \tfrac{1}{2}\left(|1\ 1\rangle + \sqrt{2}|1\ 0\rangle + |1\ -1\rangle\right)$$

$$|1, m_x = 0\rangle = \tfrac{1}{\sqrt{2}}\left(|1\ 1\rangle - |1\ -1\rangle\right)$$

$$|1, m_x = -1\rangle = \tfrac{1}{2}\left(|1\ 1\rangle - \sqrt{2}|1\ 0\rangle + |1\ -1\rangle\right)$$

The inverse expressions for the L_z eigenstates are

$$|1\ 1\rangle = \tfrac{1}{\sqrt{2}}|1, m_x = 0\rangle + \tfrac{1}{2}\left(|1\ m_x = 1\rangle + |1, m_x = -1\rangle\right)$$

$$|1\ 0\rangle = \tfrac{1}{\sqrt{2}}\left(|1\ m_x = 1\rangle - |1\ m_x = -1\rangle\right)$$

$$|1\ -1\rangle = -\tfrac{1}{\sqrt{2}}|1\ m_x = 0\rangle + \tfrac{1}{2}\left(|1\ m_x = 1\rangle + |1\ m_x = -1\rangle\right)$$

From these relations we can read off the probabilities:

$$\mathcal{P}_{L_x=\pm\hbar} = \tfrac{1}{4}, \qquad \mathcal{P}_{L_x=0} = \tfrac{1}{2}$$

(c) Using standard formulae for the spherical harmonics we obtain for the eigenfunctions of L_x

$$\mathcal{Y}_{\pm1} = \sqrt{\frac{3}{8\pi}}\,(\pm\cos\theta - i\sin\phi\sin\theta), \qquad \mathcal{Y}_0 = -\sqrt{\frac{3}{4\pi}}\,\sin\theta\cos\phi$$

Problem 5.23 Consider a system of two non-identical fermions, each with spin $1/2$. One is in a state with $S_{1x} = \hbar/2$ while the other is in a state with $S_{2y} = -\hbar/2$. What is the probability of finding the system in a state with total spin quantum numbers $s = 1$, $m_s = 0$, where m_s refers to the z-component of the total spin?

Solution

From the Pauli representation we can immediately see that

$$|S_{1x} = \hbar/2\rangle = \tfrac{1}{\sqrt{2}}(|\uparrow\rangle_1 + |\downarrow\rangle_1), \qquad |S_{1x} = -\hbar/2\rangle = \tfrac{1}{\sqrt{2}}(|\uparrow\rangle_1 - |\downarrow\rangle_1)$$

where $|\uparrow\rangle$ and $|\downarrow\rangle$ are the S_z eigenfunctions. The inverse relations are

$$|\uparrow\rangle_1 = \tfrac{1}{\sqrt{2}}(|S_{1x} = \hbar/2\rangle + |S_{1x} = -\hbar/2\rangle)$$

$$|\downarrow\rangle_1 = \tfrac{1}{\sqrt{2}}(|S_{1x} = \hbar/2\rangle - |S_{1x} = -\hbar/2\rangle)$$

In an analogous fashion, we can also see that

$$|S_{2y} = \hbar/2\rangle = \tfrac{1}{\sqrt{2}} (| \uparrow\rangle_2 + i| \downarrow\rangle_2), \qquad |S_{2y} = -\hbar/2\rangle = \tfrac{1}{\sqrt{2}} (| \uparrow\rangle_2 - i| \downarrow\rangle_2)$$

Thus we have

$$| \uparrow\rangle_2 = \tfrac{1}{\sqrt{2}} \left(|S_{2y} = \hbar/2\rangle + |S_{2y} = -\hbar/2\rangle\right)$$
$$| \downarrow\rangle_2 = -\tfrac{1}{\sqrt{2}}i \left(|S_{2y} = \hbar/2\rangle - |S_{2y} = -\hbar/2\rangle\right)$$

The $m_s = 0$ state in the $s = 1$ triplet is[7]

$$\begin{aligned}|s = 1, m_s = 0\rangle &= \tfrac{1}{\sqrt{2}} \big(| \uparrow\rangle_1 | \downarrow\rangle_2 - | \downarrow\rangle_1 | \uparrow\rangle_2\big)\\
&= -\tfrac{1}{2}i \big(e^{-i\pi/4}|S_{1x} = \hbar/2\rangle\, |S_{2y} = \hbar/2\rangle\\
&\quad - e^{i\pi/4}|S_{1x} = \hbar/2\rangle\, |S_{2y} = -\hbar/2\rangle\\
&\quad + e^{i\pi/4}|S_{1x} = -\hbar/2\rangle\, |S_{2y} = \hbar/2\rangle\\
&\quad - e^{-i\pi/4}|S_{1x} = -\hbar/2\rangle\, |S_{2y} = -\hbar/2\rangle\big)\end{aligned}$$

From this expression we can read off the probability:

$$\mathcal{P} = \big|\langle 1\, 0|S_{1x} = +,\ S_{2y} = -\rangle\big|^2 = \tfrac{1}{4}$$

Problem 5.24 Consider a system of three electrons.

(a) Find the eigenvalues of the total spin. Find a set of eigenstates of the total spin operators S^2, S_z (treat electrons 1 and 2 as a subsystem and combine the spin $\mathbf{S}_{12}$ of the latter with that of electron 3).

(b) Ignoring spatial degrees of freedom, assume that the approximate Hamiltonian of the system is

$$H = \alpha \left(S_{1x} S_{2x} + S_{1y} S_{2y} + S_{2x} S_{3x} + S_{2y} S_{3y} + S_{1x} S_{3x} + S_{1y} S_{3y}\right)$$

Show that H can be expressed in terms of S^2 and S_z^2. Find the energy eigenvalues.

Solution

(a) As suggested, we will think of the system as consisting of two subsystems, one of two electrons and one of a single electron. The first subsystem has total angular momentum values $s_{12} = 0,\ 1$, corresponding respectively to the singlet

$$|s_{12} = 0, m_{12} = 0\rangle = \tfrac{1}{\sqrt{2}} \left(| \uparrow\rangle_1 | \downarrow\rangle_2 - | \downarrow\rangle_1 | \uparrow\rangle_2\right)$$

where m_{12} specifies the z-component of the total spin s_{12} of the first subsystem,

---

[7] The factor $e^{i\pi/4}$ is a compact way of writing $\tfrac{1}{\sqrt{2}}(1 + i)$.

and to the triplet

$$|s_{12} = 1 \; m_{12} = 1\rangle = |\uparrow\rangle_1 |\uparrow\rangle_2$$
$$|s_{12} = 1 \; m_{12} = 0\rangle = \frac{1}{\sqrt{2}} (|\uparrow\rangle_1 |\downarrow\rangle_2 + |\downarrow\rangle_1 |\uparrow\rangle_2)$$
$$|s_{12} = 1 \; m_{12} = -1\rangle = |\downarrow\rangle_1 |\downarrow\rangle_2$$

Combining $\mathbf{S}_{12}$ with $\mathbf{S}_3$ gives the allowed values

$$\left| s_{12} - \tfrac{1}{2} \right| \le s_{123} \le s_{12} + \tfrac{1}{2} \qquad \Longrightarrow \qquad s_{123} = \tfrac{3}{2}, \; \tfrac{1}{2}$$

The resulting eigenstates of total spin form a *quadruplet*:

$$\left| \tfrac{3}{2} \; \tfrac{3}{2} \right\rangle = |1 \; 1\rangle |\uparrow\rangle$$
$$\left| \tfrac{3}{2} \; \tfrac{1}{2} \right\rangle = a|1 \; 0\rangle |\uparrow\rangle + b|1 \; 1\rangle |\downarrow\rangle$$
$$\left| \tfrac{3}{2} \; -\tfrac{1}{2} \right\rangle = c|1 \; -1\rangle |\uparrow\rangle + d|1 \; 0\rangle |\downarrow\rangle$$
$$\left| \tfrac{3}{2} \; -\tfrac{3}{2} \right\rangle = |1 \; -1\rangle |\downarrow\rangle$$

and a *doublet*

$$\left| \tfrac{1}{2} \; \tfrac{1}{2} \right\rangle = |0 \; 0\rangle |\uparrow\rangle$$
$$\left| \tfrac{1}{2} \; -\tfrac{1}{2} \right\rangle = |0 \; 0\rangle |\downarrow\rangle$$

The coefficients are obtained as follows:

$$S_+ \left| \tfrac{3}{2} \; \tfrac{1}{2} \right\rangle = \hbar \sqrt{3} \left| \tfrac{3}{2} \; \tfrac{3}{2} \right\rangle \qquad \Longrightarrow \qquad a = \sqrt{\tfrac{2}{3}}, \qquad b = \tfrac{1}{\sqrt{3}}$$

or

$$\left| \tfrac{3}{2} \; \tfrac{1}{2} \right\rangle = \frac{1}{\sqrt{3}} (|\uparrow\rangle_1 |\uparrow\rangle_2 |\downarrow\rangle_3 + |\uparrow\rangle_1 |\downarrow\rangle_2 |\uparrow\rangle_3 + |\downarrow\rangle_1 |\uparrow\rangle_2 |\uparrow\rangle_3)$$

Similarly,

$$S_- \left| \tfrac{3}{2} \; -\tfrac{1}{2} \right\rangle = \hbar \sqrt{3} \left| \tfrac{3}{2} \; \tfrac{3}{2} \right\rangle \qquad \Longrightarrow \qquad c = \tfrac{1}{\sqrt{3}}, \qquad d = \sqrt{\tfrac{2}{3}}$$

or

$$\left| \tfrac{3}{2} \; -\tfrac{1}{2} \right\rangle = \frac{1}{\sqrt{3}} (|\downarrow\rangle_1 |\downarrow\rangle_2 |\uparrow\rangle_3 + |\downarrow\rangle_1 |\uparrow\rangle_2 |\downarrow\rangle_3 + |\uparrow\rangle_1 |\downarrow\rangle_2 |\downarrow\rangle_3)$$

The other two states of the quadruplet are

$$\left| \tfrac{3}{2} \; \tfrac{3}{2} \right\rangle = |\uparrow\rangle_1 |\uparrow\rangle_2 |\uparrow\rangle_3, \qquad \left| \tfrac{3}{2} \; -\tfrac{3}{2} \right\rangle = |\downarrow\rangle_1 |\downarrow\rangle_2 |\downarrow\rangle_3$$

Similarly, we get for the doublet

$$\left| \tfrac{1}{2} \; \tfrac{1}{2} \right\rangle = \frac{1}{\sqrt{2}} (|\uparrow\rangle_1 |\downarrow\rangle_2 |\uparrow\rangle_3 - |\downarrow\rangle_1 |\uparrow\rangle_2 |\uparrow\rangle_3)$$
$$\left| \tfrac{1}{2} \; -\tfrac{1}{2} \right\rangle = \frac{1}{\sqrt{2}} (|\uparrow\rangle_1 |\downarrow\rangle_2 |\downarrow\rangle_3 - |\downarrow\rangle_1 |\uparrow\rangle_2 |\downarrow\rangle_3)$$

(b) The given Hamiltonian is equal to

$$H = \frac{\alpha}{2}(2\mathbf{S}_1 \cdot \mathbf{S}_2 + 2\mathbf{S}_2 \cdot \mathbf{S}_3 + 2\mathbf{S}_1 \cdot \mathbf{S}_3 - 2S_{1z}S_{2z} - 2S_{2z}S_{3z} - 2S_{1z}S_{3z})$$

$$= \frac{\alpha}{2}\left[(\mathbf{S}_1 + \mathbf{S}_2 + \mathbf{S}_3)^2 - S_1^2 - S_2^2 - S_3^2 - 2S_{1z}S_{2z} - 2S_{2z}S_{3z} - 2S_{1z}S_{3z}\right]$$

$$= \frac{\alpha}{2}\left[(\mathbf{S}_1 + \mathbf{S}_2 + \mathbf{S}_3)^2 - \frac{9\hbar^2}{4} - (S_{1z} + S_{2z} + S_{3z})^2 + S_{1z}^2 + S_{2z}^2 + S_{2z}^2\right]$$

Finally,

$$H = \frac{\alpha}{2}\left(S^2 - S_z^2 - \frac{3\hbar^2}{2}\right)$$

The energy eigenvalues are, in terms of the total spin quantum numbers s and m_s,

$$E = \frac{\alpha}{2}\left[\hbar^2 s(s+1) - \hbar^2 m_s^2 - \frac{3\hbar^2}{2}\right]$$

The possible values of s are $\frac{3}{2}, \frac{1}{2}$. Thus we have

$$\begin{array}{lll}
\text{for} & s = \frac{3}{2}, \ m_s = \pm\frac{3}{2}, & E = 0 \\
\text{for} & s = \frac{3}{2}, \ m_s = \pm\frac{1}{2}, & E = \alpha\hbar^2 \\
\text{for} & s = \frac{1}{2}, \ m_s = \pm\frac{1}{2}, & E = -\frac{1}{2}\alpha\hbar^2
\end{array}$$

Problem 5.25 Consider two spin-1 particles that occupy the state

$$|s_1 = 1, \ m_1 = 1; \ s_2 = 1, \ m_2 = 0\rangle$$

What is the probability of finding the system in an eigenstate of the total spin $\mathbf{S}^2$ with quantum number $s = 1$? What is the probability for $s = 2$?

Solution

It is not difficult to construct the eigenstates of S^2, S_z. We shall show them using boldface numbers. They are a *singlet*

$$|\mathbf{0\ 0}\rangle = \frac{1}{\sqrt{2}}|1, 0\rangle|1, 0\rangle - \frac{1}{2}(|1, 1\rangle|1, -1\rangle - |1, -1\rangle|1, 1\rangle)$$

a *triplet*

$$|\mathbf{1\ 1}\rangle = \frac{1}{\sqrt{2}}(|1, 0\rangle|1, 1\rangle - |1, 1\rangle|1, 0\rangle)$$

$$|\mathbf{1\ 0}\rangle = -\frac{1}{\sqrt{2}}|1, 0\rangle|1, 0\rangle - \frac{1}{2}(|1, 1\rangle|1, -1\rangle - |1, -1\rangle|1, 1\rangle)$$

$$|\mathbf{1\ -1}\rangle = \frac{1}{\sqrt{2}}(|1, 0\rangle|1, -1\rangle - |1, -1\rangle|1, 0\rangle)$$

and a *quintet*

$$|\mathbf{2}\ \mathbf{2}\rangle = |1, 1\rangle|1, 1\rangle$$

$$|\mathbf{2}\ \mathbf{1}\rangle = \frac{1}{\sqrt{2}}\big(|1, 0\rangle|1, 1\rangle + |1, 1\rangle|1, 0\rangle\big)$$

$$|\mathbf{2}\ \mathbf{0}\rangle = \frac{1}{\sqrt{2}}\big(|1, 1\rangle|1, -1\rangle + |1, -1\rangle|1, 1\rangle\big)$$

$$|\mathbf{2}\ -\mathbf{1}\rangle = \frac{1}{\sqrt{2}}\big(|1, 0\rangle|1, -1\rangle + |1, -1\rangle|1, 0\rangle\big)$$

$$|\mathbf{2}\ -\mathbf{2}\rangle = |1, -1\rangle|1, -1\rangle$$

From the above relations we obtain

$$|1, 1\rangle|1, 0\rangle = \frac{1}{\sqrt{2}}\big(|\mathbf{2}\ \mathbf{1}\rangle - |\mathbf{1}\ \mathbf{1}\rangle\big)$$

Thus, the probability of finding the system in either of the states shown on the right-hand side is $1/2$.

6

Quantum behaviour

Problem 6.1 A quantum system has only two energy eigenstates, $|1\rangle$, $|2\rangle$, corresponding to the energy eigenvalues E_1, E_2. Apart from the energy, the system is also characterized by a physical observable whose operator $\mathcal{P}$ acts on the energy eigenstates as follows:

$$\mathcal{P}|1\rangle = |2\rangle, \quad \mathcal{P}|2\rangle = |1\rangle$$

The operator $\mathcal{P}$ can be regarded as a type of parity operator.

(a) Assuming that the system is initially in a positive-parity eigenstate, find the state of the system at any time.

(b) At a particular time t a parity measurement is made on the system. What is the probability of finding the system with positive parity?

(c) Imagine that you make a series of parity measurements at the times $\Delta t, 2\Delta t, \ldots, N\Delta t = T$. What is the probability of finding the system with positive parity at time T?

(d) Assume that the parity measurements performed in (c) are not instantaneous but each take a minimal time $\delta\tau$. What is the survival probability of a state of positive parity if the above measurement process is carried out in the time interval T?

Solution

(a) It is clear that

$$\mathcal{P}(|1\rangle \pm |2\rangle) = \pm(|1\rangle \pm |2\rangle)$$

Thus, the parity eigenstates are

$$|\pm\rangle = \tfrac{1}{\sqrt{2}}(|1\rangle \pm |2\rangle)$$

The inverse relations are

$$|1\rangle = \tfrac{1}{\sqrt{2}}(|+\rangle + |-\rangle), \quad |1\rangle = \tfrac{1}{\sqrt{2}}(|+\rangle - |-\rangle)$$

155

The evolved state of the system is

$$|\psi(t)\rangle = e^{i\alpha t}\,(\cos\omega t\,|+\rangle + i\,\sin\omega t\,|-\rangle)$$

with

$$\omega \equiv \frac{E_2 - E_1}{2\hbar}, \qquad \alpha \equiv -\frac{E_2 + E_1}{2\hbar}$$

(b) The probability of finding the system in the state $|+\rangle$ of positive parity is $\cos^2\omega t$.

(c) At time Δt the probability of finding the system in a state of positive parity is $\cos^2\omega\Delta t$. The probability of finding the system in a positive-parity state at time $2\Delta t$ will be $\cos^2\omega\Delta t\,\cos^2\omega\Delta t$. Continuing like this, at time T the probability will be

$$\left(\cos^2\omega\Delta t\right)^N = \left(1 - \sin^2\omega T/N\right)^N$$

For $N \gg 1$ but finite, this is

$$\left[1 - \left(\frac{\omega T}{N}\right)^2\right]^N \approx \exp\left[-\frac{(\omega T)^2}{N}\right]$$

Note that for any finite N the above survival probability of the positive-parity state is always smaller than unity.

In contrast, for $N \to \infty$ the probability

$$\left[1 - \left(\frac{\omega T}{N}\right)^2\right]^N = \left(1 - \frac{\omega T}{N}\right)^N \left(1 + \frac{\omega T}{N}\right)^N = 1$$

In the limit of infinite instantaneous measurements the survival probability of a positive-parity state is unity.[1]

(d) Since there is an appreciable minimum time required for each measurement, N cannot be infinite but its maximum value will be the large but finite number $T/\delta\tau$. Also, since for $T/N > \delta\tau$ we have $\cos(\omega T/N) < \cos\omega\delta\tau$, the corresponding probability is

$$\left[\cos^2(\omega T/N)\right]^N = \exp\left[2\left(\frac{T}{\delta\tau}\right)\ln\cos\omega\delta\tau\right]$$

Problem 6.2 A pair of particles moving in one dimension is in a state characterised by the wave function

$$\Psi(x_1, x_2) = N\,\exp\left[-\frac{1}{2\alpha}(x_1 - x_2 + a)^2\right]\exp\left[-\frac{1}{2\beta}(x_1 + x_2)^2\right]$$

---

[1] The 'freezing' of the system in the initial state for a repeated series of measurements has been called the quantum Zeno effect.

(a) Discuss the behaviour of $\Psi(x_1, x_2)$ in the limit $\alpha \to 0$.

(b) Calculate the momentum-space wave function and discuss its properties in the above limit.

(c) Consider a simultaneous measurement of the positions x_1, x_2 of the two particles when the system is in the above state. What are the expected position values? What are the values resulting from a simultaneous measurement of the momenta p_1, p_2 of the two particles?

Solution

(a) The normalization constant is $N = (\alpha\beta)^{-1/4}\sqrt{2/\pi}$. In the limit $\alpha \to 0$,

$$\Psi(x_1, x_2) \sim \frac{1}{\pi} \delta(x_1 - x_2 + a) \left(\frac{\alpha}{\beta}\right)^{1/4} \exp\left[-\frac{1}{2\beta}(x_1 + x_2)^2\right]$$

This is a sharply localized amplitude describing the situation where the two particles are at a distance $x_2 - x_1 = a$. Keeping α as small as we wish but non-zero and β as large as we wish but not infinite, we have a normalizable amplitude for the two particles to be at a distance a, to any desired order of approximation.

(b) The momentum-space wave function is

$$\Phi(p_1, p_2) = \frac{(\alpha\beta)^{1/4}}{\sqrt{2\pi}} \exp\left[\frac{i}{2}(p_1 - p_2)a - \frac{\alpha}{8}(p_1 - p_2)^2 - \frac{\beta}{8}(p_1 + p_2)^2\right]$$

In the limit $\beta \to \infty$ we have

$$\Phi(p_1, p_2) \sim \delta(p_1 + p_2) 2 \left(\frac{\alpha}{\beta}\right)^{1/4} \exp\left[\frac{i}{2}(p_1 - p_2)a\right] \exp\left[-\frac{\alpha}{8}(p_1 - p_2)^2\right]$$

This is a sharply localized momentum-space amplitude for the two particles to have opposite momenta. Keeping β as large as we wish but finite and α as small as we wish, we have a normalizable amplitude for the two particles to have opposite momenta, to any desired order of approximation.

(c) A measurement of x_1 and x_2 will yield values related by $x_2 - x_1 = a$: the measurement of the position of particle 1 is sufficient to determine the position of particle 2, or vice versa. A measurement of the momenta p_1 and p_2 will give values related by $p_2 = -p_1$: measurement of the momentum of particle 2 is sufficient to determine the momentum of particle 1, or vice versa.[2] This is an example of *entanglement*.

Problem 6.3 A pair of spin-1/2 particles is produced by a source. The spin state of each particle can be measured using a Stern–Gerlach apparatus (see the schematic diagram shown in Fig. 24).

---

[2] The above wave function is very close to the wave function proposed by Einstein, Podolsky and Rosen in 1935 as a 'paradox' suggesting that both the position and momentum of a particle can have definite values independently of whether they are actually measured. The EPR controversy has been now resolved with Bell's inequality.

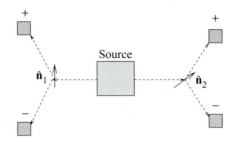

Fig. 24 Einstein–Podolsky–Rosen set-up for two spin-1/2 particles emitted by a source. The Stern–Gerlach apparatuses are represented by arrows showing their field directions. The small squares show the observed positions of spin-up and spin-down particles.

(a) Let $\hat{\mathbf{n}}_1$ and $\hat{\mathbf{n}}_2$ be the field directions of the Stern–Gerlach magnets. Consider the commuting observables

$$\sigma^{(1)} \equiv \frac{2}{\hbar} \hat{\mathbf{n}}_1 \cdot \mathbf{S}_1, \qquad \sigma^{(2)} \equiv \frac{2}{\hbar} \hat{\mathbf{n}}_2 \cdot \mathbf{S}_2$$

corresponding to the spin component of each particle along the direction of the Stern–Gerlach apparatus associated with it. What are the possible values resulting from measurement of these observables and what are the corresponding eigenstates?

(b) Consider the observable

$$\sigma^{(12)} \equiv \sigma^{(1)} \otimes \sigma^{(2)}$$

and write down its eigenvectors and eigenvalues. Assume that the pair of particles is produced in the singlet state

$$|0, 0\rangle = \frac{1}{\sqrt{2}} \left(|S_z +\rangle^{(1)} |S_z -\rangle^{(2)} - |S_z -\rangle^{(1)} |S_z +\rangle^{(2)} \right)$$

What is the expectation value of $\sigma^{(12)}$?

(c) Make the assumption that it is meaningful to assign a definite value to the spin of a particle even when it is not being measured. Assume also that the only possible results of the measurement of a spin component are $\pm\hbar/2$. Then show that the probability of finding the spins pointing in two given directions will be proportional to the overlap of the hemispheres that these two directions define. Quantify this criterion and calculate the expectation value of $\sigma^{(12)}$.

(d) Assume that the spin variables depend on a *hidden variable* λ. The expectation value of the spin observable $\sigma^{(12)}$ is determined in terms of a distribution function[3] $f(\lambda)$:

$$\langle \sigma^{(12)} \rangle = \frac{4}{\hbar^2} \int d\lambda \, f(\lambda) \, S_z^{(1)}(\lambda) \, S_\phi^{(2)}(\lambda)$$

Prove *Bell's inequality*

$$\left| \langle \sigma^{(12)}(\phi) \rangle - \langle \sigma^{(12)}(\phi') \rangle \right| \le 1 + \langle \sigma^{(12)}(\phi - \phi') \rangle$$

---

[3] Assumed to be normalized by $\int d\lambda \, f(\lambda) = 1$.

(e) Consider Bell's inequality for $\phi' = 2\phi$ and show that it is not true when applied in the context of quantum mechanics.

Solution

(a) The eigenstates and the corresponding quantum numbers relating to $\sigma^{(1)}$ and $\sigma^{(2)}$ are[4]

$$
\begin{aligned}
|S_{n_1} +\rangle^{(1)}\,|S_{n_2} +\rangle^{(2)}, & \quad +1, +1 \\
|S_{n_1} -\rangle^{(1)}\,|S_{n_2} +\rangle^{(2)}, & \quad -1, +1 \\
|S_{n_1} +\rangle^{(1)}\,|S_{n_2} -\rangle^{(2)}, & \quad +1, -1 \\
|S_{n_1} -\rangle^{(1)}\,|S_{n_2} -\rangle^{(2)}, & \quad -1, -1
\end{aligned}
$$

(b) The eigenvectors of $\sigma^{(12)}$ are the same as those of $\sigma^{(1)}$ and $\sigma^{(2)}$. The corresponding quantum numbers are the products of the pairs of quantum numbers in (a):

$$
\begin{aligned}
|S_{n_1} +\rangle^{(1)}\,|S_{n_2} +\rangle^{(2)}, & \quad +1, & \quad |S_{n_1} -\rangle^{(1)}\,|S_{n_2} +\rangle^{(2)}, & \quad -1 \\
|S_{n_1} +\rangle^{(1)}\,|S_{n_2} -\rangle^{(2)}, & \quad -1, & \quad |S_{n_1} -\rangle^{(1)}\,|S_{n_2} -\rangle^{(2)}, & \quad +1
\end{aligned}
$$

The state of the pair is given as the singlet

$$
|0\ 0\rangle = \tfrac{1}{\sqrt{2}}\left(|S_z +\rangle^{(1)}\,|S_z -\rangle^{(2)} - |S_z -\rangle^{(1)}\,|S_z +\rangle^{(2)} \right)
$$

Therefore the expectation value of $\sigma^{(12)}$ will be[5]

$$
\langle 0\ 0|\sigma^{(12)}|0\ 0\rangle = -\cos(\phi_1 - \phi_2) = -\hat{\mathbf{n}}_1 \cdot \hat{\mathbf{n}}_2
$$

(c) The expectation value of $\sigma^{(12)}$ in this scheme also would be

$$
\langle \sigma^{(12)} \rangle = \mathcal{P}_{++} + \mathcal{P}_{--} - \mathcal{P}_{+-} - \mathcal{P}_{-+}
$$

where $\mathcal{P}_{++}$ is the probability that both particles have their spin 'up' with respect to the field direction in the corresponding Stern–Gerlach apparatus. Similarly for the rest of the probabilities. Each particle, by assumption, has a well-defined spin

---

[4] The spin eigenstates for a general direction can be expressed in terms of the *standard* eigenstates with respect to z as

$$
\begin{aligned}
|S_n +\rangle &= \cos(\phi/2)|S_z +\rangle + i\sin(\phi/2)|S_z -\rangle \\
|S_n -\rangle &= i\sin(\phi/2)|S_z +\rangle + \cos(\phi/2)|S_z -\rangle
\end{aligned}
$$

The inverse relations are

$$
\begin{aligned}
|S_z +\rangle &= \cos(\phi/2)|S_n +\rangle - i\sin(\phi/2)|S_n -\rangle \\
|S_z -\rangle &= -i\sin(\phi/2)|S_n +\rangle + \cos(\phi/2)|S_n -\rangle
\end{aligned}
$$

The angle ϕ equals $\cos^{-1}(\hat{\mathbf{n}} \cdot \hat{\mathbf{z}})$.

[5] We have

$$
\langle S_z \pm |\sigma^{(i)}|S_z \pm \rangle = \pm\cos\phi_i, \qquad \langle S_z \pm |\sigma^{(i)}|S_z \mp \rangle = \mp i\sin\phi_i
$$

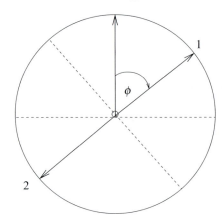

Fig. 25 Spin directions.

independently of observation. If a spin component of a particle has been measured as $+\hbar/2$ then we can conclude that its spin will lie somewhere in the hemisphere defined by this 'up' direction. Obviously the spin of the other particle will lie in the 'down' hemisphere (see Fig. 25).

Consider the spin components of the two particles along directions that make an angle ϕ. The probability of finding the spin components of both particles to be $+\hbar/2$ will be proportional to the overlap of their corresponding 'up' hemispheres, $\mathcal{P}_{++} = \phi/\pi$. The proportionality coefficient is determined by the fact that this probability must be equal to unity when $\phi = \pi$.

The probability of finding the two spin components pointing in opposite directions can be written $\mathcal{P}_{+-} = a\phi + b$. This probability would be unity if the two hemispheres coincided ($\phi = 0$); thus, $b = 1$. It vanishes if $\phi = \pi$, however. Thus, $a = -1/\pi$. Therefore $\mathcal{P}_{+-} = \mathcal{P}_{-+} = 1 - \phi/\pi$. Finally, we get

$$\langle \sigma^{(12)} \rangle = \frac{4}{\pi} \left(\phi - \frac{\pi}{2} \right)$$

(d) We have

$$\langle \sigma^{(12)}(\phi) \rangle - \langle \sigma^{(12)}(\phi') \rangle = \frac{4}{\hbar^2} \int d\lambda \, f(\lambda) \, S_z^{(1)}(\lambda) \left[S_\phi^{(2)}(\lambda) - S_{\phi'}^{(2)}(\lambda) \right]$$

$$= -\frac{4}{\hbar^2} \int d\lambda \, f(\lambda) \, S_z^{(1)}(\lambda) \, S_\phi^{(1)}(\lambda)$$

$$+ \frac{16}{\hbar^4} \int d\lambda \, f(\lambda) \, S_z^{(1)}(\lambda) \, S_\phi^{(1)}(\lambda) \, S_\phi^{(2)}(\lambda) \, S_{\phi'}^{(2)}(\lambda)$$

$$= -\frac{4}{\hbar^2} \int d\lambda \, f(\lambda) \, S_z^{(1)}(\lambda) \, S_\phi^{(1)}(\lambda) \left[1 - \frac{4}{\hbar^2} S_\phi^{(2)}(\lambda) \, S_{\phi'}^{(2)}(\lambda) \right]$$

We have used $S_\phi^{(2)}(\lambda) = -S_\phi^{(1)}(\lambda)$ and $[S_\phi^{(i)}(\lambda)]^2 = \hbar^2/4$. The last equality gives us the inequality

$$\left| \langle \sigma^{(12)}(\phi) \rangle - \langle \sigma^{(12)}(\phi') \rangle \right|$$

$$\leq \frac{4}{\hbar^2} \int d\lambda \, f(\lambda) \left| S_z^{(1)}(\lambda) \, S_\phi^{(1)}(\lambda) \right| \left| 1 - \frac{4}{\hbar^2} S_\phi^{(2)}(\lambda) \, S_{\phi'}^{(2)}(\lambda) \right|$$

$$= \int d\lambda \, f(\lambda) \left[1 - \frac{4}{\hbar^2} S_\phi^{(2)}(\lambda) \, S_{\phi'}^{(2)}(\lambda) \right]$$

Note that the quantity in square brackets is positive. Thus, we finally have

$$\left| \langle \sigma^{(12)}(\phi) \rangle - \langle \sigma^{(12)}(\phi') \rangle \right| \leq 1 + \langle \sigma^{(12)}(\phi' - \phi) \rangle$$

(e) Although Bell's inequality was derived in the framework of a so-called *local hidden variable theory*, let us apply it in quantum mechanics in the specific case $\phi' = 2\phi$. Using the quantum mechanical result found in part (b) we get $\langle \sigma^{(12)} \rangle = -\cos\phi$; Bell's inequality then implies that

$$|\cos\phi - \cos 2\phi| \leq 1 - \cos\phi$$

This is, however, *not true* in the region $0 < \phi < \pi/2$. Try, for example, $\phi = \pi/4$. Since experiment[6] has confirmed the quantum mechanical predictions, this immediately implies that the whole framework of *local hidden variable theories* is ruled out.

Problem 6.4 Neutrinos are neutral particles that come in three flavours, namely ν_e, ν_μ and ν_τ. Until recently they were considered to be massless. The observation of *neutrino oscillations*, i.e. transitional processes such as $\nu_\mu \leftrightarrow \nu_e$, is a proof of their massiveness.

(a) Consider for simplicity two neutrino species only, namely the *flavour eigenstates* $|\nu_e\rangle$ and $|\nu_\mu\rangle$. Ignoring the spatial degrees of freedom and treating the momentum as a parameter p, the Hamiltonian is a 2×2 matrix with eigenvectors $|\nu_1\rangle$ and $|\nu_2\rangle$ and eigenvalues $E_{1,2} = \sqrt{p^2c^2 + m_{1,2}^2 c^4} \approx pc + m_{1,2}^2 c^3/2p$. If the flavour eigenstates are related to the energy eigenstates by

$$|\nu_e\rangle = \cos\theta \, |\nu_1\rangle + \sin\theta \, |\nu_2\rangle$$
$$|\nu_\mu\rangle = -\sin\theta \, |\nu_1\rangle + \cos\theta \, |\nu_2\rangle$$

where θ is a parameter, calculate the matrix representing the time evolution operator.

(b) If at time $t = 0$ the neutrino system is in an *electronic neutrino state* $|\psi(0)\rangle = |\nu_e\rangle$, find the probability $\mathcal{P}_{e \to \mu}$ for it to make a transition $e \to \mu$, as a function of time. Do

---

[6] A. Aspect, P. Grangier and G. Roger, *Phys. Rev. Lett.* **49**, 91 (1982).

the same for the probability $\mathcal{P}_{e\to e}$. Alternatively, assume that initially the system is in a *muonic neutrino* state $|\psi(0)\rangle = |\nu_\mu\rangle$ and calculate the analogous probabilities $\mathcal{P}_{\mu\to e}$ and $\mathcal{P}_{\mu\to\mu}$.

(c) Show that, up to an irrelevant overall phase, the Schroedinger equation satisfied by the neutrino state can be cast in the form, with $\Delta m^2 = m_2^2 - m_1^2$,

$$i\hbar \frac{d}{dt}|\psi(t)\rangle = -\frac{\Delta m^2 c^3}{4p}\begin{pmatrix} \cos 2\theta & -\sin 2\theta \\ -\sin 2\theta & -\cos 2\theta \end{pmatrix}|\psi(t)\rangle$$

(d) Consider a situation where electron neutrinos with estimated number density N_e are emitted from a source and their density is measured at a faraway location. Derive a relation between Δm^2 and the distance at which there is maximal conversion to muon neutrinos. Derive an expression for the mixing angle θ in terms of the ratio $N_e(L)/N_e(0)$ at a location L. The average momentum of the emitted neutrinos p is considered known.

Solution

(a) In the energy-eigenstate basis the time evolution matrix is a diagonal matrix $U(t)$ with elements

$$U_{ij} = \langle v_i | e^{-iHt/\hbar} | v_j \rangle$$

so that

$$U(t) = \begin{pmatrix} e^{-iE_1 t/\hbar} & 0 \\ 0 & e^{-iE_2 t/\hbar} \end{pmatrix}$$

In the flavour-eigenstate basis we have

$$\langle v_e | U(t) | v_e \rangle = \cos^2\theta\, e^{-iE_1 t/\hbar} + \sin^2\theta\, e^{-iE_2 t/\hbar}$$
$$\langle v_\mu | U(t) | v_\mu \rangle = \sin^2\theta\, e^{-iE_1 t/\hbar} + \cos^2\theta\, e^{-iE_2 t/\hbar}$$
$$\langle v_e | U(t) | v_\mu \rangle = \langle v_\mu | U(t) | v_e \rangle = \cos\theta \sin\theta \left(e^{-iE_2 t/\hbar} - e^{-iE_1 t/\hbar} \right)$$

and the corresponding matrix $\mathcal{U}(t)$ turns out to be

$$\begin{pmatrix} \cos^2\theta\, e^{-iE_1 t/\hbar} + \sin^2\theta\, e^{-iE_2 t/\hbar} & \cos\theta \sin\theta \left(e^{-iE_2 t/\hbar} - e^{-iE_1 t/\hbar} \right) \\ \cos\theta \sin\theta \left(e^{-iE_2 t/\hbar} - e^{-iE_1 t/\hbar} \right) & \sin^2\theta\, e^{-iE_1 t/\hbar} + \cos^2\theta\, e^{-iE_2 t/\hbar} \end{pmatrix}$$

(b) In the case where $|\psi(0)\rangle = |\nu_e\rangle$, the evolved state will be

$$|\psi(t)\rangle = \left(\cos^2\theta\, e^{-iE_1 t/\hbar} + \sin^2\theta\, e^{-iE_2 t/\hbar} \right) |\nu_e\rangle$$
$$+ \cos\theta \sin\theta \left(e^{-iE_2 t/\hbar} - e^{-iE_1 t/\hbar} \right) |\nu_\mu\rangle$$

Thus

$$\mathcal{P}_{e\to\mu} = \left| \langle v_\mu | \psi(t) \rangle \right|^2 = \left| \cos\theta \sin\theta \left(e^{-iE_2 t/\hbar} - e^{-iE_1 t/\hbar} \right) \right|^2$$

giving

$$\mathcal{P}_{e \to \mu} = \sin^2 2\theta \, \sin^2 \left(\frac{E_2 - E_1}{2\hbar} t \right)$$

The survival probability of the electronic neutrino is

$$\mathcal{P}_{e \to e} = 1 - \sin^2 2\theta \, \sin^2 \left(\frac{E_2 - E_1}{2\hbar} t \right)$$

In the case where the initial state is $|\psi(0)\rangle = |\nu_\mu\rangle$, we get

$$\mathcal{P}_{\mu \to e} = \sin^2 2\theta \, \sin^2 \left(\frac{E_2 - E_1}{2\hbar} t \right) = \mathcal{P}_{e \to \mu}$$

As expected, we also have

$$\mathcal{P}_{\mu \to \mu} = 1 - \sin^2 2\theta \, \sin^2 \left(\frac{E_2 - E_1}{2\hbar} t \right) = \mathcal{P}_{e \to e}$$

(c) Consider the matrix

$$\Theta \equiv \begin{pmatrix} \cos\theta & -\sin\theta \\ \sin\theta & \cos\theta \end{pmatrix}$$

in terms of which we can transform the state vector from the flavour-eigenstate basis to the mass-eigenstate basis:

$$\Psi_m = \Theta \Psi_f$$

Thus, we have

$$i\hbar \dot{\Psi}_m = \begin{pmatrix} E_1 & 0 \\ 0 & E_2 \end{pmatrix} \begin{pmatrix} e^{-iE_1 t/\hbar} & 0 \\ 0 & e^{-iE_2 t/\hbar} \end{pmatrix} \Psi_m(0)$$

or, equivalently,

$$i\hbar \dot{\Psi}_f = \begin{pmatrix} \cos\theta & \sin\theta \\ -\sin\theta & \cos\theta \end{pmatrix} \begin{pmatrix} E_1 & 0 \\ 0 & E_2 \end{pmatrix}$$
$$\times \begin{pmatrix} e^{-iE_1 t/\hbar} & 0 \\ 0 & e^{-iE_2 t/\hbar} \end{pmatrix} \begin{pmatrix} \cos\theta & -\sin\theta \\ \sin\theta & \cos\theta \end{pmatrix} \Psi_f(0)$$

Dropping the flavour subscript this becomes

$$i\hbar \frac{d}{dt} |\psi(t)\rangle = \begin{pmatrix} \cos\theta & \sin\theta \\ -\sin\theta & \cos\theta \end{pmatrix} \begin{pmatrix} E_1 & 0 \\ 0 & E_2 \end{pmatrix}$$
$$\times \begin{pmatrix} e^{-iE_1 t/\hbar} & 0 \\ 0 & e^{-iE_2 t/\hbar} \end{pmatrix} \begin{pmatrix} \cos\theta & -\sin\theta \\ \sin\theta & \cos\theta \end{pmatrix} |\psi(0)\rangle$$

Now note that we can write

$$\begin{pmatrix} E_1 & 0 \\ 0 & E_2 \end{pmatrix} = \left[pc + \frac{(m_1^2 + m_2^2)c^3}{4p} \right] \begin{pmatrix} 1 & 0 \\ 0 & 1 \end{pmatrix} + \frac{\Delta m^2 c^3}{4p} \begin{pmatrix} -1 & 0 \\ 0 & 1 \end{pmatrix}$$

Thus, we get

$$i\hbar \frac{d}{dt} |\psi(t)\rangle = -\frac{\Delta m^2}{4p} \begin{pmatrix} \cos\theta & \sin\theta \\ -\sin\theta & \cos\theta \end{pmatrix} \begin{pmatrix} 1 & 0 \\ 0 & -1 \end{pmatrix}$$

$$\times \begin{pmatrix} \cos\theta & -\sin\theta \\ \sin\theta & \cos\theta \end{pmatrix} |\psi(t)\rangle + \left[pc + \frac{(m_1^2 + m_2^2)c^3}{4p} \right] |\psi(t)\rangle$$

or

$$i\hbar \frac{d}{dt} |\psi(t)\rangle = -\frac{\Delta m^2 c^3}{4p} \begin{pmatrix} \cos 2\theta & -\sin 2\theta \\ -\sin 2\theta & -\cos 2\theta \end{pmatrix} |\psi(t)\rangle$$

We have dropped the irrelevant phase

$$\exp\left\{ -\frac{i}{\hbar} \left[pc + \frac{(m_1^2 + m_2^2)c^3}{4p} \right] t \right\}$$

(d) Replacing the time of travel t by the distance $L = ct$, we have

$$\frac{E_2 - E_1}{2\hbar} t = (m_2^2 - m_1^2) \frac{c^3 t}{4p\hbar} = \frac{\Delta m^2 L c^2}{4p\hbar}$$

Maximal conversion of electron neutrinos occurs when

$$\frac{\Delta m^2 L_{\max} c^2}{4p\hbar} = \frac{\pi}{2}$$

that is,

$$L_{\max} \Delta m^2 = \frac{2\pi p\hbar}{c^2}$$

The mixing angle can be expressed as

$$\sin^2 2\theta = \frac{1 - N_e(L)/N_e(0)}{\sin^2(\pi L/2L_{\max})}$$

Problem 6.5 Consider a neutron interferometer composed of three crystal slabs (see Fig. 26). A beam of neutrons is split at the first slab, reflected and redirected at the second and finally superposed at the third and final slab. A *phase shifter P S* is placed along the route of one branch, giving a phase difference δ to the neutrons with which it interacts. A *spin flipper S F* that can flip the spin of a neutron is placed along the route of the other branch. By placing a detector in one of the final beams an interference pattern dependent on δ can be observed.

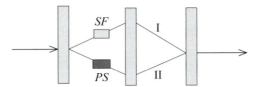

Fig. 26 Neutron interferometry.

(a) The spin flipper device is based on the operation of a static magnetic field $\mathbf{B}_0$ and a time-dependent magnetic field perpendicular to it, $\mathbf{B}_1(t) = B_1(\cos \omega t \, \hat{\mathbf{x}} + \sin \omega t \, \hat{\mathbf{y}})$. What must be the relation between the neutron magnetic dipole moment and the rest of the parameters in order to have maximum spin-flip probability? The neutron time of flight τ_N in the device is given.

(b) Let the initial state of the system be

$$|\psi_i\rangle = \tfrac{1}{\sqrt{2}} (|\phi_1\rangle \, |\uparrow\rangle + |\phi_2\rangle \, |\uparrow\rangle)$$

This is the state before the beam encounters either SF or PS. The spatial parts $|\phi_{1,2}\rangle$ correspond to known wave packets that propagate along the first and the second route correspondingly. Determine the final state of the system and calculate the spin expectation value in the final state, $\langle \psi_f | \mathbf{S} | \psi_f \rangle$.

Solution

(a) A physical system mathematically analogous to the *spin flipper* is the system of an electron in a magnetic field that has a uniform component and, perpendicular to it, a time-dependent component; this was solved in problem 5.16. Making the modification $\omega_1 \rightarrow -\omega_1$ and redefining the parameters as

$$\omega_0 \equiv \tfrac{1}{2}\mu_n B_0, \qquad \omega_1 \equiv \tfrac{1}{2}\mu_n B_1$$

we can carry over the solution (for $|\psi(0)\rangle = |\uparrow\rangle$):

$$|\psi(t)\rangle = e^{-i\omega t/2} \left[\left(\cos \frac{\gamma t}{2} + i\frac{\omega - 2\omega_0}{\gamma} \sin \frac{\gamma t}{2} \right) |\uparrow\rangle - 2i\frac{\omega_1}{\gamma} \sin \frac{\gamma t}{2} |\downarrow\rangle \right]$$

where $\gamma = \sqrt{4\omega_1^2 + (\omega - 2\omega_0)^2}$. Maximal spin-flipping occurs when $\omega = 2\omega_0$. Then the state is

$$|\psi(t)\rangle = e^{-i\omega_0 t} \cos \omega_1 t \, |\uparrow\rangle - ie^{i\omega_0 t} \sin \omega_1 t \, |\downarrow\rangle$$

If the value of the oscillating magnetic field is adjusted so that the time of flight of a neutron in the device, i.e. in the magnetic field region, satisfies $\omega_1 \tau_N = \pi/2 \implies \mu_N B_1 \tau_N = \pi$, the neutrons will be subject to a complete spin-flip. Their state will be $|\psi(\tau_N)\rangle = -ie^{i\alpha} |\downarrow\rangle$ with $\alpha = \omega_0 \tau_N = \pi B_0/2B_1$.

(b) Starting with the given initial state, after the action of the *phase shifter* and the *spin flipper* neutrons will be in the final state

$$|\psi_f\rangle = \tfrac{1}{\sqrt{2}}\left(-i|\phi_1\rangle|\downarrow\rangle + e^{i\delta}|\phi_2\rangle|\uparrow\rangle\right)$$

where α has been absorbed into δ.

It is straightforward to calculate that $\langle\uparrow|\boldsymbol{\sigma}|\uparrow\rangle = \hat{\mathbf{z}}$, $\langle\downarrow|\boldsymbol{\sigma}|\downarrow\rangle = -\hat{\mathbf{z}}$, $\langle\uparrow|\boldsymbol{\sigma}|\downarrow\rangle = \hat{\mathbf{x}} - i\hat{\mathbf{y}}$ and $\langle\downarrow|\boldsymbol{\sigma}|\uparrow\rangle = \hat{\mathbf{x}} + i\hat{\mathbf{y}}$. Thus, setting $\langle\phi_2|\phi_1\rangle = I$, we get

$$\langle\psi_f|\mathbf{S}|\psi_f\rangle = \frac{\hbar I}{2}\left(\sin\delta\,\hat{\mathbf{x}} + \cos\delta\,\hat{\mathbf{y}}\right)$$

Problem 6.6

(a) A spin-1/2 particle in the state $|S_z +\rangle$ goes through a Stern–Gerlach analyzer having orientation $\hat{\mathbf{n}} = \cos\theta\,\hat{\mathbf{z}} - \sin\theta\,\hat{\mathbf{x}}$ (see Fig. 27). What is the probability of finding the outgoing particle in the state $|S_n +\rangle$?

(b) Now consider a Stern–Gerlach device of variable orientation (Fig. 28). More specifically, assume that it can have the three different directions

$$\hat{\mathbf{n}}_1 = \hat{\mathbf{n}} = \cos\theta\,\hat{\mathbf{z}} - \sin\theta\,\hat{\mathbf{x}}$$

$$\hat{\mathbf{n}}_2 = \cos\left(\theta + \tfrac{2}{3}\pi\right)\hat{\mathbf{z}} - \sin\left(\theta + \tfrac{2}{3}\pi\right)\hat{\mathbf{x}}$$

$$\hat{\mathbf{n}}_3 = \cos\left(\theta + \tfrac{4}{3}\pi\right)\hat{\mathbf{z}} - \sin\left(\theta + \tfrac{4}{3}\pi\right)\hat{\mathbf{x}}$$

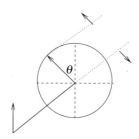

Fig. 27 Tilted Stern–Gerlach apparatus.

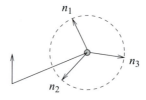

Fig. 28 Stern–Gerlach device with variable orientation.

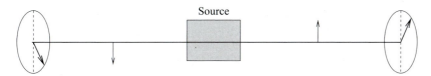

Fig. 29 A pair of spin-1/2 particles emitted in opposite directions. The arrows in the circles represent the field directions in the Stern–Gerlach analyzers.

with equal probability $(1/3)$. If a particle in the state $|S_z\,+\rangle$ enters the analyzer, what is the probability that it will come out with spin eigenvalue $+\hbar/2$?

(c) Calculate the same probability as above but now for a Stern–Gerlach analyzer that can have any orientation with equal probability.

(d) A pair of particles is emitted with the particles in opposite directions in a singlet $|0\,0\rangle$ state. Each particle goes through a Stern–Gerlach analyzer of the type introduced in (c); see Fig. 29. Calculate the probability of finding the exiting particles with opposite spin eigenvalues.

Solution

(a) The eigenvalue relation

$$(\hat{\mathbf{n}} \cdot \mathbf{S})\,|S_n\,\pm\rangle = \pm\frac{\hbar}{2}|S_n\,\pm\rangle$$

is represented in matrix form as

$$\begin{pmatrix} \cos\theta & -\sin\theta \\ -\sin\theta & -\cos\theta \end{pmatrix}\begin{pmatrix} a \\ b \end{pmatrix} = \pm\begin{pmatrix} a \\ b \end{pmatrix}$$

Solving, we obtain $a_+ = \cos(\theta/2) = b_-$ and $a_- = \sin(\theta/2) = -b_+$; this gives

$$|S_n\,+\rangle = \begin{pmatrix} \cos(\theta/2) \\ -\sin(\theta/2) \end{pmatrix}, \qquad |S_n\,-\rangle = \begin{pmatrix} \sin(\theta/2) \\ \cos(\theta/2) \end{pmatrix}$$

The probability of finding the particle in the state $|S_n\,+\rangle$ is

$$\mathcal{P} = |\langle S_z\,+|S_n\,+\rangle|^2 = \cos^2(\theta/2)$$

(b) The probability of finding a particle in the exit carrying spin eigenvalue $+\hbar/2$ is

$$\mathcal{P} = \tfrac{1}{3}\sum_{i=1,\,2,\,3}|\langle S_z\,+|S_{n_i}\,+\rangle|^2$$

$$= \tfrac{1}{3}\left[\cos^2\frac{\theta}{2} + \cos^2\left(\frac{\theta}{2}+\frac{\pi}{3}\right) + \cos^2\left(\frac{\theta}{2}+\frac{2\pi}{3}\right)\right]$$

This can be seen to give[7]

$$P = \frac{1}{3}\left(\cos^2\frac{\theta}{2} + \frac{1}{2}\cos^2\frac{\theta}{2} + \frac{3}{2}\sin^2\frac{\theta}{2}\right) = \frac{1}{2}$$

(c) In this case we should average over all orientations according to

$$\langle\cdots\rangle = \frac{1}{2\pi}\int_0^{2\pi} d\theta \ \cdots$$

Then we have

$$P = \langle|\langle S_z + |S_n +\rangle|^2\rangle = \frac{1}{2\pi}\int_0^{2\pi} d\theta \ \cos^2\frac{\theta}{2} = \frac{1}{2}$$

(d) The probability of opposite spin outcomes is

$$P = \left\langle\left\{\left|\langle 0\,0\,|\,S_n^{(1)} +\rangle\,|\,S_{n'}^{(2)} -\rangle\right|^2 + \left|\langle 0\,0\,|\,S_n^{(1)} -\rangle\,|\,S_{n'}^{(2)} +\rangle\right|^2\right\}\right\rangle$$

$$= \left\langle\cos^2\frac{1}{2}(\theta - \theta')\right\rangle = \frac{1}{2}$$

The intermediate steps involved are:

$$\langle+|\hat{n}\,+\rangle\,\langle-|\hat{n}'\,-\rangle - \langle-|\hat{n}\,+\rangle\,\langle+|\hat{n}'\,-\rangle = \cos[(\theta - \theta')/2]$$
$$\langle+|\hat{n}\,-\rangle\,\langle-|\hat{n}'\,+\rangle - \langle-|\hat{n}\,-\rangle\,\langle+|\hat{n}'\,+\rangle = -\cos[(\theta - \theta')/2]$$

Problem 6.7

(a) A system with Hamiltonian H is initially in a state $|\psi(0)\rangle \equiv |\psi_i\rangle$. Show that for small time intervals the probability of finding the system in the initial state is equal to

$$1 - (\Delta E)^2 t^2/\hbar^2 + O(t^4)$$

where ΔE is the energy uncertainty in the initial state.

(b) Consider now the transition amplitude $T_{ji}(t)$ to state $|\psi_j\rangle$. Show that $T_{ji}(t) = T_{ij}^*(-t)$. Then show that the survival probability $\mathcal{P}_{ii}$ has to be an even function of time. In a similar fashion, show that the total transition rate $\sum_j \dot{\mathcal{P}}_{ji}(0)$ at $t = 0$ vanishes.

(c) Show that the survival amplitude $T_{ii}(t)$ can be written as

$$T_{ii}(t) = \int dE \ \eta_i(E) e^{-iEt/\hbar}$$

where $\eta_i(E)$ is the so-called *spectral function* of the state $|i\rangle$. Consider the case of a

---

[7]
$$\cos(\theta/2 + \pi/3) = (1/2)\cos(\theta/2) - (\sqrt{3}/2)\sin(\theta/2)$$
$$\cos(\theta/2 + 2\pi/3) = -\cos(\theta/2 - \pi/3) = -(1/2)\cos(\theta/2) - (\sqrt{3}/2)\sin(\theta/2)$$

continuous spectrum and assume that for a particular state the spectral function has the form

$$\eta(E) = \frac{1}{2\pi} \frac{\hbar\Gamma}{(E - E_0)^2 + (\hbar\Gamma/2)^2}$$

with $\Gamma > 0$. Find the corresponding survival probability. Despite the fact that this spectral function does not correspond to a realistic spectrum, find why the short-time behaviour of the survival probability established in (a) does not apply here.

Solution

(a) The survival probability of the initial state is

$$\mathcal{P}_{ii} = \left| \langle \psi_i | e^{-iHt/\hbar} | \psi_i \rangle \right|^2$$

Expanding in time, we get

$$\mathcal{P}_{ii} \approx 1 + \frac{it}{\hbar} \left(\langle H \rangle^* - \langle H \rangle \right) + \frac{t^2}{\hbar^2} \left(|\langle H \rangle|^2 - \frac{1}{2} \langle H^2 \rangle - \frac{1}{2} \langle H^2 \rangle^* \right)$$

All the expectation values refer to the state $|\psi_i\rangle$. Obviously, due to hermiticity we have

$$\langle H \rangle^* = \langle H \rangle, \qquad \langle H^2 \rangle^* = \langle H^2 \rangle$$

and thus

$$\mathcal{P}_{ii} \approx 1 - \frac{t^2}{\hbar^2} (\Delta E)^2$$

The $O(t^3)$ term is

$$i \frac{t^3}{2\hbar^3} \left(\frac{1}{3} \langle H^3 \rangle - \frac{1}{3} \langle H^3 \rangle^* + \langle H \rangle \langle H^2 \rangle^* - \langle H \rangle^* \langle H^2 \rangle \right)$$

and it vanishes. The next non-zero correction is $O(t^4)$.

(b) From the definition of T we have

$$T_{ji}(t) = \langle j | e^{-iHt/\hbar} | i \rangle = \langle i | e^{iHt/\hbar} | j \rangle^* = T_{ij}^*(-t)$$

The survival probability is

$$\mathcal{P}_{ii}(t) = |T_{ii}(t)|^2 = T_{ii}(t) T_{ii}^*(t) = T_{ii}^*(-t) T_{ii}(-t) = |T_{ii}(-t)|^2 = \mathcal{P}_{ii}(-t)$$

We have

$$\dot{\mathcal{P}}_{ji} = \dot{T}_{ji}(t) T_{ji}^*(t) + T_{ji}(t) \dot{T}_{ji}^*(t)$$

$$= -\frac{i}{\hbar} \langle \psi_j | H e^{-iHt/\hbar} | \psi_i \rangle T_{ji}^*(t) + \frac{i}{\hbar} T_{ji}(t) \langle \psi_j | H e^{iHt/\hbar} | \psi_i \rangle$$

or

$$\dot{\mathcal{P}}_{ji}(0) = \frac{i}{\hbar} \left[T_{ji}(0) - T_{ji}^*(0) \right] \langle \psi_j | H | \psi_i \rangle$$

Summing over all states $|\psi_j\rangle$, we get

$$\sum_j \dot{\mathcal{P}}_{ji}(0) = \frac{i}{\hbar} \sum_j \langle \psi_i | H | \psi_j \rangle \langle \psi_j | \psi_i \rangle \; - \frac{i}{\hbar} \sum_j \langle \psi_i | \psi_j \rangle \langle \psi_j | H | \psi_i \rangle = 0$$

(c) Inserting the complete set of energy eigenstates into the definition of T_{ii}, we get

$$T_{ii}(t) = \sum_{E_n} \langle i | E_n \rangle \langle E_n | i \rangle \, e^{-iE_n t/\hbar} = \sum_{E_n} |\langle i | E_n \rangle|^2 \, e^{-iE_n t/\hbar}$$

This can be written as an integral:

$$T_{ii}(t) = \int dE \sum_{E_n} |\langle i | E_n \rangle|^2 \, \delta(E - E_n) e^{-iEt/\hbar} = \int dE \, \eta_i(E) e^{-iEt/\hbar}$$

where

$$\eta_i(E) = \sum_{E_n} |\langle i | E_n \rangle|^2 \, \delta(E - E_n)$$

For a continuous spectrum this will be a continuous function. Considering the given spectral function, and assuming that this is so, we obtain

$$T_{ii}(t) = \frac{1}{\pi} \int dE \, \frac{\hbar\Gamma/2}{(E - E_0)^2 + (\hbar\Gamma/2)^2} e^{-iEt/\hbar} = e^{-iE_0 t/\hbar} \, e^{-\Gamma t/2}$$

We may evaluate the above integral as a contour integral (see Fig. 30). The

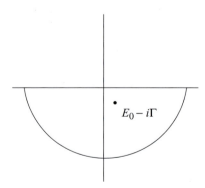

Fig. 30 The complex energy plane: evaluation of T_{ii}.

corresponding survival probability is

$$\mathcal{P}(t) = e^{-\Gamma t}$$

Note that the derived exponential decay contradicts the property shown in (a), according to which at early times we have $\mathcal{P} \propto 1 - O(t^2)$. The issue of the approximate applicability of exponential decay is broad and deep. Nevertheless, in our case, for the particular spectral function assumed it is rather clear why the early-time expansion performed in (a) would go wrong: all moments of the energy diverge. For instance,

$$\langle i|H^2|i\rangle = \sum_n E_n^2 |\langle i|E_n\rangle|^2 = \int dE\, E^2 \eta_i(E)$$

$$= \frac{\Gamma\hbar}{2\pi} \int dE\, \frac{E^2}{(E - E_0)^2 + (\hbar\Gamma/2)^2} = \infty$$

Problem 6.8 Consider a two-state system $|1\rangle$, $|2\rangle$. The Hamiltonian matrix in the orthonormal basis $\{|1\rangle, |2\rangle\}$ is a Hermitian 2×2 matrix that can be written in terms of the three Pauli matrices[8] and the unit matrix as[9] $\mathcal{H} = \frac{1}{2}(H_0 + \mathbf{H} \cdot \boldsymbol{\tau})$.

(a) Consider the *density matrix* $\rho(t) = |\psi(t)\rangle\langle\psi(t)|$ corresponding to the state $|\psi(t)\rangle$ of the system. Show that it satisfies the time-evolution equation

$$\dot{\rho} = -\frac{i}{\hbar}[H, \rho]$$

Show that the density matrix can be written as the 2×2 matrix

$$\rho = \frac{1}{2}\left(1 + \boldsymbol{\Psi} \cdot \boldsymbol{\tau}\right)$$

Show that the length of the complex vector $\boldsymbol{\Psi}$ is equal to unity.
(b) Show that the Schroedinger equation, or the corresponding time-evolution equation for ρ, is equivalent to the equation

$$\frac{d\boldsymbol{\Psi}}{dt} = \frac{1}{\hbar}\mathbf{H} \times \boldsymbol{\Psi}$$

Show that the motion of the system is periodic. Solve for $\boldsymbol{\Psi}(t)$.

[8] To avoid confusion with the spin case we shall use the symbols τ_a, $a = 1, 2, 3$:

$$\tau_1 = \begin{pmatrix} 0 & 1 \\ 1 & 0 \end{pmatrix}, \qquad \tau_2 = \begin{pmatrix} 0 & -i \\ i & 0 \end{pmatrix}, \qquad \tau_3 = \begin{pmatrix} 1 & 0 \\ 0 & -1 \end{pmatrix}$$

[9] In terms of the matrix elements $H_{ij} = \langle i|H|j\rangle$ of the original Hamiltonian, the components of the vector $\mathbf{H}$ are

$$H_1 = H_{12} + H_{12}^*, \qquad H_2 = i(H_{12} - H_{12}^*), \qquad H_3 = H_{11} - H_{22}$$

(c) How is the return probability $|\langle \psi(0)|\psi(t)\rangle|^2$ expressed in terms of the above solution? Calculate it in the simple case in which the system is initially along the '1' direction.

Solution

(a) From the Schroedinger equation and its conjugate, we can obtain the general expression

$$\dot{\rho} = \frac{d|\psi(t)\rangle}{dt}\langle\psi(t)| + |\psi(t)\rangle\frac{d\langle\psi(t)|}{dt}$$

$$= -\frac{i}{\hbar}\left(H|\psi(t)\rangle\langle\psi(t)| - |\psi(t)\rangle\langle\psi(t)|H\right) = -\frac{i}{\hbar}[H, \rho]$$

In the basis $\{|1\rangle, |2\rangle\}$ the density matrix is

$$\rho = \begin{pmatrix} |\psi_1|^2 & \psi_1\psi_2^* \\ \psi_1^*\psi_2 & |\psi_2|^2 \end{pmatrix}$$

Normalization requires $\mathrm{Tr}\,\rho = |\psi_1|^2 + |\psi_2|^2 = 1$. This is also evident from

$$\langle\psi|\rho|\psi\rangle = \langle\psi|\psi\rangle\langle\psi|\psi\rangle = |\langle\psi|\psi\rangle|^2 = 1$$

Note that the matrix representation for ρ in terms of $\boldsymbol{\Psi}$ given in the question satisfies the trace property automatically. Writing it out explicitly and equating with the above form for ρ gives

$$\frac{1}{2}\begin{pmatrix} 1 + \Psi_3 & \Psi_1 - i\Psi_2 \\ \Psi_1 + i\Psi_2 & 1 - \Psi_3 \end{pmatrix} = \begin{pmatrix} |\psi_1|^2 & \psi_1\psi_2^* \\ \psi_1^*\psi_2 & |\psi_2|^2 \end{pmatrix}$$

and

$$\Psi_1 = \psi_1\psi_2^* + \psi_1^*\psi_2, \qquad \Psi_2 = i(\psi_1\psi_2^* - \psi_1^*\psi_2), \qquad \Psi_3 = |\psi_1|^2 - |\psi_2|^2$$

(b) Substituting the vector expressions into the time-evolution equation for the density matrix, we get

$$\dot{\rho} = \frac{1}{2}\dot{\boldsymbol{\Psi}}\cdot\boldsymbol{\tau} = -\frac{i}{\hbar}\left[\frac{\tau_a}{2}, \frac{\tau_b}{2}\right]H_a\Psi_b$$

$$= \frac{1}{2\hbar}\epsilon_{abc}\tau_c\,H_a\Psi_b = \frac{1}{2\hbar}(\mathbf{H}\times\boldsymbol{\Psi})\cdot\boldsymbol{\tau}$$

Thus we can write

$$\frac{d\boldsymbol{\Psi}}{dt} = \frac{1}{\hbar}\mathbf{H}\times\boldsymbol{\Psi}$$

It is evident, by taking the inner product of this equation with $\mathbf{H}$, that the component of $\boldsymbol{\Psi}$ along the direction of the Hamiltonian vector is a constant of the motion. Similarly, Ψ^2 is a constant. Thus, we have a state vector $\boldsymbol{\Psi}$ of constant length rotating around the vector $\mathbf{H}$ (see Fig. 31).

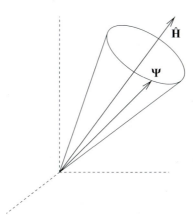

Fig. 31 State-vector rotation.

Since $\boldsymbol{\Psi}_{\parallel}(t) \equiv \left(\hat{\mathbf{H}} \cdot \boldsymbol{\Psi}(t)\right) \hat{\mathbf{H}} = \boldsymbol{\Psi}_{\parallel}(0)$, we can just write

$$\frac{d\boldsymbol{\Psi}_{\perp}}{dt} = \frac{1}{\hbar} \mathbf{H} \times \boldsymbol{\Psi}_{\perp}$$

Taking an additional time derivative we get

$$\frac{d^2\boldsymbol{\Psi}_{\perp}}{dt^2} = -\left(\frac{H^2}{\hbar^2}\right) \boldsymbol{\Psi}_{\perp}$$

The square of the Hamiltonian vector is equal to the squared difference of the energy eigenvalues:

$$H^2 = H_1^2 + H_2^2 + H_3^2 = (H_{11} - H_{22})^2 + 4|H_{12}|^2 = (E_1 - E_2)^2 \equiv \hbar^2\omega^2$$

The solution to the above equation of motion is

$$\boldsymbol{\Psi}_{\perp}(t) = \boldsymbol{\Psi}_{\perp}(0) \cos \omega t + \hat{\mathbf{H}} \times \boldsymbol{\Psi}(0) \sin \omega t$$

or

$$\boldsymbol{\Psi}(t) = \hat{\mathbf{H}} \cdot \boldsymbol{\Psi}(0) \hat{\mathbf{H}} + \left[\boldsymbol{\Psi}(0) - \hat{\mathbf{H}} \cdot \boldsymbol{\Psi}(0) \hat{\mathbf{H}}\right] \cos \omega t + \hat{\mathbf{H}} \times \boldsymbol{\Psi}(0) \sin \omega t$$

On rearrangement this becomes

$$\boldsymbol{\Psi}(t) = \hat{\mathbf{H}} \cdot \boldsymbol{\Psi}(0) \hat{\mathbf{H}}(1 - \cos \omega t) + \boldsymbol{\Psi}(0) \cos \omega t + \hat{\mathbf{H}} \times \boldsymbol{\Psi}(0) \sin \omega t$$

The period of the motion is $T = 2\pi\hbar/|E_2 - E_1|$.

(c) The return probability is

$$\begin{aligned}
\mathcal{P}(t) &= |\langle \psi(0)|\psi(t)\rangle|^2 = \langle \psi(0)|\psi(t)\rangle \langle \psi(t)|\psi(0)\rangle \\
&= \langle \psi(0)|\rho(t)|\psi(0)\rangle = \tfrac{1}{2} + \tfrac{1}{2}\boldsymbol{\Psi}(0) \cdot \boldsymbol{\Psi}(t) \\
&= \tfrac{1}{2} + \tfrac{1}{2}[\boldsymbol{\Psi}(0) \cdot \hat{\mathbf{H}}]^2 + \tfrac{1}{2}\boldsymbol{\Psi}(0) \cdot \left[\boldsymbol{\Psi}_{\perp}(0) \cos \omega t + \hat{\mathbf{H}} \times \boldsymbol{\Psi}(0) \sin \omega t\right]
\end{aligned}$$

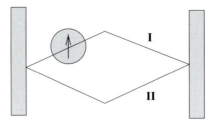

Fig. 32 Split neutron beam.

or

$$\mathcal{P}(t) = \cos^2(\omega t/2) + [\boldsymbol{\Psi}(0) \cdot \hat{\mathbf{H}}]^2 \sin^2(\omega t/2)$$

In the case where $\psi_1(0) = 1$ and $\psi_2(0) = 0$, we have $\Psi_i(0) = \delta_{i3}$ and

$$\boldsymbol{\Psi}(0) \cdot \hat{\mathbf{H}} = \frac{H_{11} - H_{22}}{\sqrt{(H_{11} - H_{22})^2 + 4|H_{12}|^2}}$$

Note also that $(\Delta E)_0^2 = |H_{12}|^2$. Thus, we may write

$$\mathcal{P}(t) = 1 - 4\frac{(\Delta E)_0^2}{(\hbar\omega)^2} \sin^2\frac{\omega t}{2}$$

Problem 6.9 A neutron beam passing through a neutron interferometer is split into two parts, one of which passes through a magnetic field (see Fig. 32). The magnetic field can be considered uniform throughout the extent of the path of the neutron branch that passes through it. Its magnitude is adjustable. Consider the interference pattern created by the existence of two branches and determine the possible magnitude differences of two magnetic fields that give the same interference pattern.

Solution

The evolution of the neutron spin is determined by the Hamiltonian

$$H_s = -\boldsymbol{\mu}_n \cdot \mathbf{B} = \tfrac{1}{2}\hbar\omega\sigma_3 \qquad (\omega \equiv 2\mu_n B/\hbar)$$

The magnetic moment is $\mu_n = g_n(|e|\hbar/2m_n c)$. The evolved state will be

$$\psi(t) = e^{-i\omega t\sigma_3/2} \psi(0) = \begin{pmatrix} e^{-i\omega t/2} & 0 \\ 0 & e^{i\omega t/2} \end{pmatrix} \psi(0)$$

Starting from an initial spin state $|\psi(0)\rangle = |\uparrow\rangle$ we end up at time τ with just a phase change, $|\psi(\tau)\rangle = e^{-i\omega\tau/2}|\uparrow\rangle$.

The state describing the interfering beams is[10]

$$|\Psi\rangle = N\left(|\phi_{\mathrm{I}}\rangle\, e^{-i\omega\tau/2}|\uparrow\rangle + |\phi_{\mathrm{II}}\rangle\,|\uparrow\rangle\right)$$

---

[10] If $\langle\phi_{\mathrm{I}}|\phi_{\mathrm{II}}\rangle = Ie^{i\delta}$, the normalization factor is
$$N = 2^{-1/2}[1 + I\cos(\delta + \omega\tau/2)]^{-1/2}$$

where τ is the time spent by the first branch in the magnetic field region and $|\phi_{\text{I}}\rangle$ and $|\phi_{\text{II}}\rangle$ correspond to the spatial wave functions of each branch. The probability of finding a neutron in the spin-up state $|\uparrow\rangle|\phi_{\text{II}}\rangle$ is

$$\mathcal{P} = \frac{1}{2}\left[\frac{1 + I^2 + 2I\cos(\delta + \omega\tau/2)}{1 + I\cos(\delta + \omega\tau/2)}\right]$$

The interference pattern is not modified for a frequency change

$$\omega' - \omega = \frac{4n\pi}{\tau} \qquad (n = 1, 2, \ldots)$$

i.e. a magnetic field change

$$\Delta B = n\left(\frac{4\pi}{\tau}\right)\left(\frac{\hbar}{2\mu_{\text{n}}}\right) = n\left(\frac{4\pi v}{L}\right)\left(\frac{\hbar}{2\mu_{\text{n}}}\right)$$
$$= n\left(\frac{8\pi^2\hbar}{L\lambda}\right)\left(\frac{\hbar}{2\mu_{\text{n}}m_{\text{n}}}\right) = n\left(\frac{8\pi^2\hbar c}{g_{\text{n}}|e|L\lambda}\right)$$

where L is the extent of the magnetic field region and λ is the average wavelength of the neutrons.

Problem 6.10

(a) Consider a system of two spin-1/2 particles in the triplet state

$$|1\ 0\rangle = \tfrac{1}{\sqrt{2}}\left(|\uparrow\rangle_{(1)}\ |\downarrow\rangle_{(2)} + |\downarrow\rangle_{(1)}\ |\uparrow\rangle_{(2)}\right)$$

and perform a measurement of $S_z^{(1)}$. Comment on the fact that a simultaneous measurement of $S_z^{(2)}$ gives an outcome that can always be predicted from the first-mentioned measurement. Show that this property, entanglement, is not shared by states that are tensor products. Is the state

$$|\Psi\rangle = \tfrac{1}{2}\left(|1\ 1\rangle + \sqrt{2}|1\ 0\rangle + |1\ -1\rangle\right)$$

entangled, i.e. is it a tensor product?

(b) Consider now the set of four states $|a\rangle$, $a = 1, 2, 3, 4$:

$$|0\rangle = \tfrac{1}{\sqrt{2}}\left(|1\ 1\rangle + i|1\ -1\rangle\right)$$
$$|1\rangle = \tfrac{1}{\sqrt{2}}\left(|1\ -1\rangle + i|1\ 1\rangle\right)$$
$$|2\rangle = \tfrac{1}{\sqrt{2}}\left(e^{-i\pi/4}|1\ 0\rangle - e^{i\pi/4}|0\ 0\rangle\right)$$
$$|3\rangle = \tfrac{1}{\sqrt{2}}\left(e^{-i\pi/4}|1\ 0\rangle + e^{i\pi/4}|0\ 0\rangle\right)$$

Show that these states are entangled and find the unitary matrix $\mathcal{U}$ such that[11]

$$|a\rangle = \mathcal{U}_{a\alpha}|\alpha\rangle$$

[11] $\{|\alpha\rangle\} = |1\ 1\rangle, |1\ -1\rangle, |1\ 0\rangle, |0\ 0\rangle$

(c) Consider a one-particle state $|\psi\rangle = C_+|\uparrow\rangle + C_-|\downarrow\rangle$ and one of the entangled states considered in (b), for example $|0\rangle$. Show that the product state can be written as

$$|\psi\rangle|0\rangle = \tfrac{1}{2}\left(|0\rangle|\psi\rangle + |1\rangle|\psi'\rangle + |2\rangle|\psi''\rangle + |3\rangle|\psi'''\rangle\right)$$

where the states $|\psi'\rangle$, $|\psi''\rangle$, $|\psi'''\rangle$ are related to $|\psi\rangle$ through a unitary transformation.

Solution

(a) A measurement of $S_z^{(1)}$ when the system is in the state $|1,\,0\rangle$ will give $\pm\hbar/2$ with equal probability. If the value $+\hbar/2$ occurs, implying that particle 1 is in the state $|\uparrow\rangle$, a measurement of $S_z^{(2)}$ will yield the value $-\hbar/2$, particle 2 being necessarily in the state $|\downarrow\rangle$. Inversely, if a measurement on the particle '1' projects it into the state $|\downarrow\rangle$, a simultaneous measurement on the other particle will necessarily give $+\hbar/2$. The two particles are in an *entangled* state. In contrast, if the state of the two particles can be written as a product then there is no entanglement. For the given state $|\Psi\rangle$, we have

$$|\Psi\rangle = \tfrac{1}{2}(|\uparrow\rangle|\uparrow\rangle + |\downarrow\rangle|\downarrow\rangle + |\uparrow\rangle|\downarrow\rangle + |\downarrow\rangle|\uparrow\rangle)$$

or

$$|\Psi\rangle = \left[\tfrac{1}{\sqrt{2}}(|\uparrow\rangle + |\downarrow\rangle)\right] \otimes \left[\tfrac{1}{\sqrt{2}}(|\uparrow\rangle + |\downarrow\rangle)\right]$$

A spin measurement of particle 1 with outcome $\pm\hbar/2$ can be accompanied by a spin measurement of particle '2' with either of the outcomes $\pm\hbar/2$. The two particles are completely disentangled.

(b) The given states can be written in terms of the product states as

$$|0\rangle = \tfrac{1}{\sqrt{2}}(|\uparrow\rangle|\uparrow\rangle + i|\downarrow\rangle|\downarrow\rangle)$$
$$|1\rangle = \tfrac{1}{\sqrt{2}}(|\downarrow\rangle|\downarrow\rangle + i|\uparrow\rangle|\uparrow\rangle)$$
$$|2\rangle = \tfrac{1}{\sqrt{2}}(|\downarrow\rangle|\uparrow\rangle - i|\uparrow\rangle|\downarrow\rangle)$$
$$|3\rangle = \tfrac{1}{\sqrt{2}}(|\uparrow\rangle|\downarrow\rangle - i|\downarrow\rangle|\uparrow\rangle)$$

A spin measurement of particle 1 in either of the first two states results in particle 2 having a spin of the same sign. Similarly, in each of the other two states a spin measurement of the first particle results in an antiparallel spin for the other.

The matrix $\mathcal{U}$ is given by

$$\mathcal{U} = \frac{1}{\sqrt{2}}\begin{pmatrix} 1 & i & 0 & 0 \\ i & 1 & 0 & 0 \\ 0 & 0 & e^{-i\pi/4} & -e^{i\pi/4} \\ 0 & 0 & e^{-i\pi/4} & e^{i\pi/4} \end{pmatrix}$$

and it is straightforward to see that it is unitary.

(c) Setting

$$|\psi'\rangle = D_+|\uparrow\rangle + D_-|\downarrow\rangle, \qquad |\psi''\rangle = E_+|\uparrow\rangle + E_-|\downarrow\rangle$$
$$|\psi'''\rangle = F_+|\uparrow\rangle + F_-|\downarrow\rangle$$

and substituting into the expression to be proved, we obtain

$$\frac{1}{2}\Big(|0\rangle|\psi\rangle + |1\rangle|\psi'\rangle + |2\rangle|\psi''\rangle + |3\rangle|\psi'''\rangle\Big)$$
$$= \frac{1}{2\sqrt{2}}\Big[(C_+ + iD_+)|\uparrow\rangle|\uparrow\rangle|\uparrow\rangle + (iC_+ + D_+)|\downarrow\rangle|\downarrow\rangle|\uparrow\rangle$$
$$+ (C_- + iD_-)|\uparrow\rangle|\uparrow\rangle|\downarrow\rangle + (iC_- + D_-)|\downarrow\rangle|\downarrow\rangle|\downarrow\rangle$$
$$+ (E_+ - iF_+)|\downarrow\rangle|\uparrow\rangle|\uparrow\rangle + (-iE_+ + F_+)|\uparrow\rangle|\downarrow\rangle|\uparrow\rangle$$
$$+ (E_- - iF_-)|\downarrow\rangle|\uparrow\rangle|\downarrow\rangle + (-iE_- + F_-)|\uparrow\rangle|\downarrow\rangle|\downarrow\rangle\Big]$$

Comparing with

$$|\psi\rangle|0\rangle$$
$$= \frac{1}{\sqrt{2}}\Big(C_+|\uparrow\rangle|\uparrow\rangle|\uparrow\rangle + iC_+|\uparrow\rangle|\downarrow\rangle|\downarrow\rangle + C_-|\downarrow\rangle|\uparrow\rangle|\uparrow\rangle + iC_-|\downarrow\rangle|\downarrow\rangle|\downarrow\rangle\Big)$$

we get

$$D_+ = -iC_+, \qquad D_- = iC_-, \qquad E_+ = C_-, \qquad E_- = -C_+$$
$$F_+ = iC_-, \qquad F_- = iC_+$$

and

$$|\psi'\rangle = -iC_+|\uparrow\rangle + iC_-|\downarrow\rangle = \begin{pmatrix} -i & 0 \\ 0 & i \end{pmatrix}|\psi\rangle$$

$$|\psi''\rangle = C_-|\uparrow\rangle - C_+|\downarrow\rangle = \begin{pmatrix} 0 & -1 \\ 1 & 0 \end{pmatrix}|\psi\rangle$$

$$|\psi'''\rangle = iC_-|\uparrow\rangle + iC_+|\downarrow\rangle = \begin{pmatrix} 0 & i \\ i & 0 \end{pmatrix}|\psi\rangle$$

According to the above, a system consisting of a spin-1/2 particle in a state $|\psi\rangle$ and a pair of spin-1/2 particles in any of the entangled states $|a\rangle$ will produce, if subject to a measurement on the latter, a particle in one of the states $|\psi\rangle$, $|\psi'\rangle$, $|\psi''\rangle$, $|\psi'''\rangle$. A complete *teleportation* of, say, state $|\psi\rangle$ would be achieved if we were to transfer classically the information contained in the specification of the unitary matrix.

7

General motion

Problem 7.1 A particle of mass m is bound in a spherical potential well

$$V(r) = \begin{cases} \infty, & 0 < r < a \\ 0, & a < r < b \\ V_0, & r > b \end{cases}$$

Find the energy eigenfunctions. Consider the case of vanishing angular momentum and find the condition that determines the energy eigenvalues. Does this condition always have a solution? What is the minimum value of the potential V_0 necessary for a bound state to exist? Is there a difference in respect to a one-dimensional square well?

Solution

The radial part of the energy eigenfunctions

$$\psi_{E\ell m} = R_{E\ell}(r) Y_{\ell m}(\theta, \phi)$$

satisfies the *radial Schroedinger equation*

$$\left[\frac{d^2}{dr^2} + \frac{2}{r} \frac{d}{dr} - \frac{\ell(\ell+1)}{r^2} - \frac{2m}{\hbar^2} [V(r) - E] \right] R_{E\ell}(r) = 0$$

Introducing

$$E \equiv \frac{\hbar^2 k^2}{2m}, \qquad V_0 - E \equiv \frac{\hbar^2 q^2}{2m}$$

we obtain

$$\left[\frac{d^2}{dr^2} + \frac{2}{r} \frac{d}{dr} - \frac{\ell(\ell+1)}{r^2} + k^2 \right] R_{E\ell}(r) = 0, \qquad a \le r < b$$

$$\left[\frac{d^2}{dr^2} + \frac{2}{r} \frac{d}{dr} - \frac{\ell(\ell+1)}{r^2} - q^2 \right] R_{E\ell}(r) = 0, \qquad r \ge b$$

with solution $(h_\ell^{(\pm)} = n_\ell \pm i j_\ell)$

$$R_{E\ell}(r) = \begin{cases} A j_\ell(kr) + B n_\ell(kr), & a \leq r < b \\ D h_\ell^{(-)}(iqr), & r \geq b \end{cases}$$

The continuity conditions read

$$R_\ell(a) = 0 \implies A = C n_\ell(ka), \quad B = -C j_\ell(ka)$$
$$C\,[n_\ell(ka) j_\ell(kb) - j_\ell(ka) n_\ell(kb)] = D h_\ell^{(-)}(iqb)$$

or

$$\frac{n_\ell(ka) j_\ell'(kb) - j_\ell(ka) n_\ell'(kb)}{n_\ell(ka) j_\ell(kb) - j_\ell(ka) n_\ell(kb)} = \frac{h_\ell^{(-)'}(iqb)}{h_\ell^{(-)}(iqb)}$$

where the prime denotes a derivative with respect to r. The last condition determines the allowed (discrete) energy eigenvalues.

In the case $\ell = 0$, after some algebra the last condition simplifies to

$$-q = k \cot k(b - a)$$

Defining $\xi \equiv k(b - a)$, $2m V_0(b - a)^2/\hbar^2 \equiv \beta^2$, we can write the eigenvalue equation in the form

$$\tan \xi = -\frac{\xi}{\sqrt{\beta^2 - \xi^2}}$$

This equation can be solved graphically. The right-hand side blows up at $\xi = \beta$, while the left-hand side blows up at $\pi/2$. Thus, in order to have at least one solution, the following must hold:

$$\beta > \frac{\pi}{2} \implies V_0 > \frac{\hbar^2 \pi^2}{8m(b - a)^2}$$

In contrast, a one-dimensional square well always has at least one (even) solution. Note that the bound-state condition we have obtained here corresponds to the one-dimensional square-well condition for odd bound states, which is met only for a sufficiently deep square well.

Problem 7.2 A hydrogen-like atom with atomic number $\mathcal{Z}$ is in its ground state when, due to nuclear processes (operating at a time scale much shorter than the characteristic time scale of the atom $\tau \sim \hbar a_0/e^2$), its nucleus is modified to have the atomic number increased by one unit, i.e. to $\mathcal{Z} + 1$. The electronic state of the atom does not change during this process. What is the probability of finding the atom in the new ground state at a later time? Answer the same question for the new first excited state.

Solution

The wave function of the atom at $t = 0$ is

$$\psi_{100}(r) = \frac{Z^{3/2}}{\left(\pi a_0^3\right)^{1/2}} e^{-Zr/a_0}$$

The energy eigenfunctions of the modified system $\overline{\psi}_{n\ell m}(\mathbf{r})$ are obtained by the substitution $Z \to Z + 1$ to the original eigenfunctions. For example,[1]

$$\overline{\psi}_{100}(r) = \frac{(Z+1)^{3/2}}{\left(\pi a_0^3\right)^{1/2}} e^{-(Z+1)r/a_0}$$

$$\overline{\psi}_{200}(r) = \frac{(Z+1)^{3/2}}{\left(32\pi a_0^3\right)^{1/2}} \left(2 - \frac{Z+1}{a_0}r\right) e^{-(Z+1)r/2a_0}$$

The evolved wave function of the system will be

$$\psi(\mathbf{r}, t) = \sum_{n=1}^{\infty} \sum_{\ell=0}^{\infty} \sum_{m=-\ell}^{+\ell} C_{n\ell m} e^{-i\overline{E}_n t/\hbar} \overline{\psi}_{n\ell m}(\mathbf{r})$$

where the energy eigenvalues are

$$\overline{E}_n = \frac{Z+1}{Z} E_n = -\frac{(Z+1)e^2}{2a_0 n^2}$$

The coefficients $C_{n\ell m}$ can be obtained by taking the inner product with one of the orthogonal eigenfunctions. We get

$$\int d^3r\, \overline{\psi}^*_{n'\ell'm'}\psi(t) = \sum_{n\ell m} C_{n\ell m} e^{-i\overline{E}_n t/\hbar} \int d^3r\, \overline{\psi}^*_{n'\ell'm'}\overline{\psi}_{n\ell m} = C_{n'\ell m'} e^{-i\overline{E}_{n'}t/\hbar}$$

At $t = 0$, we have

$$C_{n\ell m} = \int d^3r\, \overline{\psi}^*_{n\ell m}(\mathbf{r})\psi(\mathbf{r}, 0) = \int d^3\, \overline{\psi}^*_{n\ell m}(\mathbf{r})\psi_{100}(r)$$

$$= \delta_{\ell 0}\delta_{m0} \int d^3r\, \overline{\psi}_{n00}(r)\psi_{100}(r)$$

The probability of finding the system in the new ground state is

$$\mathcal{P}_0 = \left|C_{100}e^{-i\overline{E}_0 t/\hbar}\right|^2 = \left|\frac{[Z(Z+1)]^{3/2}}{\pi a_0^3} 4\pi \int_0^{\infty} dr\, r^2 \exp\left[-\frac{Z}{a_0}r - \frac{Z+1}{a_0}r\right]\right|^2$$

$$= \frac{[Z(Z+1)]^3}{\left(Z+\frac{1}{2}\right)^6}$$

---

[1] The Bohr radius is defined as $a_0 = \hbar^2/m_e e^2$.

Similarly, the probability of finding the system in the new first excited state is

$$\mathcal{P}_1 = \left| C_{200} e^{-i\bar{E}_2 t/\hbar} \right|^2 = \frac{2^{11}}{3^8} \frac{[\mathcal{Z}(\mathcal{Z}+1)]^3}{\left(\mathcal{Z}+\frac{1}{3}\right)^8}$$

Note that for $\mathcal{Z} \gg 1$, $\mathcal{P}_0 \to 1$ and $\mathcal{P}_1 \to 0$.

Problem 7.3 A particle of mass m is bound in a central potential $V(r)$. The particle is in an eigenstate $|\psi_{E\ell m}\rangle$ of the energy and the angular momentum.

(a) Consider the operator

$$G = \mathbf{r} \cdot \mathbf{p} + \mathbf{p} \cdot \mathbf{r}$$

Show that the expectation value $\langle \psi_{E\ell m} | G | \psi_{E\ell m} \rangle$ vanishes at all times.

(b) Show that for any state

$$\frac{d}{dt} \langle G \rangle = 4 \left\langle \frac{p^2}{2m} \right\rangle - 2 \langle [\mathbf{r} \cdot \nabla V(r)] \rangle$$

In particular, show that for the above bound state of the system the left-hand side vanishes and

$$\left\langle \psi_{E\ell m} \left| \frac{p^2}{2m} \right| \psi_{E\ell m} \right\rangle = \frac{1}{2} \langle \psi_{E\ell m} | [\mathbf{r} \cdot \nabla V(r)] | \psi_{E\ell m} \rangle$$

(c) Consider the above relation (known as the *virial theorem*) for the case of the hydrogen atom and compute the matrix elements $\langle n \, \ell \, m | V(r) | n \, \ell \, m \rangle$, $\langle n \, \ell \, m | p^2/2m | n \, \ell \, m \rangle$.

Solution

(a) Using the canonical commutation relations we can write

$$G = 2\hbar \mathbf{r} \cdot \mathbf{p} - i\hbar (\nabla \cdot \mathbf{r}) = 2\hbar \mathbf{r} \cdot \mathbf{p} - 3i\hbar$$

$$\implies \quad -2i\hbar \mathbf{r} \cdot \nabla - 3i\hbar = -2i\hbar \, r \frac{\partial}{\partial r} - 3i\hbar$$

Thus, the expectation value of the above operator in the state

$$\psi_{E\ell m}(\mathbf{r}) = R_{E\ell}(r) Y_{\ell m}(\Omega) e^{-iEt/\hbar}$$

will be

$$\langle \psi_{E\ell m}(t) | G | \psi_{E\ell m}(t) \rangle = \int_0^\infty dr \, r^2 R_{E\ell}(r) \left(-2i\hbar \, r \frac{\partial}{\partial r} - 3i\hbar \right) R_{E\ell}(r)$$

$$= -i\hbar \int_0^\infty dr \, R_{E\ell}(r) \left[2r^3 R'_{E\ell}(r) + 3r^2 R_{E\ell}(r) \right]$$

$$= -i\hbar \int_0^\infty dr \left\{ 2r^3 \left[R_{E\ell}^2(r) \right]' + 3r^2 R_{E\ell}^2(r) \right\}$$

$$= -i\hbar \int_0^\infty dr \left[r^3 R_{E\ell}^2(r) \right]' = 0$$

We have used the fact that for a bound state the radial wave function is real.

(b) For any state, the time evolution of the expectation value of the operator G satisfies

$$\frac{d}{dt}\langle G \rangle = \frac{i}{\hbar}\langle [H, G] \rangle$$

The commutator can be computed as follows:

$$[H, G] = \frac{1}{2m}[p^2, G] + [V(r), G]$$

$$= \frac{1}{2m}\left(2p_j[p_j, x_k]p_k + 2[p_j, x_k]p_k p_j\right) + 2i\hbar \mathbf{r} \cdot \nabla V(r)$$

$$= -\frac{2i\hbar}{m}p^2 + 2i\hbar r V'(r)$$

Thus, we get

$$\frac{d}{dt}\langle G \rangle = 4\left\langle \frac{p^2}{2m} \right\rangle - 2\left\langle r\frac{dV(r)}{dr} \right\rangle$$

In the particular case of the state $\psi_{E\ell m}$ the expectation value $\langle G \rangle$ vanishes at all times and we get

$$\left\langle \psi_{E\ell m} \left| \frac{p^2}{2m} \right| \psi_{E\ell m} \right\rangle = \frac{1}{2}\left\langle \psi_{E\ell m} \left| r\frac{dV(r)}{dr} \right| \psi_{E\ell m} \right\rangle$$

(c) In the case of a hydrogen atom $V(r) = -e^2/r$ and the above relation gives

$$\left\langle n\,\ell\,m \left| \frac{p^2}{2m} \right| n\,\ell\,m \right\rangle \equiv \langle T \rangle_{n\ell m} = \frac{1}{2}\langle rV'(r) \rangle_{n\ell m} = -\frac{1}{2}\langle V \rangle_{n\ell m}$$

Also, we have

$$\langle T \rangle_{n\ell m} + \langle V \rangle_{n\ell m} = E_n$$

Thus

$$\langle V \rangle_{n\ell m} = 2E_n = -\frac{me^4}{\hbar^2 n^2}, \qquad \langle T \rangle_{n\ell m} = -E_n = \frac{me^4}{2\hbar^2 n^2}$$

Problem 7.4 A hydrogen atom is in an energy and angular momentum eigenstate $\psi_{n\ell m}$. The energy eigenvalues $E_n = -m_e^2 e^4/2n^2\hbar^2$ are considered known. Consider the expectation values of the kinetic and the potential energy $\langle T \rangle_{n\ell m}$, $\langle V \rangle_{n\ell m}$ as functions of the parameters m_e and e^2 and show that

$$\langle T \rangle_{n\ell m} = -E_n, \qquad \langle V \rangle_{n\ell m} = 2E_n$$

Solution

If we vary the expectation value of the kinetic energy with respect to the mass parameter, we get

$$m_e \frac{\partial}{\partial m_e} \left\langle n \ell m \left| \frac{p^2}{2m_e} \right| n \ell m \right\rangle = -\left\langle n \ell m \left| \frac{p^2}{2m_e} \right| n \ell m \right\rangle = -\langle T \rangle_{n\ell m}$$

Note that the normalized state $|n \ell m\rangle$ is dimensionless and does not carry any mass dependence.[2] Similarly, we get

$$e^2 \frac{\partial}{\partial e^2} \left\langle n \ell m \left| -\frac{e^2}{r} \right| n \ell m \right\rangle = \left\langle n \ell m \left| -\frac{e^2}{r} \right| n \ell m \right\rangle = \langle V \rangle_{n\ell m}$$

If, however, we vary the expectation value of the total energy,

$$\langle H \rangle = \langle T \rangle + \langle V \rangle = E_n = -\frac{m_e e^4}{2\hbar^2 n^2}$$

we get

$$m_e \frac{\partial}{\partial m_e} \langle T \rangle = E_n, \qquad e^2 \frac{\partial}{\partial e^2} \langle V \rangle = 2E_n$$

Combining these results, we arrive at the desired relations

$$\langle T \rangle_{n\ell m} = -E_n, \qquad \langle V \rangle_{n\ell m} = 2E_n$$

Problem 7.5 A particle of mass μ is bound in a central potential $V(r)$. The particle is in an eigenstate of the energy and angular momentum $|E \ell m\rangle$.

(a) Prove that the following relationship between expectation values is true:

$$\frac{\mu}{\hbar^2} \left[2sE \left\langle r^{s-1} \right\rangle - 2s \left\langle r^{s-1} V(r) \right\rangle - \left\langle r^s V'(r) \right\rangle \right]$$

$$+ \frac{1}{4} \left[s(s-1)(s-2) - 4(s-1)\ell(\ell+1) \right] \left\langle r^{s-3} \right\rangle + 2\pi |\psi(0)|^2 \delta_{s,0} = 0$$

where s is a non-negative integer.[3]

(b) Apply this formula for the hydrogen atom ($V(r) = -e^2/r$) and calculate[4] the expectation value of the radius $\langle r \rangle$ and the expectation value of the square of the radius $\langle r^2 \rangle$ for any energy eigenstate. Consider the uncertainty Δr and show that it becomes smallest for the maximal value of angular momentum. Show that for $\ell = n - 1$ the ratio $\Delta r / \langle r \rangle$ tends to zero for very large values of the principal quantum number.

---

[2] Equivalently, in the expression $\int d^3r \, \psi^*_{n\ell m}(\mathbf{r}) \{\cdots\} \psi_{n\ell m}(\mathbf{r})$, the wave function has the dimension length to the power $-3/2$ and, thus, can be expressed as $L^{-3/2} f(\mathbf{r}/L)$ in terms of a dimensionless function. The length L is buried in the redefinition of the integration variable.

[3] Hint: Consider the radial equation in terms of the *one-dimensional* radial wave function $u(r)$, multiply it by $r^s u'(r)$ and integrate.

[4] The hydrogen energy eigenfunction at the origin is $\psi_{n\ell m}(0) = \delta_{\ell,0}(\pi a_0^3 n^3)^{-1/2}$, where $a_0 = \hbar^2/\mu e^2$ is the Bohr radius.

Solution

(a) We consider the radial Schroedinger equation

$$u''_{E\ell}(r) = \left\{ \frac{\ell(\ell+1)}{r^2} + \frac{2\mu}{\hbar^2}[V(r) - E] \right\} u_{E\ell}(r)$$

multiply by $r^s u'_{E\ell}(r)$ and integrate by parts. The left-hand side becomes

$$-\frac{1}{2}\delta_{s,0}R^2(0) + \frac{s}{2}\int_0^\infty dr\, u\left[(s-1)r^{s-2}u' + r^{s-1}u''\right]$$

$$= -\frac{1}{2}\delta_{s,0}R^2(0) + \frac{s(s-1)}{4}\int_0^\infty dr\, r^{s-2}(u^2)'$$

$$+ \frac{s}{2}\int_0^\infty dr\, r^{s-1}u^2\left\{\frac{\ell(\ell+1)}{r^2} + \frac{2\mu}{\hbar^2}[V(r) - E]\right\}$$

$$= -\frac{1}{2}\delta_{s,0}R^2(0) - \frac{1}{4}[-2s\ell(\ell+1) + s(s-1)(s-2)]\langle r^{s-3}\rangle$$

$$+ s\frac{m}{\hbar^2}\langle r^{s-1}(V - E)\rangle$$

The right-hand side is

$$-\frac{1}{2}\ell(\ell+1)(s-2)\langle r^{s-3}\rangle + s\frac{\mu E}{\hbar^2}\langle r^{s-1}\rangle - s\frac{\mu}{\hbar^2}\langle r^{s-1}V\rangle - \frac{\mu}{\hbar^2}\langle r^s V'\rangle$$

Combining them, we arrive at the relation[5]

$$\frac{\mu}{\hbar^2}\left[2sE\langle r^{s-1}\rangle - 2s\langle r^{s-1}V(r)\rangle - \langle r^s V'(r)\rangle\right]$$

$$+ \frac{1}{4}[s(s-1)(s-2) - 4(s-1)\ell(\ell+1)]\langle r^{s-3}\rangle$$

$$= 2\pi[\ell(\ell+1) - 1]|\psi(0)|^2 \delta_{s,0}$$

An alternative way to arrive at this relation is to consider first the commutator

$$[H,\, [H,\, r^{s+1}]]$$

with

$$H = \frac{p_r^2}{2\mu} + \frac{\ell(\ell+1)}{\mu r^2} + V(r)$$

and compute it in terms of the radial commutator $[p_r,\, f(r)] = -i\hbar f'(r)$. The radial-momentum operator is

$$p_r = \frac{-i\hbar}{r}\frac{\partial}{\partial r}r = -i\hbar\left(\frac{\partial}{\partial r} + \frac{1}{r}\right)$$

5                    $\psi(0) = Y_{\ell m}(0)R_{E\ell}(0) = (4\pi)^{-1/2}R_{E\ell}(0)$

Taking the matrix element of the resulting operator equation between eigenstates $|E\ \ell\ m\rangle$ gives the above formula.

(b) In the case of the hydrogen atom ($V = -e^2/r$, $E = -e^2/2n^2 a_0$), the above relation becomes

$$-\frac{1}{4}\left[s(s-1)(s-2) - 4(s-1)\ell(\ell+1)\right]\langle r^{s-3}\rangle$$

$$+\frac{s}{a_0^2 n^2}\langle r^{s-1}\rangle - \frac{2s-1}{a_0}\langle r^{s-2}\rangle = \frac{2}{a_0^3 n^3}\delta_{s,0}\delta_{\ell,0}$$

For $s = 2$, we get

$$\langle r\rangle = \tfrac{1}{2}a_0 n^2\left[3 - \ell(\ell+1)a_0\langle r^{-1}\rangle\right] = \tfrac{1}{2}a_0\left[3n^2 - \ell(\ell+1)\right]$$

For the last step we have used the relation $\langle V\rangle = 2E$ implied by the virial theorem.
For $s = 3$ we get

$$\langle r^2\rangle = \tfrac{1}{2}a_0^2 n^2\left[5n^2 + 1 - 3\ell(\ell+1)\right]$$

It is straightforward to see that

$$(\Delta r)^2 = \langle r^2\rangle - (\langle r\rangle)^2 = \tfrac{1}{4}a_0^2\left[n^4 + 2n^2 - \ell^2(\ell+1)^2\right]$$

The uncertainty Δr becomes smallest for

$$\ell_{\max} = n - 1 \qquad \Longrightarrow \qquad (\Delta r)^2 = \tfrac{1}{4}a_0^2 n^2(2n+1)$$

In this case, the relative uncertainty is

$$\frac{(\Delta r)^2}{\langle r\rangle^2} = \frac{\tfrac{1}{4}a_0^2 n^2(2n+1)}{\tfrac{1}{4}a_0^2 n^2(2n+1)^2} = \frac{1}{2n+1}$$

and tends to zero as $n \to \infty$, as expected from the correspondence principle.

Problem 7.6 An isotropic harmonic oscillator is in an eigenstate of energy and angular momentum $|n\ \ell\ m\rangle$ with energy eigenvalue $E = \hbar\omega(n + 3/2)$.

(a) By considering the dependence of the expectation values of the kinetic and potential energy on the parameters μ (mass) and ω, prove that

$$\langle T\rangle = \langle V\rangle = \frac{E}{2} = \frac{\hbar\omega}{2}\left(n + \frac{3}{2}\right)$$

This result is the *virial theorem* for the isotropic harmonic oscillator.

(b) Calculate the position and momentum uncertainty product $(\Delta \mathbf{r})^2(\Delta \mathbf{p})^2$ for the oscillator when in the given eigenstate and check the Uncertainty Principle.

(c) Prove that the following relation between expectation values is true:

$$\frac{2\mu E}{\hbar^2} s \langle r^{s-1} \rangle - \left(\frac{\mu\omega}{\hbar}\right)^2 (s+1)\langle r^{s+1} \rangle + \tfrac{1}{4}[s(s-1)(s-2) - 4(s-1)\ell(\ell+1)]\langle r^{s-3} \rangle = 0$$

where s is a positive integer.[6]

(d) From the relation in part (c) obtain $\langle r^4 \rangle$ and $(\Delta r^2)^2$. Show that for the maximal value of ℓ the uncertainty Δr^2 achieves its minimal value. In this case show that the ratio $(\Delta r^2)^2/(\langle r^2 \rangle)^2$ tends to zero in the limit of very large energies.

Solution

(a) We have

$$\mu\frac{\partial}{\partial\mu}\langle T \rangle = \mu\frac{\partial}{\partial\mu}\left\langle \frac{p^2}{2\mu} \right\rangle = -\langle T \rangle$$

$$\mu\frac{\partial}{\partial\mu}\langle V \rangle = \mu\frac{\partial}{\partial\mu}\left\langle \frac{\mu\omega^2}{2}r^2 \right\rangle = \langle V \rangle$$

$$\omega\frac{\partial}{\partial\omega}\langle V \rangle = \omega\frac{\partial}{\partial\omega}\left\langle \frac{\mu\omega^2}{2}r^2 \right\rangle = 2\langle V \rangle$$

$$\omega\frac{\partial}{\partial\omega}E = E = 2\langle V \rangle$$

$$\mu\frac{\partial}{\partial\mu}E = 0 = -\langle T \rangle + \langle V \rangle$$

Thus, we get

$$\langle T \rangle = \langle V \rangle = \frac{E}{2} = \frac{\hbar\omega}{2}\left(n + \frac{3}{2}\right)$$

(b) Since $\langle \mathbf{p} \rangle = \langle \mathbf{r} \rangle = 0$ from symmetry, the corresponding uncertainty product squared is

$$(\Delta\mathbf{r})^2(\Delta\mathbf{p})^2 = \langle r^2 \rangle \langle p^2 \rangle = \frac{4}{\omega^2}\langle V \rangle \langle T \rangle = \frac{E^2}{\omega^2}$$

$$= \hbar^2\left(n + \frac{3}{2}\right)^2 \geq \frac{1}{4}|\langle [x_i, p_j] \rangle|^2 = \frac{3\hbar^2}{4}$$

(c) See problem 7.5.

(d) For $s = 3$ we obtain

$$\frac{6\mu E}{\hbar^2}\langle r^2 \rangle - 4\left(\frac{\mu\omega}{\hbar}\right)^2\langle r^4 \rangle + \frac{3}{2} - 2\ell(\ell+1) = 0$$

---

[6] Hint: Consider the radial equation in terms of the *one-dimensional* radial wave function $u(r)$, multiply it by $r^s u'(r)$ and integrate.

or

$$\langle r^4 \rangle = \frac{1}{4} \left(\frac{\hbar}{\mu\omega} \right)^2 \left[6 \left(n + \frac{3}{2} \right)^2 + \frac{3}{2} - 2\ell(\ell + 1) \right]$$

The corresponding uncertainty squared is

$$(\Delta r^2)^2 = \langle r^4 \rangle - \langle r^2 \rangle^2 = \frac{1}{2} \left(\frac{\hbar}{\mu\omega} \right)^2 \left[\left(n + \frac{3}{2} \right)^2 + \frac{3}{4} - \ell(\ell + 1) \right]$$

For $\ell = n$, this gets its minimum value:

$$(\Delta r^2)^2 = \left(\frac{\hbar}{\mu\omega} \right)^2 \left(n + \frac{3}{2} \right)$$

The relative uncertainty squared is

$$\frac{(\Delta r^2)^2}{(\langle r^2 \rangle)^2} = \left(n + \frac{3}{2} \right)^{-1}$$

and tends to zero for very large n.

Problem 7.7 Consider an isotropic two-dimensional harmonic oscillator, with Hamiltonian

$$H = \frac{p_x^2}{2\mu} + \frac{p_y^2}{2\mu} + \frac{\mu\omega^2}{2} \left(x^2 + y^2 \right)$$

(a) Introduce polar coordinates ρ, ϕ and show that the energy and angular momentum eigenfunctions are of the form

$$\frac{1}{\sqrt{2\pi}} e^{i\nu\phi} F_{E\nu}(\rho)$$

where the radial part satisfies

$$\left[-\frac{d^2}{d\rho^2} - \frac{1}{\rho} \frac{d}{d\rho} + \frac{\nu^2}{\rho^2} + \frac{\rho^2}{\alpha^4} - \frac{2(N+1)}{\alpha^2} \right] F_{N\ell}(\rho) = 0$$

The length α is the usual harmonic-oscillator characteristic length $\alpha = (\hbar/\mu\omega)^{1/2}$. The number N takes non-negative integer values. The angular momentum takes values $\nu = 0, 1, \ldots, N$.

(b) Consider the ground state, $N = \nu = 0$, and find the corresponding normalized wave function. Calculate the expectation value $\langle \rho^2 \rangle$ for the ground state. Do the same for the excited state for which $N = \nu = 1$.

(c) Introduce into the above radial equation the transformation

$$\rho = \xi \, r^{1/2}, \qquad F(\rho) \propto r^{1/2} \, R(r)$$

where ξ is a parameter to be determined; the function $R(r)$ satisfies the *three-dimensional radial Schroedinger equation for the hydrogen atom*. For what values of the angular momentum ν and the oscillator quantum number is this correspondence possible?

(d) Show that the following relation between expectation values is true for any eigenstate corresponding to ν, N:

$$\langle \rho^2 \rangle \left\langle \frac{1}{r} \right\rangle = \xi^2$$

Using this relation show that

$$\langle \rho^2 \rangle_{N\nu} = \alpha^2 (N+1)$$

Solution

(b) It is straightforward to check that in the case $\nu = N = 0$ the radial Schroedinger equation has the solution

$$F_{00}(\rho) = Ce^{-\rho^2/2\alpha^2}$$

The normalization constant is $C = \sqrt{2}/\alpha$. The expectation value of ρ^2 in this state is

$$\langle \rho^2 \rangle_0 = \frac{2}{\alpha^2} \int_0^\infty d\rho \, \rho \rho^2 e^{-\rho^2/\alpha^2} = \alpha^2$$

Similarly the wave function of the excited state $N = \nu = 1$ can be found by substitution to be

$$F_{11}(\rho) = \frac{\sqrt{2}}{\alpha^2} \rho \, e^{-\rho^2/2\alpha^2}$$

The corresponding expectation value of the square of the radius is

$$\langle \rho^2 \rangle_{11} = 2\alpha^2$$

(c) Performing the change of variables, we get

$$\left\{ \frac{d^2}{dr^2} + \frac{2}{r}\frac{d}{dr} - \frac{\nu^2 - 1}{4r^2} - \frac{\xi^4}{4\alpha^4} + \left[\frac{(N+1)\xi^2}{2\alpha^2} \right] \frac{1}{r} \right\} R(r) = 0$$

This is the radial Schroedinger equation for a standard (three-dimensional) hydrogen atom, provided that

$$\frac{\nu^2 - 1}{4} = \ell(\ell+1) \qquad \Longrightarrow \qquad \nu = 2\ell + 1 = 1, 3, \ldots, N$$

$$N + 1 = 2n = 2, 4, \ldots \qquad \Longrightarrow \qquad N = 1, 3, \ldots$$

$$\frac{\xi^2}{2\alpha^2} = \frac{1}{a_0 n}$$

Thus, the correspondence is possible for odd values of N and ν. The last relation can be considered as a definition of the Bohr radius a_0 for this hydrogen-atom equivalent.

(d) The expectation value of the square of the oscillator radius, for any eigenstate, is

$$\langle \rho^2 \rangle_{N\nu} = \frac{\int d\rho \, \rho \rho^2 F^2(\rho)}{\int d\rho \, \rho F^2(\rho)} = \frac{\int dr \, \rho^2 r R^2(r)}{\int dr \, r R^2(r)}$$

$$= \xi^2 \int dr \, r^2 R^2(r) \Big/ \int dr \, r^2 \left(\frac{1}{r}\right) R^2(r) = \xi^2 \Big/ \left\langle \frac{1}{r}\right\rangle$$

The expectation value of the inverse radius for the hydrogen atom is known, by virtue of the virial theorem $\langle T \rangle = \frac{1}{2}\langle r V' \rangle$, to be given by

$$\left\langle \frac{1}{r} \right\rangle_{n\ell} = -\frac{1}{e^2}\langle V \rangle_{n\ell} = \frac{2|E_n|}{e^2} = \frac{1}{a_0 n^2}$$

Thus, we obtain

$$\langle \rho^2 \rangle_{N\nu} = \xi^2 a_0 n^2 = \left(\frac{2\alpha^2}{a_0 n}\right) a_0 n^2 = \alpha^2 \, 2n = \alpha^2 (N + 1)$$

Note that this result is true for the $N = \nu = 1$ case as well as the ground-state case.

Problem 7.8 A particle of mass μ is confined to a two-dimensional plane and is subject to harmonic forces, which, to a good approximation, can be put in the form of the following Hamiltonian:

$$H = \frac{p_x^2}{2\mu} + \frac{p_y^2}{2\mu} + \frac{\mu\omega^2}{2}\left(x^2 + y^2\right) + \nu\mu\omega^2 xy$$

The only constraint that the parameter ν satisfies is $\nu < 1$. Note that ν is not necessarily small.

(a) Consider a rotation of the variables that leaves the kinetic energy and $x^2 + y^2$ invariant:

$$x_1 = \cos\alpha x + \sin\alpha \, y, \qquad x_2 = -\sin\alpha \, x + \cos\alpha y$$
$$p_1 = \cos\alpha p_x + \sin\alpha \, p_y, \qquad p_2 = -\sin\alpha \, p_x + \cos\alpha p_y$$

Show that it preserves the canonical commutation relations. Choose the parameter angle α in such a way that the full Hamiltonian becomes diagonal.

(b) Find the eigenvalues and eigenfunctions of the energy.

Solution

(a) Expressing the potential energy in terms of the new canonical variables, we get

$$\tfrac{1}{2}\mu\omega^2\left(x_1^2 + x_2^2\right) + \nu\mu\omega^2\left[x_1x_2\cos 2\alpha + \tfrac{1}{2}\sin 2\alpha\left(x_2^2 - x_1^2\right)\right]$$

Taking $\alpha = \pi/4$, we get a diagonal potential energy

$$V(x_1, x_2) = \tfrac{1}{2}\mu\omega^2\left[(1 - \nu)x_1^2 + (1 + \nu)x_2^2\right]$$

This choice corresponds to $x_1 = \frac{1}{\sqrt{2}}(x + y),\ x_2 = \frac{1}{\sqrt{2}}(y - x)$.

(b) The system is the sum of two independent harmonic oscillators with frequencies $\Omega_1^2 = \omega^2(1 - \nu)$ and $\Omega_2^2 = \omega^2(1 + \nu)$. The energy eigenvalues will be

$$E_{n_1 n_2} = \hbar\Omega_1\left(n_1 + \tfrac{1}{2}\right) + \hbar\Omega_2\left(n_2 + \tfrac{1}{2}\right)$$

The two quantum numbers take on the standard one-dimensional harmonic oscillator values $n_1,\ n_2 = 0, 1, \ldots$. The energy eigenfunctions will be $\Psi_{n_1 n_2}(x_1, x_2) = \psi_{n_1}(x_1)\,\psi_{n_2}(x_2)$, where $\psi_n(x)$ are the one-dimensional harmonic oscillator energy eigenfunctions.

Problem 7.9 The *delta-shell* potential is a very simple, although crude and somewhat artificial, model of the force experienced by a neutron interacting with a nucleus. Consider an attractive delta-shell potential of radius a, the strength of which is parametrized in terms of a parameter g^2:

$$V(r) = -\frac{\hbar^2 g^2}{2\mu}\,\delta(r - a)$$

(a) Investigate the existence of bound states in the case of negative energy.
(b) Consider the case of positive energy and find the corresponding energy and angular momentum eigenfunctions.

Solution

(a) For negative energies $E = -\hbar^2\kappa^2/2\mu$, the radial Schroedinger equation is

$$u''_{E\ell}(r) = \left[\frac{\ell(\ell + 1)}{r^2} - g^2\delta(r - a) + \kappa^2\right]u_{E\ell}(r)$$

The energy eigenfunctions are[7]

$$u(r) = \begin{cases} r < a, & u_<(r) = Ar j_\ell(i\kappa r) \\ r > a, & u_>(r) = r B h_\ell^{(-)}(i\kappa r) \end{cases}$$

---

[7] The spherical Hankel functions are defined as

$$h_\ell^{(\pm)}(x) = n_\ell(x) \pm i j_\ell(x)$$

in terms of the spherical Neumann (n_ℓ) and Bessel (j_ℓ) functions.

Continuity of the wave function at $r = a$ implies

$$A\, j_\ell(i\kappa a) = B\, h_\ell^{(-)}(i\kappa a)$$

Integrating around the point $r = a$, we obtain the *discontinuity condition* for the derivative of the eigenfunctions:

$$u'(+a) - u'(-a) = -g^2 u(a)$$

In terms of the above eigenfunction forms, this is

$$B h_\ell^{(-)\prime}(i\kappa a) - A j_\ell'(i\kappa a) = -g^2 A j_\ell(i\kappa a)$$

The condition on the energy in order for there to be a bound-state solution reads

$$\frac{1}{g^2} = \left[\frac{j_\ell'(i\kappa a)}{j_\ell(i\kappa a)} - \frac{h_\ell^{(-)\prime}(i\kappa a)}{h_\ell^{(-)}(i\kappa a)} \right]^{-1}$$

This equation has one solution for sufficiently strong coupling g^2; as an illustration, we can consider the case $\ell = 0$, for which it simplifies to

$$\frac{1 - e^{-\xi}}{\xi} = \frac{1}{g^2 a}$$

with $\xi \equiv 2\kappa a$. Since the left-hand side is always smaller than unity, there will be a solution only if

$$g^2 a > 1$$

(b) In the case of positive energies $E = \hbar^2 k^2 / 2\mu > 0$, the radial Schroedinger equation is

$$u''_{E\ell}(r) = \left[\frac{\ell(\ell + 1)}{r^2} - g^2 \delta(r - a) - k^2 \right] u_{E\ell}(r)$$

The energy eigenfunctions are

$$u(r) = \begin{cases} r < a, & u_<(r) = A r j_\ell(kr) \\ r > a, & u_>(r) = r B h_\ell^{(-)}(kr) + r C h_\ell^{(+)}(kr) \end{cases}$$

Continuity of the wave function at $r = a$ implies that

$$A j_\ell(ka) = B h_\ell^{(-)}(ka) + C h_\ell^{(+)}(ka)$$

Integrating around the point $r = a$, we obtain the *discontinuity condition* for the derivative of the eigenfunctions,

$$u'(+a) - u'(-a) = -g^2 u(a)$$

which, in terms of the above wave-function forms, becomes

$$Bh_\ell^{(-)'}(ka) + Ch_\ell^{(+)'}(ka) - Aj_\ell'(ka) = -g^2 Aj_\ell(ka)$$

From these two independent homogeneous equations, we can obtain the ratio

$$\frac{C}{B} = -\frac{g^2 + \dfrac{h_\ell^{(-)'}(ka)}{h_\ell^{(-)}(ka)} - \dfrac{j_\ell'(ka)}{j_\ell(ka)}}{g^2 + \dfrac{h_\ell^{(+)'}(ka)}{h_\ell^{(+)}(ka)} - \dfrac{j_\ell'(ka)}{j_\ell(ka)}}$$

An equivalent but more illuminating expression is

$$\frac{C}{B} = -\frac{1 - iX}{1 + iX}$$

where

$$X \equiv \frac{g^2 j_\ell^2(ka)}{n_\ell'(ka)j_\ell(ka) - j_\ell'(ka)n_\ell(ka) + g^2 j_\ell(ka)n_\ell(ka)}$$

It is clear that $|C/B|^2 = 1$, as expected from the conservation of probability. It is more appropriate to introduce the *phase shift* $\delta_\ell(k)$, defined via

$$\frac{C}{B} = -e^{-2i\delta_\ell}$$

with $\delta_\ell = \arctan X$. Note that, using the asymptotic form of the Hankel functions $h_\ell^{(\pm)} \sim -(\pm i)^\ell x^{-1} e^{\mp ix}$, we obtain for the asymptotic form of the outside eigenfunction the expression

$$u_>(r) \sim -2ik^{-1} B e^{-i\delta_\ell} \sin(kr - \ell\pi/2 + \delta_\ell)$$

This should be contrasted with the asymptotic form of the solution in the absence of the potential, namely $u(r) \sim \sin(kr - \ell\pi/2)$.

Problem 7.10 Consider a repulsive *delta-shell* potential

$$V(r) = \frac{\hbar^2 g^2}{2\mu} \delta(r - a)$$

(a) Find the energy eigenfunctions.
(b) Consider the limit $g^2 \to \infty$ corresponding to an impenetrable shell and discuss the possible solutions.

Solution

(a) The radial Schroedinger equation

$$u''_{E\ell}(r) = \left[\frac{\ell(\ell+1)}{r^2} + g^2\delta(r-a) - k^2\right]u_{E\ell}(r)$$

leads to the energy eigenfunctions:

$$u(r) = \begin{cases} r < a, & u_<(r) = Arj_\ell(kr) \\ r > a, & u_>(r) = rBh_\ell^{(-)}(kr) + rCh_\ell^{(+)}(kr) \end{cases}$$

Continuity of the wave function at $r = a$ implies

$$Aj_\ell(ka) = Bh_\ell^{(-)}(ka) + Ch_\ell^{(+)}(ka)$$

Integrating around the point $r = a$, we obtain the *discontinuity condition* for the derivative of the eigenfunctions:

$$u'(+a) - u'(-a) = g^2 u(a)$$

which, in terms of the above wave function forms, becomes

$$Bh_\ell^{(-)'}(ka) + Ch_\ell^{(+)'}(ka) - Aj'_\ell(ka) = g^2 Aj_\ell(ka)$$

From these two independent homogeneous equations, we can obtain the ratio

$$\frac{C}{B} = -\frac{-g^2 + \dfrac{h_\ell^{(-)'}(ka)}{h_\ell^{(-)}(ka)} - \dfrac{j'_\ell(ka)}{j_\ell(ka)}}{-g^2 + \dfrac{h_\ell^{(+)'}(ka)}{h_\ell^{(+)}(ka)} - \dfrac{j'_\ell(ka)}{j_\ell(ka)}}$$

As in the previous problem, an equivalent expression is

$$\frac{C}{B} = -\frac{1 - iX}{1 + iX}$$

where

$$X \equiv \frac{-g^2 j_\ell^2(ka)}{n'_\ell(ka)j_\ell(ka) - j'_\ell(ka)n_\ell(ka) - g^2 j_\ell(ka)n_\ell(ka)}$$

It is clear that $|C/B|^2 = 1$, as required from the conservation of probability. Again it is useful to introduce the *phase shift* $\delta_\ell(k)$, defined via

$$\frac{C}{B} = -e^{-2i\delta_\ell}$$

with $\delta_\ell = \arctan X$. Note that, using the asymptotic form of the Hankel functions $h_\ell^{(\pm)} \sim -(\pm i)^\ell x^{-1} e^{\mp ix}$, we obtain for the asymptotic form of the outer eigenfunction the expression

$$u_>(r) \sim -2ik^{-1}Be^{-i\delta_\ell}\sin(kr - \ell\pi/2 + \delta_\ell)$$

This should be contrasted with the asymptotic form of the solution in the absence of the potential, namely $u(r) \sim \sin(kr - \ell\pi/2)$.

(b) In the limit $g^2 \to \infty$, we get

$$u_>(r) = \tilde{B}r\,[j_\ell(kr)n_\ell(ka) - j_\ell(ka)n_\ell(kr)]$$

which vanishes at the shell. The phase shift is the same as that for an infinitely hard sphere, namely

$$\delta_\ell = \arctan\frac{j_\ell(ka)}{n_\ell(ka)}$$

The outer solution is completely independent of the internal solution,

$$u_<(r) = Arj_\ell(kr)$$

which should vanish on the shell, namely

$$j_\ell(ka) = 0$$

The possible energies corresponding to this condition coincide with the (infinite) zeros of the spherical Bessel function. In the simplest case, $\ell = 0$, we just get

$$\ell = 0 \quad \implies \quad k_n = \frac{n\pi}{a}, \quad n = 1, 2, \ldots$$

Similarly, in the $\ell = 1$ case we get the condition

$$\ell = 1 \quad \implies \quad \tan ka = ka$$

which has also an infinity of solutions.

Note that the two branches of the solution are independent. Thus, we could have one of the following physical situations: a particle bound in the interior of the impenetrable shell with the above special energy values E_n ($A \neq 0$, $B = 0$), or a particle always in the outer region ($A = 0$, $B \neq 0$) for any energy $E > 0$.

Problem 7.11 A neutron is bound in a nucleus. The system is approximated by a spherical 'square well' of radius a and depth $-V_0$ and it is in its ground state. Find the momentum uncertainty of the system. Consider also the hypothetical case in which the depth of the well has the special value $V_0 \approx 9\hbar^2\pi^2/16m_na^2$. Show that in this case there is a unique bound state for zero angular momentum and find its energy.

Solution

The lowest-energy state has to have $\ell = 0$ and $E < 0$. The corresponding radial eigenfunction is

$$u_0(r) = \begin{cases} 0 < r < a, & A \sin qr \\ a < r < \infty, & Be^{-\kappa r} \end{cases}$$

with

$$q^2 \equiv \frac{2m_n}{\hbar^2}(V_0 - |E|), \qquad \kappa^2 \equiv \frac{2m_n}{\hbar^2}|E| \qquad \left(q^2 + \kappa^2 = \frac{2m_n V_0}{\hbar^2} \equiv \frac{\gamma^2}{a^2}\right)$$

Continuity implies that

$$B = Ae^{\kappa a} \sin qa = -\frac{q}{\kappa} Ae^{\kappa a} \cos qa$$

or

$$qa \cot qa = -\kappa a$$

The last equation gives the condition determining the existing bound-state eigenvalues. Introducing $\xi \equiv qa$, we can write it as

$$\frac{\xi \cot \xi}{\sqrt{\gamma^2 - \xi^2}} = -1$$

In order to have at least one solution γ has to be greater than $\pi/2$. This is equivalent to

$$V_0 \geq \frac{\pi^2 \hbar^2}{8m_n a^2} \equiv V_{min}$$

Our system occupies the ground state, whose energy $E_0 = -\hbar^2 \kappa^2 / 2m_n$ corresponds to values κ and q satisfying the above condition.

The normalization constant N of the ground-state wave function,

$$u_0(r) = N \left[\Theta(a - r) \sin qr + \sin qa \Theta(r - a) e^{-\kappa(r-a)}\right]$$

is given by

$$N^{-2} = \frac{1}{2}\left(a - \frac{\sin 2qa}{2q}\right) + \frac{\sin^2 qa}{2\kappa}$$

Using the eigenvalue condition, this simplifies to $N^2 = 2\kappa/(1 + \kappa a)$.

The expectation value of the momentum vanishes, owing to the spherical symmetry of the ground state. The expectation value of the square of the momentum is proportional to the expectation value of the kinetic energy:

$$\langle p^2 \rangle = 2m_{\mathrm{n}} \langle T \rangle = 2m_{\mathrm{n}} (E_0 - \langle V \rangle) = -2m_{\mathrm{n}}|E_0| + 2m_{\mathrm{n}} V_0 \int_0^a dr\, u_0^2(r)$$

the integral is given by

$$\int_0^a dr\, u_0^2(r) = N^2 \int_0^a dr\, \sin^2 qr = \frac{N^2}{2} \left(a - \frac{\sin 2qa}{2q} \right)$$

Using the eigenvalue condition, we finally arrive at the uncertainty expression

$$(\Delta p)^2 = (\hbar q)^2 \frac{\kappa a}{1 + \kappa a}$$

The special case $V_0 = 9\hbar^2 \pi^2 / 16 m_{\mathrm{n}} a^2$ corresponds to

$$\gamma_0 = \frac{3\pi}{2\sqrt{2}} \qquad \Longrightarrow \qquad \frac{\pi}{2} < \gamma_0 < \frac{3\pi}{2}$$

Note that the upper limit is the minimum value necessary for at least two bound states, as can be easily seen from a graphical solution of the eigenvalue equation. Thus, we have a unique bound state with $\ell = 0$ corresponding to

$$\frac{\xi \cot \xi}{\sqrt{9\pi^2/8 - \xi^2}} = -1 \quad \Longrightarrow \quad \xi = \frac{3\pi}{4} \quad \Longrightarrow \quad E_0 = -\frac{9\hbar^2 \pi^2}{32 m_{\mathrm{n}} a^2}$$

Problem 7.12 Derive the *Thomas–Reiche–Kuhn sum rule* in the form

$$\frac{2\mu}{\hbar^2} \sum_\nu (E_\nu - E_0)\, |\langle \nu | \hat{\mathbf{n}} \cdot \mathbf{r} | 0 \rangle|^2 = 1$$

for a particle of mass μ. The direction $\hat{\mathbf{n}}$ is arbitrary. In the particular case of the hydrogen atom ($|\nu\rangle \rightarrow |n\,\ell\,m\rangle$), calculate the contribution of the $n = 2$ states to this sum. Take $\hat{\mathbf{n}} = \hat{\mathbf{z}}$.

Solution

Consider the easily provable commutator identity

$$[x_i, [H, x_j]] = -\frac{i\hbar}{\mu}[x_i, p_j] = \frac{\hbar^2}{\mu}\delta_{ij}$$

This can be written in the form

$$x_i H x_j - x_i x_j H - H x_j x_i + x_j H x_i = \frac{\hbar^2}{\mu}\delta_{ij}$$

Let us find its matrix element with respect to the ground state $|0\rangle$ and introduce a complete sum of states. We obtain

$$\sum_{\nu} \left[\langle 0|x_i|\nu \rangle E_{\nu} \langle \nu|x_j|0 \rangle - \langle 0|x_i|\nu \rangle \langle \nu|x_j|0 \rangle E_0 \right.$$

$$\left. - E_0 \langle \nu|x_j|\nu \rangle \langle \nu|x_i|0 \rangle + \langle 0|x_j|\nu \rangle E_{\nu} \langle \nu|x_i|0 \rangle \right] = \frac{\hbar^2}{\mu} \delta_{ij}$$

This is equivalent to

$$\sum_{\nu} (E_{\nu} - E_0) \left[\langle 0|x_j|\nu \rangle \langle \nu|x_i|0 \rangle + \langle 0|x_i|\nu \rangle \langle \nu|x_j|0 \rangle \right] = \frac{\hbar^2}{\mu} \delta_{ij}$$

Introducing a unit vector $\hat{\mathbf{n}}$ and multiplying both sides by $\hat{n}_i \hat{n}_j$, we obtain

$$\sum_{\nu} (E_{\nu} - E_0) |\langle 0|\hat{\mathbf{n}} \cdot \mathbf{r}|\nu \rangle|^2 = \frac{\hbar^2}{2\mu}$$

In the case of the hydrogen atom, we must also include in the above sum rule the contribution of the continuum of the scattering states in the form of an integral. The discrete part reads

$$\sum_{n=1}^{\infty} \sum_{\ell=0}^{n-1} \sum_{m=-\ell}^{+\ell} (E_n - E_1) |\langle 1\ 0\ 0|\hat{\mathbf{n}} \cdot \mathbf{r}|n\ \ell\ m\rangle|^2$$

The contribution of the first excited state $n = 2$ is

$$(E_2 - E_1) \sum_{\ell=0,1} \sum_{m=-\ell}^{\ell} |\langle 2\ \ell\ m|z|1\ 0\ 0\rangle|^2$$

$$= (E_2 - E_1) \left[|\langle 2\ 0\ 0|z|1\ 0\ 0\rangle|^2 + |\langle 2\ 1\ 1|z|1\ 0\ 0\rangle|^2 \right.$$

$$\left. + |\langle 2\ 1\ -1|z|1\ 0\ 0\rangle|^2 + |\langle 2\ 1\ 0|z|1\ 0\ 0\rangle|^2 \right]$$

It is clear that, owing to the odd parity of z, the matrix element $\langle 2\ 0\ 0|z|1\ 0\ 0\rangle$ will vanish. Similarly, from the fact that $[L_z, z] = 0$, we get

$$\langle 2\ 1\ \pm 1|[L_z, z]|1\ 0\ 0\rangle = \pm \langle 2\ 1\ \pm 1|z|1\ 0\ 0\rangle = 0$$

Thus, only the matrix element $\langle 2\ 1\ 0|z|1\ 0\ 0\rangle$ contributes. Using the known forms of the hydrogen eigenfunctions

$$\psi_{100} = (\pi a_0)^{-1/2} e^{-r/a_0}$$

and

$$\psi_{210} = (32\pi a_0)^{-1/2} \frac{r}{a_0} e^{-r/2a_0} \cos\theta$$

we obtain

$$\langle 2\,1\,0|z|1\,0\,0\rangle = a_0 \left(\frac{2\sqrt{2}}{3}\right)^5$$

The corresponding contribution to the Thomas–Reiche–Kuhn sum is

$$\frac{2\mu}{\hbar^2}\frac{3e^2}{8a_0}\left(\frac{8}{9}\right)^5 a_0^2 = \frac{3}{4}\left(\frac{8}{9}\right)^5 = 0.416 \approx 40\%$$

Problem 7.13 Consider a repulsive delta-shell potential

$$V(r) = \frac{\hbar^2 g^2}{2\mu}\,\delta(r-a)$$

Take the limit $g^2 \rightarrow \infty$.

(a) Obtain the energy eigenvalues and eigenfunctions that correspond to a particle captured in the interior of this impenetrable shell. Discuss the ground state as well as the first few excited states and give a rough estimate of their energies.

(b) Calculate the matrix elements

$$\langle E\,\ell\,m|z|E_0\,0\,0\rangle$$

between the ground state and any other energy and angular momentum eigenstate.[8]

(c) Prove the Thomas–Reiche–Kuhn sum rule

$$\sum_E (E - E_0)|\langle E|z|E_0\rangle|^2 = \frac{\hbar^2}{2\mu}$$

for this case. Then show that it reduces to the following sum:

$$\sum_n \frac{(ak_n)^2}{[(ak_n)^2 - \pi^2]^3} = \frac{3}{16\pi^2}$$

where the wave numbers k_n correspond to the $\ell = 1$ energy levels. Compare numerically the first and second terms in this sum.

---

[8] Take the following integrals as given:

$$\int_0^a dr\, r^2\, j_1^2(kr) = \frac{1}{4k^4 a}\left(-4 + 2k^2 a^2 + 2\cos 2ka + ka\sin 2ka\right)$$

$$\int_0^a dr\, r^3\, j_1(kr) j_0\left(\frac{\pi r}{a}\right) = -\left(\frac{a^3}{k}\right)\left\{\frac{\cos ak}{\pi^2 - (ak)^2} + \frac{2ak\sin ak}{[\pi^2 - (ak)^2]^2}\right\}$$

$$+ \left(\frac{a^2}{k^2}\right)\frac{\sin ak}{\pi^2 - (ak)^2}$$

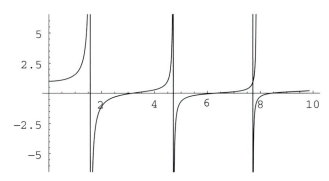

Fig. 33 Plot of $(\tan x)/x$.

Solution

(a) The delta-shell potential in the infinite strength ($g^2 \to \infty$) limit (see problem 7.10) leads to the energy eigenfunctions

$$R_{E\ell}(r) = \begin{cases} r < a, & N j_\ell(kr) \\ r > a, & 0 \end{cases}$$

The corresponding eigenvalues are given by $j_\ell(ka) = 0$. For $\ell = 0$, the eigenvalues are

$$E_n = \frac{\hbar^2 n^2 \pi^2}{2\mu a^2} \qquad (n = 1,\ 2,\ \ldots)$$

For $\ell = 1$, the eigenvalue condition has the form

$$\tan ka = ka$$

As it can be immediately inferred from the graphical solution to this equation (cf. Fig. 33), the smallest root occurs in the interval $(\pi, 3\pi/2)$. Thus, it corresponds to the first excited state. The numerical value of this root is $ka \approx 4.493$ or 1.43π. The next root occurs at $ka \approx 7.730$ or 2.46π. A very rough estimate of the position of these roots reads

$$k_n a \approx \left(n + \tfrac{1}{2}\right)\pi - \frac{1}{\pi}\frac{1}{\left(n + \tfrac{1}{2}\right)}$$

For $n = 1$ and $n = 2$ we get $ka \sim 2.5$ and $ka \approx 7.72$, which are roughly equal to the previously stated values.

(b) The matrix element $\langle E\ \ell\ m|z|E_0\ 0\ 0\rangle$ can be calculated, starting from

$$\int d\Omega\, Y_{\ell m}^*(\Omega) \cos\theta\, Y_{00}(\Omega)\, N_{E\ell} N_0 \int_0^a dr\, r^2 j_\ell(kr) r j_0(k_0 r)$$

The angular integral is

$$\frac{1}{\sqrt{4\pi}} \int d\Omega\, Y_{\ell m}^*(\Omega) \cos\theta = \frac{1}{\sqrt{3}} \int d\Omega\, Y_{\ell m}^*(\Omega) Y_{10}(\Omega) = \frac{1}{\sqrt{3}} \delta_{\ell,1} \delta_{m,0}$$

Thus the matrix element will be

$$\langle E \ \ell \ m|z|E_0 \ 0 \ 0\rangle = \frac{1}{\sqrt{3}}\delta_{\ell,1}\delta_{m,0} \ N_{E1}N_0 \int_0^a dr \ r^2 \ j_1(kr)r j_0(k_0 r)$$

The ground-state wave number is $k_0 = \pi/a$, while the wave number k corresponds to the solutions of $j_1(ka) = 0$ or $\tan ka = ka$. The normalization constants are given by

$$N_{E1}^{-2} = \int_0^a dr \ r^2 \ j_1^2(kr) = \frac{a^3}{2[1 + (ka)^2]}$$

$$N_0^{-2} = \int_0^a dr \ r^2 \ j_0^2(k_0 r) = \frac{a^3}{2\pi^2}$$

The integral appearing in the expression for the matrix element is

$$\int_0^a dr \ r^3 \ j_1(kr)j_0(k_0 r) = \frac{2ka \cos ka}{(k^2 - k_0^2)^2}$$

We have used the eigenvalue condition $\tan ka = ka$. Note that we can also use $\cos^2 ka = (1 + \tan^2 ka)^{-1} = [1 + (ka)^2]^{-1}$. Substituting this in the above, we get

$$\langle E \ \ell \ m|z|E_0 \ 0 \ 0\rangle = \frac{4\pi}{\sqrt{3}a^3}\delta_{\ell,1}\delta_{m,0} \frac{ka}{(k^2 - k_0^2)^2}$$

(c) Starting from the Thomas–Reiche–Kuhn sum rule as proved in problem 7.12,

$$\sum_{n,\ell,m}(k^2 - k_0^2)|\langle E_n \ \ell \ m|z|E_0 \ 0 \ 0\rangle|^2 = 1$$

and substituting the calculated matrix element, we obtain

$$\frac{16\pi^2}{3}\sum_n \frac{(k_n a)^2}{[(k_n a)^2 - \pi^2]^4} = 1$$

Now, the sum runs over the $\ell = 1$ eigenstates, for which

$$k_1 \approx 1.43\left(\frac{\pi}{a}\right), \qquad k_2 \approx 2.46\left(\frac{\pi}{a}\right), \qquad \cdots$$

corresponding to the first, third, etc. excited levels. A rough numerical estimate of the first two terms gives

$$0.967 + 0.02 + \cdots \sim 1$$

which shows that the sum rule is very quickly saturated by the first excited state.

Problem 7.14 Consider a hydrogen atom in its ground state. What is the expectation value of the kinetic energy? What is the probability of finding the atom with a kinetic energy larger than the expectation value? Calculate the uncertainty in the position variable. Do the same for the uncertainty in the momentum and check the validity of the Heisenberg inequality.

Solution

With the help of the known ground-state wave function we can calculate the expectation value of the kinetic energy:[9]

$$\langle T \rangle_1 = \frac{\hbar^2}{2\mu\pi a_0^3} \int d^3r \, e^{-r/2a_0} \, \nabla^2 \, e^{-r/2a_0} = -\frac{\hbar^2}{\mu a_0^4} \int_0^\infty dr \, r^2 \left(-\frac{1}{r} + \frac{1}{4a_0} \right) e^{-r/a_0}$$

$$= \frac{\hbar^2}{2\mu a_0^2} = \frac{e^2}{2a_0} = -E_1$$

Alternatively, we could have used the virial theorem

$$\langle T \rangle = -\tfrac{1}{2}\langle r V' \rangle = -\tfrac{1}{2}\langle V \rangle$$

which, together with

$$\langle T \rangle_1 + \langle V \rangle_1 = E_1$$

implies that $\langle T \rangle_1 = -E_1$ and $\langle V \rangle_1 = 2E_1$.

A measurement of kinetic energy is equivalent to a measurement of momentum. The probability amplitude density for finding the system with momentum **p** will be

$$\phi(p) = \int d^3r \, \frac{e^{i\mathbf{p}\cdot\mathbf{r}/\hbar}}{(2\pi\hbar)^{3/2}} \, \psi_{100}(r)$$

$$= \frac{2\pi}{(2\pi\hbar)^{3/2}(\pi a_0^3)^{1/2}} \int_0^\infty dr \, r^2 \int_{-1}^1 d\cos\theta \, e^{ipr\cos\theta/\hbar} \, e^{-r/a_0}$$

$$= -\frac{i}{p(2\pi^2 a_0^3)^{1/2}} \int_0^\infty dr \, r \left\{ \exp\left[-\left(\frac{1}{a_0} - \frac{i}{\hbar}p \right) r \right] - \text{c.c.} \right\}$$

$$= -\frac{i a_0^2}{p(2\pi^2\hbar a_0^3)^{1/2}} \frac{\partial}{\partial a_0} \left(\int_0^\infty dr \left\{ \exp\left[-\left(\frac{1}{a_0} - \frac{i}{\hbar}p \right) r \right] - \text{c.c.} \right\} \right)$$

$$= -\frac{i a_0^2}{p(2\pi^2\hbar a_0^3)^{1/2}} \frac{\partial}{\partial a_0} \left[\left(\frac{1}{a_0} - \frac{i}{\hbar}p \right)^{-1} - \text{c.c.} \right]$$

$$= \frac{2a_0^2}{(2\pi^2\hbar^3 a_0^3)^{1/2}} \frac{\partial}{\partial a_0} \left(\frac{1}{a_0^2} + \frac{p^2}{\hbar^2} \right)^{-1}$$

$$= \frac{4a_0^{-1}}{(2\pi^2\hbar^3 a_0^3)^{1/2}} \left(\frac{1}{a_0^2} + \frac{p^2}{\hbar^2} \right)^{-2} = \frac{4}{\pi\sqrt{2}} \left(\frac{a_0}{\hbar} \right)^{3/2} \left(1 + \frac{a_0^2 p^2}{\hbar^2} \right)^{-2}$$

---

[9] The Bohr radius is $a_0 = \hbar^2/\mu e^2$ and the hydrogen energy levels are $E_n = -e^4/2n^2 a_0$.

The corresponding probability density will be

$$\mathcal{P}(p) = \frac{8}{\pi^2} \left(\frac{a_0}{\hbar}\right)^3 \frac{1}{\left(\frac{1}{a_0^2} + \frac{p^2}{\hbar^2}\right)^4}$$

The probability of finding the system with momentum

$$p \geq p_0 \equiv \sqrt{2\mu\langle T\rangle_1} = \sqrt{\frac{\mu e^2}{a_0}} = \frac{\hbar}{a_0}$$

is given by

$$P = 4\pi \int_{p_0}^{\infty} dp\, p^2 \mathcal{P}(p) = \frac{32}{\pi} \int_1^{\infty} d\xi\, \frac{\xi^2}{(1+\xi^2)^4}$$

The integral at hand can be easily computed:[10]

$$\int_1^{\infty} d\xi\, \frac{\xi^2}{(1+\xi^2)^4} = \frac{\pi}{32} - \frac{4+3\pi}{192}$$

Thus, the probability of finding the system with momentum larger than the expectation value is

$$P = \frac{1}{2} - \frac{2}{3\pi} \approx 0.2878$$

The expectation value of the position $\langle 1\,0\,0|\mathbf{x}|1\,0\,0\rangle$ vanishes owing to spherical symmetry. The square of the position has a non-vanishing expectation value and it is given by

$$(\Delta \mathbf{x})^2 = \langle r^2\rangle - \langle \mathbf{x}\rangle^2 = \frac{4\pi}{(\pi a_0^3)} \int_0^{\infty} dr\, r^4 e^{-2r/a_0} = \frac{4}{a_0^3} \left(\frac{2}{a_0}\right)^5 \Gamma(5) = 3a_0^2$$

The square root of this value should be comparable to the expectation value of the radius, which is

$$\langle r\rangle = \frac{4}{a_0^3} \int_0^{\infty} dr\, r^3 e^{-2r/a_0} = \frac{3a_0}{2}$$

Either of these lengths gives the approximate size of the atom in its ground state. The corresponding uncertainty is

$$(\Delta r)^2 = \langle r^2\rangle - \langle r\rangle^2 = \left(3 - \frac{9}{4}\right) a_0^2 = \frac{3a_0^2}{4}$$

---

[10] Consider the change of variable $x = \tan\varphi$.

The uncertainty in the momentum is known from the expectation value of the kinetic energy:

$$(\Delta \mathbf{p})^2 = \langle p^2 \rangle - \langle \mathbf{p} \rangle^2 = 2\mu \langle T \rangle = \frac{\hbar^2}{a_0^2}$$

The Uncertainty Principle is clearly satisfied since

$$(\Delta \mathbf{p})^2 (\Delta \mathbf{x})^2 = 3\hbar^2$$

Problem 7.15 Some hadrons can be successfully described as bound states of heavy quarks in a non-relativistic spin-independent potential. The most common example is a pair comprising a *charmed quark* and *antiquark* ($m_c = 1.5 \, \text{GeV}/c^2$) described in its centre-of-mass frame by a linear interaction potential

$$V(r) = g^2 r + V_0$$

(a) Calculate the expectation values of the radius and the square of the radius for any energy eigenstate with $\ell = 0$. Show that both are expressible in terms of one dimensionful parameter a, with the dimensions of length, times a dimensionless factor. Show that this is generally true for any expectation value $\langle r^n \rangle$ in any energy eigenstate.

(b) Consider the square of the momentum operator and show that the expectation value of any power of it scales according to the corresponding power of a, for any energy eigenstate with $\ell = 0$. Write down the momentum–position uncertainty product and derive an inequality for the dimensionless coefficient in the energy eigenvalues.

(c) Assume now that the same interaction potential is valid not only for the charmed-quark ground state but also for the *bottom*-quark–antiquark bound state ($m_b = 4.5 \, \text{GeV}/c^2$). Calculate the corresponding expectation values of the radius and the square of the radius in any $\ell = 0$ eigenstate. Calculate the relative ratio $\langle r^\nu \rangle_c / \langle r^\nu \rangle_b$, where ν is an integer, of the radius expectation values for the charmed-quark case and the bottom-quark case.

(d) The energy eigenvalues of the ground state ($n = \ell = 0$) and the first excited state ($n = 1$, $\ell = 0$) of the charmed-quark system are $E_0^{(c)} \approx 3.1 \, \text{GeV}/c^2$ and $E_1^{(c)} \approx 3.7 \, \text{GeV}/c^2$. Assume also that the ground-state energy of the bottom-quark system is $E_0^{(b)} \approx 9.5 \, \text{GeV}/c^2$. Can the energy eigenvalue $E_1^{(b)}$ be predicted?

Solution

(a) Following problem 7.5, we can prove that

$$\frac{\mu}{\hbar^2} \left[2s E \langle r^{s-1} \rangle - 2s \langle r^{s-1} V(r) \rangle - \langle r^s V'(r) \rangle \right]$$

$$+ \frac{1}{4} \left[s(s-1)(s-2) - 4(s-1)\ell(\ell+1) \right] \langle r^{s-3} \rangle + 2\pi |\psi(0)|^2 \delta_{s,0} = 0$$

In our case we have, for any eigenstate with $\ell = 0$,

$$\frac{\mu_c}{\hbar^2} \left[2s(E_n - V_0)\langle r^{s-1}\rangle - (2s+1)g^2\langle r^s\rangle \right]$$

$$+ \tfrac{1}{4}s(s-1)(s-2)\langle r^{s-3}\rangle + 2\pi|\psi_0(0)|^2\delta_{s,0} = 0$$

where n is the *principal quantum number* and labels the energy eigenstates E_n. The mass $\mu_c = m_c/2$ is the reduced mass of the quark–antiquark system. For $s = 0$, we get

$$|\psi_0(0)|^2 = \frac{g^2\mu_c}{2\pi\hbar^2}$$

For $s = 1$, we have

$$\langle r\rangle = \frac{2}{3g^2}(E_n - V_0)$$

Finally, for $s = 2$ we obtain

$$\langle r^2\rangle = \frac{4}{5g^2}(E_n - V_0)\langle r\rangle = \frac{8}{15g^4}(E_n - V_0)^2$$

The radial Schroedinger equation has the form

$$\left[-\frac{1}{r}\frac{d^2}{dr^2}r + \frac{\ell(\ell+1)}{r^2} + \frac{2\mu_c}{\hbar^2}(g^2 r + V_0) \right] R_{n\ell}(r) = \frac{2\mu_c}{\hbar^2} E_n R_{n\ell}(r)$$

Equivalently, we can write

$$\left[-\frac{1}{r}\frac{d^2}{dr^2}r + \frac{\ell(\ell+1)}{r^2} + 2a^{-3}r \right] R_{n\ell}(r) = \frac{2\mu_c}{\hbar^2}(E_n - V_0) R_{n\ell}(r)$$

where we have introduced the parameter

$$a = \left(\frac{\hbar^2}{\mu_c g^2} \right)^{1/3}$$

The parameter combination that multiplies $R_{n\ell}(r)$ on the right-hand side has the dimensions of inverse length squared. Thus, we can always write

$$\frac{2\mu_c}{\hbar^2}(E_n - V_0) = a^{-2}2\epsilon_n \qquad \Longrightarrow \qquad E_n = V_0 + \epsilon_n \left(\frac{\hbar^2 g^4}{\mu_c} \right)^{1/3}$$

where ϵ_n is a dimensionless function of the quantum number n. Equivalently, we may write

$$E_n = V_0 + \epsilon_n \frac{\hbar^2}{\mu_c a^2} = V_0 + \epsilon_n a g^2$$

The radial Schroedinger equation takes the form

$$a^2 \left[-\frac{1}{r} \frac{d^2}{dr^2} r + \frac{\ell(\ell+1)}{r^2} + \frac{2r}{a^3} \right] R_{n\ell}(r) = \epsilon_n R_{n\ell}(r)$$

Therefore, its solutions will be functions of the dimensionless variable $\bar{r} = r/a$. Their normalization constants will equal a dimensionless number times $a^{-3/2}$. Any radius expectation value will be given by

$$\langle r^\nu \rangle_{n\ell} = a^\nu I_{n\ell}$$

in terms of the dimensionless number[11] $I_{n\ell}$.

For the calculated radius expectation values, we get

$$\langle r \rangle = \frac{2}{3} \epsilon_n a, \qquad \langle r^2 \rangle = \frac{8}{15} \epsilon_n^2 a^2$$

Note that the ratio

$$\frac{\langle r^2 \rangle}{\langle r \rangle^2} = \frac{6}{5}$$

is independent of ϵ_n.

(b) From the *virial theorem*[12] $2\langle T \rangle = \langle r V'(r) \rangle$ we have, for any eigenstate of the energy with $\ell = 0$,

$$\langle p^2 \rangle = 2\mu \langle T \rangle = \mu g^2 \langle r \rangle = \frac{2}{3} \mu g^2 a \epsilon_n = \frac{2}{3} \frac{\hbar^2 \epsilon_n}{a^2}$$

Since owing to spherical symmetry $\langle \mathbf{p} \rangle = \langle \mathbf{x} \rangle = 0$, the Heisenberg uncertainty product will be

$$(\Delta \mathbf{x})^2 (\Delta \mathbf{p})^2 = \frac{16}{45} \hbar^2 \epsilon_n^3$$

Heisenberg's inequality implies that

$$\epsilon_n \geq \left(\frac{3}{4} \times \frac{45}{16} \right)^{1/3} = \frac{3}{4} 5^{1/3}$$

(c) The only thing that changes is the mass. This modifies the characteristic length according to

$$a_c = \left(\frac{\hbar^2}{\mu_c g^2} \right)^{1/3} \qquad \Longrightarrow \qquad a_b = \left(\frac{\hbar^2}{\mu_b g^2} \right)^{1/3} = a_c \left(\frac{\mu_c}{\mu_b} \right)^{1/3}$$

---

[11] $I_{n\ell} = \int_0^\infty d\bar{r}\, \bar{r}^2 R_{n\ell}^2(\bar{r})$.

[12] It can be derived from the general expectation-value relation used in (a).

The corresponding expectation values are

$$\langle r \rangle_b = \frac{2}{3}\epsilon_n a_b = \left(\frac{\mu_c}{\mu_b}\right)^{1/3} \langle r \rangle_c$$

$$\langle r^2 \rangle_b = \frac{8}{15}\epsilon_n^2 a_b^2 = \left(\frac{\mu_c}{\mu_b}\right)^{2/3} \langle r^2 \rangle_c$$

Thus, we obtain

$$\frac{\langle r \rangle_b}{\langle r \rangle_c} = \left(\frac{\mu_c}{\mu_b}\right)^{1/3} \approx \left(\frac{1}{3}\right)^{1/3}, \qquad \frac{\langle r^2 \rangle_b}{\langle r^2 \rangle_c} = \left(\frac{\mu_c}{\mu_b}\right)^{2/3} \approx \left(\frac{1}{3}\right)^{2/3}$$

Note however that, in general, for every energy eigenstate and for every radius power we may write

$$\frac{\langle r^\nu \rangle_{n\ell}^{(b)}}{\langle r^\nu \rangle_{n\ell}^{(c)}} = \frac{a_b^\nu I_{n\ell}^{(\nu)}}{a_c^\nu I_{n\ell}^{(\nu)}} = \left(\frac{a_b}{a_c}\right)^\nu = \left(\frac{\mu_c}{\mu_b}\right)^{\nu/3}$$

since the dimensionless integrals $I_{n\ell}^{(\nu)}$ are identical.

(d) In terms of the known values $E_0^{(c)}$, $E_1^{(c)}$, $E_0^{(b)}$, we get

$$E_1^{(c)} - E_0^{(c)} = (\epsilon_1 - \epsilon_0)\left(\frac{\hbar^2 g^4}{\mu_c}\right)^{1/3}, \qquad E_1^{(b)} - E_0^{(b)} = (\epsilon_1 - \epsilon_0)\left(\frac{\hbar^2 g^4}{\mu_b}\right)^{1/3}$$

or

$$E_1^{(b)} = E_0^{(b)} + \left(\frac{\mu_c}{\mu_b}\right)^{1/3}\left(E_1^{(c)} - E_0^{(c)}\right) \approx 9.9\,\mathrm{GeV}/c^2$$

Problem 7.16 Consider a spherically symmetric potential of the form

$$V(r) = \lambda^2 r^\alpha$$

A particle of mass μ is bound in it. Using dimensional analysis, determine the dependence of the energy eigenvalues on the parameters of the system up to a multiplicative dimensionless factor. Consider the case where this potential describes the centre-of-mass interaction between two particles, of masses M and m, and the first particle is replaced by a similar particle but of mass M'. How would the energy eigenvalues change?

Solution

The parameters that appear in the Hamiltonian are $\hbar^2/\mu$ and λ^2. Thus

$$E_n = \epsilon_n \left(\frac{\hbar^2}{\mu}\right)^x (\lambda^2)^y$$

where ϵ_n is a dimensionless factor depending on the radial quantum number that labels the energy eigenvalues. The parameters have dimensions as follows:

$$\frac{\hbar^2}{\mu} \implies ML^4T^{-2}, \qquad \lambda^2 \implies ML^{2-\alpha}T^{-2}$$

Thus, in order to match the left-hand side of the expression for E_n we must have $1 = x + y$ and $2 = 4x + (2 - \alpha)y$, or $x = \alpha/(2 + \alpha)$ and $y = 2/(2 + \alpha)$. Then the energy can be written as

$$E_n = \epsilon_n \left(\frac{\hbar^2}{\mu}\right)^{\alpha/(\alpha+2)} (\lambda^2)^{2/(\alpha+2)}$$

An equivalent parametrisation makes use of a characteristic length $d \equiv (\hbar^2/\mu\lambda^2)^{1/\alpha+2}$. In terms of d, we have $E_n = \epsilon_n \hbar^2/\mu d^2$.

For the application considered, the eigenvalues would change to E'_n, where

$$\frac{E'_n}{E_n} = \left(\frac{\mu}{\mu'}\right)^{\alpha/(\alpha+2)} = \left[\frac{Mm/(M+m)}{M'm/(M'+m)}\right]^{\alpha/(\alpha+2)} = \left[\frac{M(M'+m)}{M'(M+m)}\right]^{\alpha/(\alpha+2)}$$

Problem 7.17 Consider the *electric dipole and quadrupole moment* operators for a particle of electric charge e:

$$d_i \equiv e\, x_i, \qquad Q_{ij} \equiv e\left(x_i x_j - \tfrac{1}{3}\delta_{ij}r^2\right)$$

Calculate the expectation values of these operators when the particle occupies the ground state ($n = 1$, $\ell = 0$) or the first excited state ($n = 2$, $\ell = 1$) of the hydrogen atom.

Solution

The diagonal matrix elements of the electric dipole moment vanish since these states are parity eigenstates and the position operator is odd under parity:

$$\mathcal{P}\mathbf{x}\mathcal{P} = -\mathbf{x}$$
$$\implies \quad \langle n\,\ell\,m|\mathbf{x}|n\,\ell\,m\rangle = -\langle n\,\ell\,m|\mathcal{P}\mathbf{x}\mathcal{P}|n\,\ell\,m\rangle$$
$$= -(-1)^\ell(-1)^\ell \langle n\,\ell\,m|\mathbf{x}|n\,\ell\,m\rangle = -\langle n\,\ell\,m|\mathbf{x}|n\,\ell\,m\rangle = 0$$

Since for any spherically symmetric eigenfunction it is true that

$$\langle\psi|x_i x_j|\psi\rangle = \delta_{ij} J \qquad \implies \qquad J = \tfrac{1}{3}\langle\psi|r^2|\psi\rangle$$

we should have

$$\langle n\,\ell\,0|Q_{ij}|n\,\ell\,0\rangle = e\left(\langle n\,\ell\,0|x_i x_j|n\,\ell\,0\rangle - \tfrac{1}{3}\delta_{ij}\langle n\,\ell\,0|r^2|n\,\ell\,0\rangle\right) = 0$$

This takes care of all states but $|2\ 1\ \pm 1\rangle$. For this state, we have

$$\mathcal{Q}_{ij} = \frac{e}{64\pi a_0^5} \int_0^\infty dr\, r^6 e^{-r/a_0} \int d\Omega\, \sin^2\theta\, \mathcal{M}_{ij}$$

where

$$\mathcal{M}_{ij} = \begin{pmatrix} \sin^2\theta \cos^2\phi - \frac{1}{3} & \sin^2\theta \cos\phi \sin\phi & \sin\theta \cos\theta \cos\phi \\ \sin^2\theta \cos\phi \sin\phi & \sin^2\theta \sin^2\phi - \frac{1}{3} & \sin\theta \cos\theta \sin\phi \\ \sin\theta \cos\theta \cos\phi & \sin\theta \cos\theta \sin\phi & \cos^2\theta - \frac{1}{3} \end{pmatrix}$$

Performing the integrals, we obtain the diagonal electric quadrupole moment

$$\mathcal{Q}_{ij} = \frac{ea_0^2}{64\pi}(6!)\left(\frac{8\pi}{45}\right)\begin{pmatrix} 1 & 0 & 0 \\ 0 & 1 & 0 \\ 0 & 0 & -2 \end{pmatrix} = 2ea_0^2 \begin{pmatrix} 1 & 0 & 0 \\ 0 & 1 & 0 \\ 0 & 0 & -2 \end{pmatrix}$$

Problem 7.18 An electron is moving under the influence of a uniform electrostatic field $\mathcal{E}$. Consider the probability amplitude that the electron, having momentum $\mathbf{p}_i$ at time $t = 0$, will be found at time $t > 0$ with momentum $\mathbf{p}_f$.

(a) Show that this amplitude vanishes unless

$$\mathbf{p}_f = \mathbf{p}_i + e\mathcal{E}t$$

This is what would be expected classically from a force constant in space and time.

(b) Denote by $\mathcal{K}(\mathbf{r}_f, \mathbf{r}_i; t)$ the amplitude for finding the electron at the position $\mathbf{r}_f$ at time t, when initially (at $t = 0$) it is at the position $\mathbf{r}_i$, and see whether this amplitude (*propagator*) satisfies the *reflection invariance* satisfied by the free propagator in the absence of an external field:

$$\mathcal{K}_0(\mathbf{r}_f, \mathbf{r}_i; t) = \mathcal{K}_0(-\mathbf{r}_f, -\mathbf{r}_i; t)$$

(c) How is $\mathcal{K}(\mathbf{r}_f, \mathbf{r}_i; t)$ related to $\mathcal{K}(\mathbf{r}_i, \mathbf{r}_f; t)$?

(d) Calculate the propagator explicitly and check all the above results.

Solution

(a) The electron Hamiltonian in the presence of the uniform electric field is

$$H = \frac{p^2}{2m} - e\mathcal{E} \cdot \mathbf{r}$$

The amplitude for finding the electron with momentum $\mathbf{p}_f$ at time $t > 0$ when it had momentum $\mathbf{p}_i$ initially is

$$\tilde{\mathcal{K}}(\mathbf{p}_f, \mathbf{p}_i; t) = \langle \mathbf{p}_f | e^{-iHt/\hbar} | \mathbf{p}_i \rangle$$

Consider the commutator

$$[\mathbf{p},\ e^{-iHt/\hbar}] = e^{-iHt/\hbar}\left(e^{iHt/\hbar}\mathbf{p}e^{-iHt/\hbar}\right) - e^{-iHt/\hbar}\mathbf{p} = e^{-iHt/\hbar}\left[\mathbf{p}_H(t) - \mathbf{p}\right]$$

The Schroedinger momentum operator coincides with the initial Heisenberg momentum operator. Taking the matrix element of the above operator relation between momentum eigenstates, we obtain

$$\langle\mathbf{p}_f|[\mathbf{p},\ e^{-iHt/\hbar}]|\mathbf{p}_i\rangle = (\mathbf{p}_f - \mathbf{p}_i)\langle\mathbf{p}_f|e^{-iHt/\hbar}|\mathbf{p}_i\rangle = \langle\mathbf{p}_f|e^{-iHt/\hbar}\left[\mathbf{p}_H(t) - \mathbf{p}\right]|\mathbf{p}_i\rangle$$

The Heisenberg operator appearing in the right-hand side of the last relation can be obtained from the solution of the Heisenberg equation as follows:

$$\frac{d\mathbf{p}_H}{dt} = \frac{i}{\hbar}[H,\ \mathbf{p}_H] = -\frac{ie}{\hbar}\mathcal{E}_j[x_j,\ \mathbf{p}] = e\mathcal{E}$$

Thus, integrating with respect to time, we get

$$\mathbf{p}_H(t) = \mathbf{p} + e\mathcal{E}t$$

Returning to the above momentum matrix element, we can write

$$(\mathbf{p}_f - \mathbf{p}_i)\langle\mathbf{p}_f|e^{-iHt/\hbar}|\mathbf{p}_i\rangle = e\mathcal{E}t\ \langle\mathbf{p}_f|e^{-iHt/\hbar}|\mathbf{p}_i\rangle$$

or, equivalently,

$$(\mathbf{p}_f - \mathbf{p}_i - e\mathcal{E}t)\tilde{\mathcal{K}}(\mathbf{p}_f,\ \mathbf{p}_i\ ;\ t) = 0$$

which proves that this amplitude vanishes unless

$$\mathbf{p}_f = \mathbf{p}_i + et\mathcal{E}$$

This is something expected classically, since the momentum transfer of a force that is constant in space and time should be $\mathcal{F}t = e\mathcal{E}t$. Note that this amplitude is related to the spatial *propagator* as

$$\mathcal{K}(\mathbf{r}',\ \mathbf{r}\ ;\ t) \equiv \langle\mathbf{r}'|e^{-iHt/\hbar}|\mathbf{r}\rangle = \int\int\frac{d^3p'd^3p}{(2\pi\hbar)^3}\ e^{i\mathbf{p}'\cdot\mathbf{r}'/\hbar}\ e^{-i\mathbf{p}\cdot\mathbf{r}/\hbar}\ \tilde{\mathcal{K}}(\mathbf{p}',\ \mathbf{p}\ ;\ t)$$

(b) Consider the space-reflected propagator

$$\mathcal{K}(-\mathbf{r}',\ -\mathbf{r}\ ;\ t) = \langle-\mathbf{r}'|e^{-iHt/\hbar}|-\mathbf{r}\rangle = \langle\mathbf{r}'|\mathcal{P}e^{-iHt/\hbar}\mathcal{P}|\mathbf{r}\rangle$$

where $\mathcal{P}$ is the parity operator. This is further equal to

$$\langle\mathbf{r}'|e^{-iHt/\hbar}|\mathbf{r}\rangle + \langle\mathbf{r}'|[\mathcal{P},\ e^{-iHt/\hbar}]\mathcal{P}|\mathbf{r}\rangle = \mathcal{K}(\mathbf{r}',\ \mathbf{r}\ ;\ t) + \langle\mathbf{r}'|[\mathcal{P},\ e^{-iHt/\hbar}]\mathcal{P}|\mathbf{r}\rangle$$

Since the last commutator is non-zero as long as the electric field is present, we conclude that

$$\mathcal{K}(-\mathbf{r}',\ -\mathbf{r}\ ;\ t) \neq \mathcal{K}(\mathbf{r}',\ \mathbf{r}\ ;\ t)$$

(c) It is useful here to consider the expression for the propagator in terms of energy eigenstates,

$$\mathcal{K}(\mathbf{r}', \mathbf{r}; t) = \sum_n \psi_n(\mathbf{r})\psi_n^*(\mathbf{r}')\, e^{-iE_n t/\hbar}$$

Because of the hermiticity and reality of the Hamiltonian, we have

$$\langle \mathbf{r}'|e^{-iHt/\hbar}|\mathbf{r}\rangle = \langle \mathbf{r}|e^{iHt/\hbar}|\mathbf{r}'\rangle^* = \sum_n \psi_n^*(\mathbf{r})\psi_n(\mathbf{r}')\, e^{-iE_n t/\hbar} = \mathcal{K}(\mathbf{r}, \mathbf{r}'; t)$$

Thus, despite the fact that

$$\mathcal{K}(-\mathbf{r}', -\mathbf{r}; t) \neq \mathcal{K}(\mathbf{r}', \mathbf{r}; t)$$

we still have

$$\mathcal{K}(\mathbf{r}, \mathbf{r}'; t) = \mathcal{K}(\mathbf{r}', \mathbf{r}; t)$$

as in the free case.

(d) As above, it is convenient to go to the Heisenberg picture, which can be solved explicitly. The solved Heisenberg momentum and position operators are

$$\mathbf{p}_{\mathrm{H}} = \mathbf{p} + e\mathcal{E}t, \qquad \mathbf{r}_{\mathrm{H}} = \mathbf{r} + \frac{t}{m}\mathbf{p} + \frac{et^2}{2m}\mathcal{E}$$

Consider now

$$\mathbf{r}'\mathcal{K}(\mathbf{r}', \mathbf{r}; t) = \mathbf{r}'\langle \mathbf{r}'|e^{-iHt/\hbar}|\mathbf{r}\rangle = \langle \mathbf{r}'|\mathbf{r}e^{-iHt/\hbar}|\mathbf{r}\rangle$$

$$= \langle \mathbf{r}'|e^{-iHt/\hbar}\mathbf{r}_{\mathrm{H}}|\mathbf{r}\rangle = \left\langle \mathbf{r}'\left|e^{-iHt/\hbar}\left(\mathbf{r} + \frac{t}{m}\mathbf{p} + \frac{et^2}{2m}\mathcal{E}\right)\right|\mathbf{r}\right\rangle$$

$$= \left(\mathbf{r} + \frac{et^2}{2m}\mathcal{E}\right)\mathcal{K}(\mathbf{r}', \mathbf{r}; t) + \frac{t}{m}\langle \mathbf{r}'|e^{-iHt/\hbar}\mathbf{p}|\mathbf{r}\rangle$$

The last term, being of the form

$$\langle \psi|p_j|\mathbf{r}\rangle = \langle \mathbf{r}|p_j|\psi\rangle^* = +i\hbar \frac{\partial}{\partial x_j}\langle \mathbf{r}|\psi\rangle^* = +i\hbar \frac{\partial}{\partial x_j}\langle \psi|\mathbf{r}\rangle$$

modifies the above expression into a differential equation,

$$i\frac{\hbar t}{m}\nabla \mathcal{K}(\mathbf{r}', \mathbf{r}; t) = \left(\mathbf{r}' - \mathbf{r} - \frac{et^2}{2m}\mathcal{E}\right)\mathcal{K}(\mathbf{r}', \mathbf{r}; t)$$

Integrating with respect to $\mathbf{r}$, we get

$$\ln \mathcal{K}(\mathbf{r}', \mathbf{r}; t) - \ln \mathcal{K}(\mathbf{r}', 0; t) = i\frac{m}{\hbar t}\left(\frac{1}{2}r^2 - \mathbf{r}\cdot\mathbf{r}' + \frac{et^2}{2m}\mathcal{E}\cdot\mathbf{r}\right)$$

With an exactly analogous series of manipulations, we can obtain the symmetric expression

$$\ln \mathcal{K}(\mathbf{r'}, \mathbf{r}; t) - \ln \mathcal{K}(0, \mathbf{r}; t) = i \frac{m}{\hbar t} \left(\frac{1}{2}(r')^2 - \mathbf{r} \cdot \mathbf{r'} + \frac{et^2}{2m} \boldsymbol{\mathcal{E}} \cdot \mathbf{r'} \right)$$

Taking $\mathbf{r'} = 0$ in the first expression, we get

$$\ln \mathcal{K}(0, \mathbf{r}; t) - \ln \mathcal{K}(0, 0; t) = i \frac{m}{\hbar t} \left(\frac{1}{2} r^2 + \frac{et^2}{2m} \boldsymbol{\mathcal{E}} \cdot \mathbf{r} \right)$$

Substituting this into the second expression, we arrive at

$$\ln \mathcal{K}(\mathbf{r'}, \mathbf{r}; t) - \ln \mathcal{K}(0, 0; t) = \frac{im}{2\hbar t} \left[(\mathbf{r} - \mathbf{r'})^2 + \frac{et^2}{m} \boldsymbol{\mathcal{E}} \cdot (\mathbf{r} + \mathbf{r'}) \right]$$

or

$$\mathcal{K}(\mathbf{r'}, \mathbf{r}; t) = \mathcal{K}(0, 0; t) \exp \left\{ \frac{im}{2\hbar t} \left[(\mathbf{r} - \mathbf{r'})^2 + \frac{et^2}{m} \boldsymbol{\mathcal{E}} \cdot (\mathbf{r} + \mathbf{r'}) \right] \right\}$$

In order to calculate the time-dependent coefficient $\mathcal{K}(0, 0; t)$, we can use the identity

$$\mathcal{K}(0, 0; t) = \int d^3 r \, \mathcal{K}(0, \mathbf{r}; t - t') \mathcal{K}(\mathbf{r}, 0; t')$$

which holds for any intermediate time t' and take $t' = t/2$. Then we have

$$\mathcal{K}(0, 0; t) = \int d^3 r \, [\mathcal{K}(\mathbf{r}, 0; t/2)]^2$$

$$= [\mathcal{K}(0, 0; t/2)]^2 \int d^3 r \, \exp \left(i \frac{2m}{\hbar t} r^2 + i \frac{et}{2\hbar} \boldsymbol{\mathcal{E}} \cdot \mathbf{r} \right)$$

or

$$\mathcal{K}(0, 0; t) = [\mathcal{K}(0, 0; t/2)]^2 \left(\frac{i\hbar t}{2m} \right)^{3/2} \exp \left(-i \frac{e^2 \mathcal{E}^2}{32 m \hbar} t^3 \right)$$

Note that in the free case ($\mathcal{E} = 0$) this relation is immediately satisfied by the known solution $\mathcal{K}_0(0, 0; t) = (m/2\pi i\hbar t)^{3/2}$. It is clear that the appropriate trial ansatz is $\mathcal{K}(0, 0; t) = \mathcal{K}_0(0, 0; t) \, e^{i\alpha t^3}$, which, when substituted, yields $\alpha = -e^2 \mathcal{E}^2 / 24 m\hbar$ and

$$\mathcal{K}(0, 0; t) = \mathcal{K}_0(0, 0; t) \exp \left(-i \frac{e^2 \mathcal{E}^2}{24 m\hbar} t^3 \right)$$

Thus, the full propagator is

$$\mathcal{K}(\mathbf{r'}, \mathbf{r}; t) = \mathcal{K}_0(\mathbf{r'}, \mathbf{r}; t) \exp \left\{ \frac{iet}{2\hbar} \left[\boldsymbol{\mathcal{E}} \cdot (\mathbf{r} + \mathbf{r'}) \right] \right\} \exp \left(-i \frac{e^2 \mathcal{E}^2}{24 m\hbar} t^3 \right)$$

where $\mathcal{K}_0$ is the free propagator. The non-invariance under spatial reflection and the $\mathbf{r}' \leftrightarrow \mathbf{r}$ symmetry are now manifest.

The momentum-space amplitude that we encountered in part (a) is

$$\tilde{\mathcal{K}}(\mathbf{p}_f, \, \mathbf{p}_i \, ; \, t)$$

$$= \mathcal{K}(0, 0; t) \int \int \frac{d^3 r' d^3 r}{(2\pi)^3} \, e^{i\mathbf{p}_f \cdot \mathbf{r}'/\hbar} e^{-i\mathbf{p}_i \cdot \mathbf{r}/\hbar}$$

$$\times \exp\left[\frac{im}{2\hbar t} (\mathbf{r}' - \mathbf{r})^2\right] \exp\left\{\frac{iet}{2\hbar} \left[\boldsymbol{\mathcal{E}} \cdot (\mathbf{r} + \mathbf{r}')\right]\right\}$$

$$= \mathcal{K}(0, 0; t) \int \int \frac{d^3 r' d^3 r}{(2\pi)^3} \, e^{i\mathbf{q}_f \cdot \mathbf{r}'/\hbar} e^{-i\mathbf{q}_i \cdot \mathbf{r}/\hbar} \exp\left[\frac{im}{2\hbar t} (\mathbf{r}' - \mathbf{r})^2\right]$$

$$= \tilde{\mathcal{K}}_0(\mathbf{q}_f, \, \mathbf{q}_i; \, t) \exp\left(-i\frac{e^2 \mathcal{E}^2}{24m\hbar} t^3\right)$$

where

$$\mathbf{q}_f = \mathbf{p}_f + \frac{et}{2}\boldsymbol{\mathcal{E}}, \qquad \mathbf{q}_i = \mathbf{p}_i - \frac{et}{2}\boldsymbol{\mathcal{E}}$$

However, the free amplitude is particularly simple. Thus, we have

$$\tilde{\mathcal{K}}(\mathbf{p}_f, \, \mathbf{p}_i \, ; \, t) = \tilde{\mathcal{K}}_0(\mathbf{q}_f, \, \mathbf{q}_i; \, t) \exp\left(-i\frac{e^2 \mathcal{E}^2}{24m\hbar} t^3\right)$$

$$= \langle \mathbf{q}_f | e^{-i H_0 t/\hbar} | \mathbf{q}_i \rangle \exp\left(-i\frac{e^2 \mathcal{E}^2}{24m\hbar} t^3\right)$$

$$= \exp\left(-i\frac{e^2 \mathcal{E}^2}{24m\hbar} t^3\right) \exp\left(-\frac{i q_i^2 t}{2m\hbar}\right) \delta(\mathbf{q}_f - \mathbf{q}_i)$$

$$= \exp\left(-i\frac{e^2 \mathcal{E}^2}{24m\hbar} t^3\right) \exp\left(-\frac{i q_i^2 t}{2m\hbar}\right) \delta(\mathbf{p}_f - \mathbf{p}_i - et\boldsymbol{\mathcal{E}})$$

Problem 7.19 Consider the Hermitian operator

$$D = \tfrac{1}{2} \left(\mathbf{x} \cdot \mathbf{p} + \mathbf{p} \cdot \mathbf{x}\right)$$

This operator is associated with *dilatations*, i.e. rescalings of the position coordinates.

(a) Establish the commutation relations

$$[D, \, x_j] = -i\hbar x_j, \qquad [D, \, p_j] = i\hbar p_j, \qquad [D, \, L_j] = 0$$

(b) Prove the finite dilatation properties

$$e^{i\alpha D/\hbar} \, x_j \, e^{-i\alpha D/\hbar} = e^\alpha \, x_j, \qquad e^{i\alpha D/\hbar} \, \psi(x) = e^{d\alpha/2} \, \psi(e^\alpha x)$$

where α is a parameter and d is the number of space dimensions.

(c) Show that for a free particle we have

$$[D, H_0] = 2i\hbar H_0$$

Taking the matrix element of this relation between energy eigenstates, we get $(E - E')\langle E|D|E'\rangle = 2i\hbar E\langle E|E'\rangle$. Verify this relation explicitly for plane and spherical waves.

(d) Consider the one-dimensional Hamiltonian

$$H = \frac{p^2}{2m} + \frac{\lambda}{x^2}$$

Show that it obeys the same dilatation commutator relation as a free Hamiltonian. Is this property compatible with a discrete energy spectrum?

Solution
(b) We can write

$$D = \mathbf{x} \cdot \mathbf{p} - \frac{i\hbar}{2} d$$

where d is the number of space dimensions. Then, we have

$$\langle x| e^{i\alpha D/\hbar} x_j |\psi\rangle = e^{\alpha d/2} e^{\alpha \mathbf{x} \cdot \nabla} \langle x|x_j |\psi\rangle = e^{\alpha d/2} e^{\alpha \mathbf{x} \cdot \nabla} x_j \langle x|\psi\rangle$$

Note, now, that the operator exponent is just the Taylor translation operator:

$$\exp(\alpha \mathbf{x} \cdot \nabla) = \exp\left(\alpha \sum_i x_i \frac{\partial}{\partial x_i}\right) = \exp\left(\alpha \sum_i \frac{\partial}{\partial \ln x_i}\right)$$

Thus, the coordinates are translated as follows:

$$\ln x_i \to \ln x_i + \alpha = \ln(e^\alpha x_i) \qquad \Longrightarrow \qquad x_i \to e^\alpha x_i$$

Therefore, we have

$$\langle x| e^{i\alpha D/\hbar} x_j |\psi\rangle = e^{\alpha(1+d/2)} x_j \langle e^\alpha x|\psi\rangle = e^\alpha x_j \langle x|e^{i\alpha D/\hbar}|\psi\rangle$$

and, finally,

$$e^{i\alpha D/\hbar} x_j e^{-i\alpha D/\hbar} = e^\alpha x_j$$

and also

$$e^{i\alpha D/\hbar} \psi(x) = e^{\alpha d/2} \psi(e^\alpha x)$$

(c) It is straightforward to see that $[D, p^2] = 2i\hbar p^2$. Taking the expectation value of the dilatation commutator of the Hamiltonian between energy eigenstates we get

$$(E - E')\langle E|D|E'\rangle = 2i\hbar E\langle E|E'\rangle$$

For plane waves (in one dimension, for simplicity), we get

$$\left[k^2 - (k')^2\right] \int \frac{dx}{2\pi} e^{-ikx} \left(-i\hbar x \frac{\partial}{\partial x} - \frac{i\hbar}{2}\right) e^{ik'x} = 2i\hbar k^2 \delta(k - k')$$

or

$$\left[k^2 - (k')^2\right] \left(i\hbar k' \frac{\partial}{\partial k'} - \frac{i\hbar}{2}\right) \delta(k - k') = 2i\hbar k^2 \delta(k - k')$$

The left-hand side is

$$-i\hbar \frac{\partial}{\partial k'} \left\{k' \left[k^2 - (k')^2\right]\right\} \delta(k - k') = -i\hbar \left[k^2 - 3(k')^2\right] \delta(k - k')$$
$$= 2i\hbar k^2 \delta(k - k')$$

and, therefore, is equal to the right-hand side.

For spherical waves it is sufficient to verify that

$$\left[k^2 - (k')^2\right] \int_0^\infty dr\, r^2 j_\ell(kr) \left(-i\hbar r \frac{\partial}{\partial r} - \frac{3i\hbar}{2}\right) j_\ell(k'r) = i\hbar\pi \delta(k - k')$$

The left-hand side is equal to

$$i\hbar \left[k^2 - (k')^2\right] k' \frac{\partial}{\partial k'} \int_0^\infty dr\, r^2 j_\ell(kr)\, j_\ell(k'r)$$

$$= \frac{i\hbar\pi}{2k^2} \left[k^2 - (k')^2\right] k' \frac{\partial}{\partial k'} \delta(k - k')$$

$$= -\frac{i\hbar\pi}{2} \frac{\partial}{\partial k'} \left[k' - \frac{(k')^3}{k^2}\right] \delta(k - k') = i\hbar\pi\, \delta(k - k')$$

which coincides with the right-hand side. We have used the orthogonality property of spherical Bessel functions,

$$\int_0^\infty dr\, r^2 j_\ell(kr) j_\ell(k'r) = \frac{\pi}{2k^2} \delta(k - k')$$

(d) This Hamiltonian, as well as an analogous higher-dimensional one with interaction $V \propto r^{-2}$, corresponds to the Schroedinger equation

$$\left[-\frac{d^2}{dx^2} + \left(\frac{2m\lambda}{\hbar^2}\right) \frac{1}{x^2}\right] \psi(x) = \frac{2mE}{\hbar^2} \psi(x) = k^2 \psi(x)$$

Note that the effective interaction parameter $\bar{\lambda} = 2m\lambda/\hbar^2$ is dimensionless. The fact that both the kinetic term and the interaction term scale with the same power of length is reflected in the commutation relation

$$[D, H] = 2i\hbar H$$

which can be proved in a straightforward fashion.

If $|E_0\rangle$ is a normalizable state (i.e. in the discrete spectrum), taking the expectation value of the above commutation relation in this state gives

$$\langle E_0|[D,\ H]|E_0\rangle = 2i\hbar\langle E_0|H|E_0\rangle$$

or

$$E_0 = 0$$

Thus, any such Hamiltonian cannot have normalizable discrete eigenstates.

Problem 7.20 Consider an infinite region in which a uniform electric field $\mathcal{E} = \mathcal{E}\hat{\mathbf{y}}$ is present. An electron, characterized by a well-localized wave function $\psi(\mathbf{r}) = Ne^{i\mathbf{k}\cdot\mathbf{r}}e^{-\alpha r^2/2}$, with $\mathbf{k} = k_x\hat{\mathbf{x}} + k_y\hat{\mathbf{y}}$ and $k_y > 0$, enters the region (at $t = 0$); see Fig. 34. Find the expectation values of the position and momentum of the electron at any subsequent time of its presence in the region. Calculate the time T at which the expectation value of the position component of the electron in the direction of the field will coincide with its initial value and write down the expectation values of its position and momentum at that time. Calculate the size of the wave packet at any time $0 < t \leq T$ and show that it is independent of the electric field. Show also that the momentum uncertainty is the same as in the free case and, therefore, constant.

Solution

The Hamiltonian of the system is

$$H = \frac{p^2}{2m} - e\mathcal{E}\cdot\mathbf{r}$$

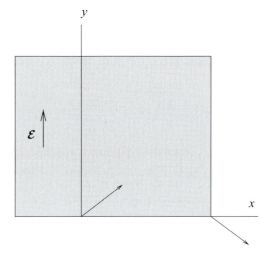

Fig. 34 Motion in a uniform electric field. The small arrows represent the particle's velocity.

Let us consider the Heisenberg equations for the position and momentum; note that they are linear in the corresponding operators, something that, as in the case of the harmonic oscillator, suggests strong similarities with the evolution of the classical system. They are

$$\frac{d\mathbf{r}}{dt} = \frac{\mathbf{p}}{m}, \qquad \frac{d\mathbf{p}}{dt} = \frac{i}{\hbar}[H, \mathbf{p}] = e\mathcal{E}$$

They can be easily integrated out to give

$$\mathbf{p}(t) = \mathbf{p}(0) + e\mathcal{E}t, \qquad \mathbf{r}(t) = \mathbf{r}(0) + \frac{\mathbf{p}(0)}{m}t + \frac{et^2}{2m}\mathcal{E}$$

The initial expectation values of the electron position and momentum are

$$\langle \mathbf{r} \rangle_0 = |N|^2 \int d^3r \, \mathbf{r} e^{-\alpha r^2} = 0$$

because of rotational symmetry, and[13]

$$\langle \mathbf{p} \rangle_0 = |N|^2 \int d^3r \, e^{-i\mathbf{k}\cdot\mathbf{r}} e^{-\alpha r^2/2} \left(-i\hbar\nabla\right) e^{i\mathbf{k}\cdot\mathbf{r}} e^{-\alpha r^2/2}$$

$$= |N|^2 \int d^3r \, e^{-\alpha r^2} (\hbar\mathbf{k} + i\alpha\mathbf{r}) = \hbar\mathbf{k}$$

The evolved expectation values are

$$\langle \mathbf{r} \rangle_t = \frac{\hbar\mathbf{k}}{m}t + \frac{et^2}{2m}\mathcal{E}, \qquad \langle \mathbf{p} \rangle_t = \hbar\mathbf{k} + e\mathcal{E}t$$

Writing the position expectation value in components, we get

$$\langle z \rangle_t = 0, \qquad \langle x \rangle_t = \frac{\hbar t}{m}k_x, \qquad \langle y \rangle_t = \frac{\hbar t}{m}k_y + \frac{et^2}{2m}\mathcal{E}$$

It is clear that the motion takes place entirely in the xy-plane. In fact, it is the parabola

$$\langle y \rangle_t = \frac{k_y}{k_x}\langle x \rangle_t + \frac{e\mathcal{E}m}{2\hbar^2 k_x^2}\langle x \rangle_t^2$$

The exit point, at $t = T$, occurs when the xz-plane is reached again and so $\langle y \rangle_T = 0$. The time T is equal to $2\hbar k_y/|e|\mathcal{E}$. The exit position and momentum expectation values are

$$\langle \mathbf{r} \rangle_T = \frac{2\hbar^2 k_x k_y}{|e|\mathcal{E}}\hat{\mathbf{x}}, \qquad \langle \mathbf{p} \rangle = \hbar\mathbf{k} + e\mathcal{E}T = k_x\hat{\mathbf{x}} - k_y\hat{\mathbf{y}}$$

---

[13] Note that $|N|^2 \int d^3r e^{-\alpha r^2} = \langle \psi | \psi \rangle = 1$.

Taking the square of the position-operator solution of the Heisenberg equation we obtain

$$r^2(t) = r^2(0) + \frac{t^2}{m^2}p^2(0) + \frac{t}{m}[\mathbf{p}(0) \cdot \mathbf{r}(0) + \mathbf{r}(0) \cdot \mathbf{p}(0)]$$

$$+ \frac{et^3}{m^2}\boldsymbol{\mathcal{E}} \cdot \mathbf{p}(0) + \frac{et^2}{m}\boldsymbol{\mathcal{E}} \cdot \mathbf{r}(0) + \frac{e^2t^4}{4m^2}\mathcal{E}^2$$

The expectation value is

$$\langle r^2 \rangle_t = \langle r^2 \rangle_0 + \frac{t^2}{m^2}\langle p^2 \rangle_0 + \frac{t}{m}\langle \mathbf{p} \cdot \mathbf{r} + \mathbf{r} \cdot \mathbf{p} \rangle_0 + \frac{et^3\hbar}{m^2}\boldsymbol{\mathcal{E}} \cdot \mathbf{k} + \frac{e^2t^4}{4m^2}\mathcal{E}^2$$

The square of the expectation value of the position is, however,

$$\langle \mathbf{r} \rangle_t^2 = \frac{t^2\hbar^2k^2}{m^2} + \frac{e^2t^4}{4m^2}\mathcal{E}^2 + \frac{\hbar t^3 e}{m^2}\mathbf{k} \cdot \boldsymbol{\mathcal{E}}$$

Subtracting, we get

$$(\Delta\mathbf{r})_t^2 = \langle r^2 \rangle_0 + \frac{t^2}{m^2}\langle p^2 \rangle_0 + \frac{t}{m}\langle \mathbf{p} \cdot \mathbf{r} + \mathbf{r} \cdot \mathbf{p} \rangle_0 - \frac{t^2\hbar^2k^2}{m^2}$$

which is completely independent of the electric field at all times $0 < t \le T$. It is not difficult to calculate that

$$\langle r^2 \rangle_0 = \frac{3}{2\alpha}, \qquad \langle p^2 \rangle_0 = \frac{3\alpha\hbar^2}{2} + k^2\hbar^2, \qquad \langle \mathbf{r} \cdot \mathbf{p} + \mathbf{p} \cdot \mathbf{r} \rangle_0 = 0$$

and, finally, obtain

$$(\Delta\mathbf{r})_t^2 = \frac{3}{2\alpha}\left[1 + \left(\frac{\alpha\hbar t}{m}\right)^2\right]$$

Similarly, we can calculate the momentum uncertainty $(\Delta\mathbf{p})_t^2 = \frac{3}{2}\alpha\hbar^2$ and see that it is independent of the electric field but also constant, as in the free case.

Problem 7.21 Consider a particle of mass m interacting with a central potential

$$V(r) = \frac{\lambda}{r^2}$$

(a) Write down the radial Schroedinger equation and show that the eigenfunctions corresponding to the continuous spectrum can be expressed in terms of the solutions[14] of the Bessel equation

$$x^2y''(x) + xy'(x) + (x^2 - n^2)y(x) = 0$$

---

[14] The solutions of the Bessel equation are the Bessel functions $J_n(x)$ and $Y_n(x)$.

which depend upon the dimensionless parameter $\gamma = 2m\lambda/\hbar^2$ and the wave number.

(b) Assuming momentarily the possibility of bound states, consider the dependence of the energy eigenvalues on the available parameters on dimensional grounds. What are the conclusions? Arrive at a similar conclusion by assuming the existence of bound states and applying the virial theorem $\langle T \rangle = \frac{1}{2}\langle r V' \rangle$.

(c) Consider now the particle under a modified potential

$$V(r) = -\left(\frac{\hbar^2|\gamma|}{2m}\right)\frac{1}{r^{2(1-\eta)}}$$

where η is a number smaller than unity. With the help of dimensional analysis determine the dependence of the discrete energy eigenvalues on the parameters of the system. Define a characteristic length that determines the size of the discrete eigenfunctions. What happens to this length in the limit $\eta \to 0$?

Solution

(a) The radial equation is

$$\left[-\frac{\hbar^2}{2m}\left(\frac{d^2}{dr^2} + \frac{2}{r}\frac{d}{dr}\right) + \frac{\hbar^2}{2m}\frac{\ell(\ell+1)}{r^2} + \frac{\lambda}{r^2}\right]R_{E\ell}(r) = ER_{E\ell}(r)$$

For $E = \hbar^2 k^2/2m > 0$ it becomes

$$R''(x) + \frac{2}{x}R'(x) + \left[1 - \frac{\ell(\ell+1)+\gamma}{x^2}\right]R(x) = 0$$

where $x \equiv kr$. Changing variable according to

$$R(x) = x^\alpha y(x) \qquad \Longrightarrow \qquad \alpha = -\frac{1}{2}$$

we get the equation

$$y''(x) + \frac{1}{x}y'(x) + \left[1 - \frac{\ell(\ell+1)+\gamma+\frac{1}{4}}{x^2}\right]y(x) = 0$$

This is the Bessel equation with

$$\nu^2 = \ell(\ell+1) + \gamma + \frac{1}{4}$$

and solutions $J_\nu(x)$, $Y_\nu(x)$. The corresponding radial eigenfunctions are

$$R_\ell(kr) = \frac{1}{\sqrt{kr}}J_\nu(kr), \quad \frac{1}{\sqrt{kr}}Y_\nu(kr)$$

(b) Since the Schroedinger equation is of the form

$$\left[-\left(\frac{d^2}{dr^2} + \frac{2}{r}\frac{d}{dr}\right) + \frac{\ell(\ell+1)}{r^2} - \frac{|\gamma|}{r^2}\right] R_{E\ell}(r) = -\frac{2m|E|}{\hbar^2} R_{E\ell}(r)$$

it is clear that there is no dimensionful parameter to express the $(\text{length})^{-2}$ quantity $2mE/\hbar^2$. This reflects the absence of normalizable discrete eigenstates. Similarly, if we assume that such a state exists and apply the virial theorem, we get

$$\langle T \rangle = \tfrac{1}{2}\langle r V'(r)\rangle = -\langle V \rangle \qquad \Longrightarrow \qquad \langle H \rangle = 0$$

(c) The parameters of the problem are $\hbar^2/2m$ and γ. The first has dimensions $ML^4 T^{-2}$, while the second has dimension $L^{-2\eta}$. If the energy eigenvalues have the dependence

$$E = \left(\frac{\hbar^2}{2m}\right)^x (|\gamma|)^y \, \hat{\epsilon}$$

where $\hat{\epsilon}$ is a dimensionless function of the appropriate radial quantum number, we obtain

$$ML^2 T^{-2} = \left(ML^4 T^{-2}\right)^x L^{-2y\eta} \qquad \Longrightarrow \qquad x = 1, \; y = \frac{1}{\eta}$$

Thus, we have

$$E = \frac{\hbar^2}{2m}(|\gamma|)^{1/\eta} \, \hat{\epsilon}$$

It is obvious that the characteristic length of the system is

$$d = \frac{1}{|\gamma|^{1/2\eta}} = \exp\left(-\frac{1}{2\eta}\ln|\gamma|\right)$$

This length determines the overall extent of the corresponding eigenfunctions. In the limit $\eta \to 0$, it tends to zero for strong coupling ($|\gamma| > 1$) and becomes infinite for weak coupling ($|\gamma| < 1$).

Problem 7.22 An electron moves in the presence of a uniform magnetic field $\mathbf{B} = \hat{z}B$. The Hamiltonian of the system is

$$H = \frac{1}{2\mu}\left[\mathbf{p} - \frac{e}{c}\mathbf{A}(\mathbf{r})\right]^2$$

(a) Write down the velocity operators $\mathbf{v}$ and calculate their mutual commutator $[v_i, \, v_j]$.
(b) Derive the Heisenberg equations of motion for the velocity operators and solve them.

(c) At a given time, taken to be $t = 0$, a measurement of the y-component of the velocity of the electron is performed, yielding the value v_y'. Almost instantaneously, a measurement is made of the z-component, giving the value v_z'. What is the state of the system at the moment that these measurements are completed?

(d) Write down the state of the system at a later time $t > 0$. At what time is this state an eigenstate of v_z with the same eigenvalue v_z'? Also, at what time is this state an eigenstate of v_y with the same eigenvalue v_y'? At which times $t > 0$ is the system in an eigenstate of v_y with eigenvalue $-v_y'$?

Solution

(a) The velocity operator is

$$\mathbf{v} \equiv \frac{d}{dt}\mathbf{r} = \frac{i}{\hbar}[H, \mathbf{r}] = \frac{1}{\mu}\left[\mathbf{p} - \frac{e}{c}\mathbf{A}(\mathbf{r})\right]$$

The commutator of the velocity components is

$$[v_i, v_j] = -\frac{e}{\mu^2 c}[p_i, A_j(\mathbf{r})] + \frac{e}{\mu^2 c}[p_j, A_i(\mathbf{r})]$$

$$= \frac{ie\hbar}{\mu^2 c}\left(\partial_i A_j - \partial_j A_i\right) = \frac{ie\hbar}{\mu^2 c}\epsilon_{ijk}B_k = i\hbar\left(\frac{eB}{\mu^2 c}\right)\epsilon_{ijz}$$

Thus, we get

$$[v_x, v_z] = [v_y, v_z] = 0, \qquad [v_x, v_y] = i\hbar\left(\frac{eB}{\mu^2 c}\right)$$

(b) The Heisenberg equations of motion for the velocity operators are

$$\dot{\mathbf{v}} = \frac{i}{\hbar}\left[\frac{\mu}{2}(\mathbf{v})^2, \mathbf{v}\right]$$

that is,

$$\dot{v}_i = \frac{i\mu}{2\hbar}v_j[v_j, v_i] + \frac{i\mu}{2\hbar}[v_i, v_j]v_j = -\frac{eB}{c\mu}\epsilon_{jiz}v_j$$

or

$$\dot{v}_z = 0, \qquad \dot{v}_x = \frac{eB}{c\mu}v_y, \qquad \dot{v}_y = -\frac{eB}{c\mu}v_x$$

The first shows that the velocity component parallel to the magnetic field is a constant of the motion, i.e. $v_z(t) = v_z(0)$. The transverse components can be decoupled by differentiating the above equations once more. We can also introduce the *cyclotron frequency* $\omega \equiv |e|B/\mu c$. We have

$$\ddot{v}_x = -\omega^2 v_x, \qquad \ddot{v}_y = -\omega^2 v_y$$

with solution

$$v_x(t) = v_x(0)\cos\omega t - v_y(0)\sin\omega t$$
$$v_y(t) = v_y(0)\cos\omega t + v_x(0)\sin\omega t$$

(c) Since the velocity components v_y and v_z commute, they can be measured simultaneously. Thus their measurement at $t = 0$, yielding the eigenvalues v_y' and v_z', determines the initial state of the system to be

$$|\psi(0)\rangle = |v_y', v_z'\rangle$$

(d) Since no subsequent measurement is made, the evolved state at times $t > 0$ will be

$$|\psi(t)\rangle = e^{-iHt/\hbar} |v_y', v_z'\rangle$$

The fact that v_z is a constant of the motion, i.e. $\dot{v}_z = [H, v_z] = 0$, means that the evolved state will continue to be an eigenstate of this operator with the same eigenvalue at all times:

$$v_z|\psi(t)\rangle = e^{-iHt/\hbar} v_z|v_y', v_z'\rangle = e^{-iHt/\hbar} v_z'|v_y', v_z'\rangle = v_z'|\psi(t)\rangle$$

Acting on the evolved state with the operator v_y, we get

$$\begin{aligned}
v_y|\psi(t)\rangle &= v_y e^{-iHt/\hbar}|\psi(0)\rangle \\
&= e^{-iHt/\hbar} \left(e^{iHt/\hbar} v_y e^{-iHt/\hbar}\right) |\psi(0)\rangle = e^{-iHt/\hbar} v_y(t)|\psi(0)\rangle \\
&= e^{-iHt/\hbar} \left(v_y \cos\omega t + v_x \sin\omega t\right) |v_y', v_z'\rangle \\
&= v_y' \cos\omega t|\psi(t)\rangle + \sin\omega t e^{-iHt/\hbar} v_x|\psi(0)\rangle
\end{aligned}$$

It is clear that at the times

$$t_n = \left(\frac{2n\pi}{\omega}\right) \qquad (n = 1, 2, \ldots)$$

the second term disappears and the evolved state becomes an eigenstate of v_y with the original eigenvalue v_y':

$$v_y|\psi(t_n)\rangle = v_y'|\psi(t_n)\rangle$$

Note that it is a simultaneous eigenstate of v_z having the original eigenvalue v_z'. It is also clear that at the times

$$\bar{t}_n = \left(\frac{n\pi}{\omega}\right) \qquad (n = 1, 2, \ldots)$$

the system occupies a simultaneous eigenstate of v_y and v_z corresponding to eigenvalues $-v_y'$ and v_z'.

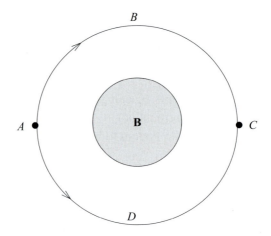

Fig. 35 Phase difference induced by the presence of magnetic flux.

Problem 7.23 Consider a uniform magnetic field that has a non-zero value $\mathbf{B} = \hat{\mathbf{z}}B$ inside an infinite cylinder parallel to its direction. A beam of particles of charge q and mass μ travelling in the plane perpendicular to the field is split into two beams that are directed towards the same point along different paths, as shown in Fig. 35. The size of the particle wave packet is much smaller than both the characteristic lengths of the system and the distances travelled, so that it makes sense to talk about a particle trajectory. Thus, the particles never travel through the region of non-zero magnetic field. Write down the vector potential in all space. Calculate the classical action for a particle travelling in the region of vanishing magnetic field. Show that there will be a phase difference between the two beams proportional to the magnetic flux.

Solution

A vector potential that gives the magnetic field $\hat{\mathbf{z}}B$ in the cylinder is

$$\mathbf{A}_< = \frac{B}{2}\left(-y\hat{\mathbf{x}} + x\hat{\mathbf{y}}\right) = \frac{B}{2}\rho\hat{\phi}$$

This is continuous at the surface of the cylinder ($\rho = R$) with the pure-gradient vector potential outside the cylinder,

$$\mathbf{A}_> = \frac{BR^2}{2}\nabla\phi = \frac{BR^2}{2}\frac{\hat{\phi}}{\rho}$$

The Lagrangian for a particle of charge q and mass μ moving in the presence of a vector potential $\mathbf{A}$ is

$$L = \frac{\mu}{2}v^2 + \frac{q}{c}\mathbf{A}\cdot\mathbf{v}$$

The classical action is

$$S[\mathbf{r};t] = \int_0^t dt' \left(\frac{\mu}{2}v^2 + \frac{q}{c}\mathbf{A}\cdot\mathbf{v}\right) = \frac{\mu}{2}\int_0^{\mathbf{r}} d\mathbf{r}\cdot\mathbf{v} + \frac{q}{c}\int_0^{\mathbf{r}} d\mathbf{r}\cdot\mathbf{A}$$

In the region of vanishing magnetic field, the Lorentz force also vanishes and the classical equation of motion gives constant $\mathbf{v}$. The vector potential term in the action, however, takes the form

$$\frac{qBR^2}{2c}\int^{\mathbf{r}} d\mathbf{r}\cdot\nabla\phi = \frac{qBR^2}{2c}\phi$$

An approximate expression for the probability amplitude for finding at the point C a particle that has followed path ABC is $\psi_{ABC} \propto e^{i\mathcal{S}_{ABC}/\hbar}$. Similarly, for the alternative path ADC we get $\psi_{ADC} \propto e^{i\mathcal{S}_{ADC}/\hbar}$. There will be a phase difference

$$\Delta\alpha = \frac{1}{\hbar}(\mathcal{S}_{ABC} - \mathcal{S}_{ADC}) = \frac{\mu}{2\hbar}\oint d\mathbf{r}\cdot\mathbf{v} + \frac{qBR^2}{2c\hbar}(\phi_{ABC} - \phi_{ADC})$$

$$= 0 + \left(\frac{qBR^2}{2c\hbar}\right)2\pi \quad\Longrightarrow\quad \Delta\alpha = \frac{q}{c\hbar}\Phi_B$$

where Φ_B is the magnetic flux.

Problem 7.24 A particle of charge q and mass μ is subject to a uniform magnetic field $\mathbf{B}$.

(a) Starting from the Hamiltonian

$$H = \frac{1}{2\mu}\left[\mathbf{p} - \frac{q}{c}\mathbf{A}(\mathbf{r})\right]^2$$

derive the velocity operator $\mathbf{v}$. Calculate the commutator $[v_i, v_j]$. Prove the operator identity

$$\left(\frac{i}{\hbar}\right)^2 [H, [H, \mathbf{v}_\perp]] = -\omega^2\mathbf{v}_\perp$$

where $\omega \equiv |q|B/\mu c$. If $|E_a\rangle$ and $|E_b\rangle$ are two eigenstates of the energy with eigenvalues E_a and E_b, show that the matrix element of the transverse velocity between these states vanishes unless the energy eigenvalues differ by $\hbar\omega$, namely

$$\langle E_a|v_\perp|E_b\rangle \neq 0 \quad\Longrightarrow\quad E_a - E_b = \pm\hbar\omega$$

(b) Show that the expectation value of the velocity component in the direction of the magnetic field $\langle v_\| \rangle$ is a constant of the motion. Show that the expectation value of the velocity in the transverse direction is given by

$$\langle \mathbf{v}_\perp\rangle_t = \langle \mathbf{v}_\perp\rangle_0 + \frac{q}{2\mu}\langle\mathbf{r}\rangle_t \times \mathbf{B} - \frac{q}{2\mu}\langle\mathbf{r}\rangle_0 \times \mathbf{B}$$

(c) Calculate the expectation value $\langle \mathbf{r}_\perp \rangle_t$ and show that its motion is a circle of radius $r_c^2 = (\langle \mathbf{v}_\perp \rangle_0)^2 / \omega^2$ around a point $\mathbf{R}_0 = \langle \mathbf{R} \rangle_0$, where

$$\mathbf{R} = \mathbf{r}_\perp + \frac{1}{\omega} \mathbf{v} \times \hat{\mathbf{B}}$$

Show also that $[H, \mathbf{R}] = 0$ and calculate the commutator $[R_i, R_j]$.

(d) Calculate the commutators

$$\left(\frac{i}{\hbar}\right)^2 [H, [H, (\mathbf{r}_\perp - \mathbf{R})]], \qquad \frac{i}{\hbar}[H, (\mathbf{r}_\perp - \mathbf{R})^2], \qquad \frac{i}{\hbar}[H, v_\perp^2]$$

Show that the matrix elements $\langle E_a | (\mathbf{r}_\perp - \mathbf{R}) | E_b \rangle$ between energy eigenstates obey the same selection rule as the matrix elements of the transverse velocity.

(e) Show that the uncertainty in the transverse velocity is independent of time, i.e.

$$\left\langle [\mathbf{v}_\perp - \langle \mathbf{v}_\perp \rangle_t]^2 \right\rangle_t = \left\langle [\mathbf{v}_\perp - \langle \mathbf{v}_\perp \rangle_0]^2 \right\rangle_0$$

or, equivalently,

$$\left\langle [\mathbf{v}_\perp - \langle \mathbf{v} \rangle_t]^2 \right\rangle_t = \left\langle [\mathbf{v}_\perp - \langle \mathbf{v} \rangle_0]^2 \right\rangle_0$$

Derive the corresponding relation for $\mathbf{r}_\perp - \mathbf{R}$.

(f) Derive the time evolution of the angular momentum $\mathbf{L}$. Show that the angular momentum component parallel to the magnetic field, $\mathbf{L}_{||}$, is conserved.

Solution

(a) As in the first part of problem 7.22, we have

$$\mathbf{v} = \frac{1}{\mu}\left[\mathbf{p} - \frac{q}{c}\mathbf{A}(\mathbf{r})\right] \qquad \Longrightarrow \qquad [v_i, v_j] = \frac{i\hbar q}{\mu^2 c}\epsilon_{ijk} B_k$$

It is straightforward to introduce velocity components parallel and transverse to the direction defined by the magnetic field:

$$\mathbf{v} = \mathbf{v}_{||} + \mathbf{v}_\perp$$

where

$$\mathbf{v}_{||} = \hat{\mathbf{B}}(\mathbf{v} \cdot \hat{\mathbf{B}})$$
$$\mathbf{v}_\perp = \hat{\mathbf{B}} \times (\mathbf{v} \times \hat{\mathbf{B}})$$

If we take $\hat{\mathbf{B}} = \hat{\mathbf{z}}$, this corresponds to

$$v_{||} = v_z, \qquad \mathbf{v}_\perp = -\hat{\mathbf{z}} \times (\hat{\mathbf{z}} \times \mathbf{v}) = v_x \hat{\mathbf{x}} + v_y \hat{\mathbf{y}}$$

In an exactly analogous way, we can decompose the position operator as

$$\mathbf{r} = \mathbf{r}_{||} + \mathbf{r}_\perp = \hat{\mathbf{B}}\left(\mathbf{r} \cdot \hat{\mathbf{B}}\right) + \hat{\mathbf{B}} \times \left(\mathbf{r} \times \hat{\mathbf{B}}\right)$$

Since the Hamiltonian is

$$H = \frac{\mu}{2}\,(\mathbf{v})^2 = \frac{\mu}{2}\,(\mathbf{v}_\perp)^2 + \frac{\mu}{2}\,(\mathbf{v}_\parallel)^2$$

and

$$[v_j,\, v_\parallel] = 0, \qquad [v_{\perp i},\, v_{\perp j}] = [v_i,\, v_j] = \frac{i\hbar q}{\mu^2 c}\,\epsilon_{ijk}\, B_k$$

we arrive at

$$[H,\, \mathbf{v}_\perp] = -\left(\frac{i\hbar q}{\mu c}\right)(\mathbf{v} \times \mathbf{B})$$

and

$$[H,\, [H,\, \mathbf{v}_\perp]] = \hbar^2 \omega^2\, \mathbf{v}_\perp$$

Rearranging, we obtain

$$\left(\frac{i}{\hbar}\right)^2 [H,\, [H,\, \mathbf{v}_\perp]] = -\omega^2\, \mathbf{v}_\perp$$

which is the required identity. Note that, starting from Heisenberg's equation of motion for the transverse velocity operator and differentiating it with respect to time, we arrive at the very simple equation of motion

$$\frac{d\mathbf{v}_\perp}{dt} = \frac{i}{\hbar}\,[H,\, \mathbf{v}_\perp] \qquad \Longrightarrow \qquad \frac{d^2\mathbf{v}_\perp}{dt^2} = -\omega^2 \mathbf{v}_\perp$$

which will be used below. Returning to the identity proved above and taking its matrix element between energy eigenstates, we get

$$\langle E_a |\,[H,\, [H,\, \mathbf{v}_\perp]]\,| E_b \rangle = (\hbar\omega)^2\, \langle E_a | \mathbf{v}_\perp | E_b \rangle$$

or

$$(E_a - E_b)^2\, \langle E_a | \mathbf{v}_\perp | E_b \rangle = (\hbar\omega)^2\, \langle E_a | \mathbf{v}_\perp | E_b \rangle$$

and finally

$$\left[(E_a - E_b)^2 - (\hbar\omega)^2\right] \langle E_a | \mathbf{v}_\perp | E_b \rangle = 0$$

This implies that the matrix element has to vanish unless

$$E_a - E_b = \pm\hbar\omega$$

(b) We proved in the solution to (a) that $[\mathbf{v}_j,\, \mathbf{v}_\parallel] = 0$. This implies that

$$[H,\, \mathbf{v}_\parallel] = 0$$

which is equivalent to stating that the parallel velocity component is a constant of the motion,

$$v_{||}(t) = v_{||}(0)$$

The parallel component of the position operator will be

$$r_{||}(t) = r_{||}(0) + v_{||}(0)\, t$$

From the Heisenberg equation for the transverse velocity,

$$\frac{d\mathbf{v}_\perp}{dt} = \frac{i}{\hbar}[H,\, \mathbf{v}_\perp]$$

we get

$$\frac{d\mathbf{v}_\perp}{dt} = \frac{q}{\mu c}\,\mathbf{v} \times \mathbf{B} = \frac{q}{\mu c}\frac{d}{dt}(\mathbf{r} \times \mathbf{B})$$

or

$$\mathbf{v}_\perp(t) + \frac{q}{\mu c}\mathbf{B} \times \mathbf{r}(t) = \mathbf{v}_\perp(0) + \frac{q}{\mu c}\mathbf{B} \times \mathbf{r}(0)$$

The corresponding relation between the expectation values is

$$\langle \mathbf{v}_\perp \rangle_t = \langle \mathbf{v}_\perp \rangle_0 + \frac{q}{c\mu}\,\langle \mathbf{r} \rangle_t \times \mathbf{B} - \frac{q}{c\mu}\,\langle \mathbf{r} \rangle_0 \times \mathbf{B}$$

We can actually do better than this and solve for the transverse velocity operator using the second-order differential equation for the latter obtained previously, namely

$$\frac{d\mathbf{v}_\perp}{dt} = \frac{i}{\hbar}[H,\, \mathbf{v}_\perp]$$

It is straightforward to deduce that

$$\mathbf{v}_\perp(t) = \mathbf{v}_\perp(0)\cos\omega t + \left[\mathbf{v}_\perp(0) \times \hat{\mathbf{B}}\right]\sin\omega t$$

and so

$$\langle \mathbf{v}_\perp \rangle_t = \langle \mathbf{v}_\perp \rangle_0 \cos\omega t + \left[\langle \mathbf{v} \rangle_0 \times \mathbf{B}\right]\sin\omega t$$

(c) Integrating the last two equations, we obtain

$$\mathbf{r}_\perp(t) = \mathbf{r}_\perp(0) + \frac{1}{\omega}\left\{\mathbf{v}(0) \times \hat{\mathbf{B}} + \mathbf{v}_\perp(0)\sin\omega t - \left[\mathbf{v}(0) \times \hat{\mathbf{B}}\right]\cos\omega t\right\}$$

and

$$\langle \mathbf{r}_\perp \rangle_t = \langle \mathbf{r}_\perp \rangle_0 + \frac{1}{\omega}\left[\langle \mathbf{v} \rangle_0 \times \hat{\mathbf{B}} + \langle \mathbf{v}_\perp \rangle_0 \sin\omega t - \left(\langle \mathbf{v} \rangle_0 \times \hat{\mathbf{B}}\right)\cos\omega t\right]$$

Introducing

$$\mathbf{R} \equiv \mathbf{r}_\perp + \frac{1}{\omega}\mathbf{v} \times \hat{\mathbf{B}}, \qquad \mathbf{R}_0 \equiv \langle \mathbf{R} \rangle_0, \qquad r_c^2 \equiv \omega^{-2} \langle \mathbf{v}_\perp \rangle_0^2$$

we can write

$$\langle \mathbf{r}_\perp \rangle_t - \mathbf{R}_0 = \frac{1}{\omega}\left[\langle \mathbf{v}_\perp \rangle_0 \sin \omega t - \left(\langle \mathbf{v} \rangle_0 \times \hat{\mathbf{B}}\right)\cos \omega t\right]$$

or

$$\left(\langle \mathbf{r}_\perp \rangle_t - \mathbf{R}_0\right)^2 = \frac{1}{\omega^2}\left[\langle \mathbf{v}_\perp \rangle_0^2 \sin^2 \omega t + \left(\langle \mathbf{v} \rangle_0 \times \hat{\mathbf{B}}\right)^2 \cos^2 \omega t\right] = r_c^2$$

Thus, $\langle \mathbf{r}_\perp \rangle_t$ describes a circle of radius r_c with its centre at $\mathbf{R}_0$. Since $\langle \mathbf{r}_{||} \rangle_t = \langle \mathbf{v}_{||} \rangle_0 t$, the particle moves in a helix in the direction of the magnetic field. It is straightforward to see that $\mathbf{R}$ is a constant of the motion:

$$[H, R_i] = \frac{1}{2\mu}\left(v_j[v_j, R_i] + [v_j, R_i]v_j\right) = 0$$

since

$$[v_j, R_i] = 0$$

Thus, there is a corresponding operator equation for the circle,

$$[\mathbf{r}_\perp(t) - \mathbf{R}]^2 = \frac{1}{\omega^2}[\mathbf{v}_\perp(0)]^2$$

It is interesting to see that the commutator of different components of $\mathbf{R}$ is non-zero: we have

$$[R_i, R_j] = -\frac{i\hbar}{\mu\omega}\epsilon_{ijk}\hat{\mathbf{B}}_k$$

(d) It is not difficult to obtain

$$\left(\frac{i}{\hbar}\right)^2 [H, [H, (\mathbf{r}_\perp - \mathbf{R})]] = \frac{i}{\hbar}\left[H, \frac{d\mathbf{r}_\perp}{dt}\right] = \frac{d^2\mathbf{r}_\perp}{dt^2}$$

$$= -\omega^2 (\mathbf{r}_\perp - \mathbf{R}_0)$$

$$\frac{i}{\hbar}\left[H, (\mathbf{r}_\perp - \mathbf{R})^2\right] = 0$$

$$\frac{i}{\hbar}\left[H, v_\perp^2\right] = -\frac{i}{\hbar}[H, v_{||}^2] = 0$$

From the first of these commutators we get

$$(E_a - E_b)^2 \langle E_a | (\mathbf{r}_\perp - \mathbf{R}_0) | E_b \rangle = (\hbar\omega)^2 \langle E_a | (\mathbf{r}_\perp - \mathbf{R}_0) | E_b \rangle$$

that is,

$$\left[(E_a - E_b)^2 - (\hbar\omega)^2\right]\langle E_a | (\mathbf{r}_\perp - \mathbf{R}_0) | E_b \rangle = 0$$

which clearly means that this matrix element vanishes unless the energy difference $E_a - E_b = \pm\hbar\omega$.

(e) We have

$$\left[\mathbf{v}_\perp(t) - \langle\mathbf{v}\rangle_t\right]^2 = \left[\mathbf{v}_\perp(t) - \langle\mathbf{v}_\perp\rangle_t\right]^2 + \langle\mathbf{v}_\|\rangle^2$$

Now substitute into the transverse term on the right-hand side the solved expression from (b) for the transverse velocity operator:

$$\left[\mathbf{v}_\perp(t) - \langle\mathbf{v}_\perp\rangle_t\right]^2 = \left[\mathbf{v}_\perp(0) - \langle\mathbf{v}_\perp\rangle_0\right]^2 \sin^2\omega t + \left\{\left[\mathbf{v}_\perp(0) - \langle\mathbf{v}_\perp\rangle_0\right] \times \hat{\mathbf{B}}\right\}^2 \cos^2\omega t$$
$$= \left[\mathbf{v}_\perp(0) - \langle\mathbf{v}_\perp\rangle_0\right]^2$$

Thus, we obtain

$$\left\langle(\mathbf{v}_\perp - \langle\mathbf{v}_\perp\rangle_t)^2\right\rangle_t = \left\langle(\mathbf{v}_\perp - \langle\mathbf{v}_\perp\rangle_0)^2\right\rangle_0$$

or, equivalently,

$$\left\langle(\mathbf{v}_\perp - \langle\mathbf{v}\rangle_t)^2\right\rangle_t = \left\langle(\mathbf{v}_\perp - \langle\mathbf{v}\rangle_0)^2\right\rangle_0$$

From the expressions for the transverse and parallel components of the position operator,

$$\mathbf{r}_\perp(t) = \mathbf{R} + \frac{1}{\omega}\mathbf{v}_\perp(0)\sin\omega t - \frac{1}{\omega}\mathbf{v}(0) \times \hat{\mathbf{B}}\cos\omega t$$
$$\mathbf{r}_\|(t) = \mathbf{r}_\|(0) + \mathbf{v}_\|(0)\,t$$

we can write

$$\left[\mathbf{r}_\perp(t) - \langle\mathbf{r}\rangle_t\right]^2 = \left[\mathbf{r}_\perp(t) - \langle\mathbf{r}_\perp\rangle_t\right]^2 + \langle\mathbf{r}_\|\rangle_t^2$$

The transverse term is

$$\left[\mathbf{r}_\perp(t) - \langle\mathbf{r}_\perp\rangle_t\right]^2$$
$$= \left(\mathbf{R} - \mathbf{R}_0 + \frac{1}{\omega}\left[\mathbf{v}_\perp(0) - \langle\mathbf{v}_\perp\rangle_0\right]\sin\omega t - \frac{1}{\omega}\left\{\left[\mathbf{v}_\perp(0) - \langle\mathbf{v}_\perp\rangle_0\right] \times \hat{\mathbf{B}}\right\}\cos\omega t\right)^2$$

or, remembering that we are considering an operator equation,

$$\left[\mathbf{r}_\perp(t) - \langle\mathbf{r}_\perp\rangle_t\right]^2 = (\mathbf{R} - \mathbf{R}_0)^2$$
$$+ \frac{1}{\omega^2}\left\{\left[\mathbf{v}_\perp(0) - \langle\mathbf{v}_\perp\rangle_0\right]\sin\omega t\right.$$
$$\left. - \left[\mathbf{v}_\perp(0) - \langle\mathbf{v}_\perp\rangle_0\right] \times \hat{\mathbf{B}}\cos\omega t\right\}^2$$
$$+ \frac{1}{\omega}(\mathbf{R} - \mathbf{R}_0) \cdot \left\{\left[\mathbf{v}_\perp(0) - \langle\mathbf{v}_\perp\rangle_0\right]\sin\omega t\right.$$
$$\left. - \left[\mathbf{v}_\perp(0) - \langle\mathbf{v}_\perp\rangle_0\right] \times \hat{\mathbf{B}}\cos\omega t\right\}$$
$$+ \frac{1}{\omega}\left\{\left[\mathbf{v}_\perp(0) - \langle\mathbf{v}_\perp\rangle_0\right]\sin\omega t\right.$$
$$\left. - \left[\mathbf{v}_\perp(0) - \langle\mathbf{v}_\perp\rangle_0\right] \times \hat{\mathbf{B}}\cos\omega t\right\} \cdot (\mathbf{R} - \mathbf{R}_0)$$

Rearranging,

$$\left[\mathbf{r}_\perp(t) - \langle\mathbf{r}_\perp\rangle_t\right]^2 = \frac{1}{\omega^2}\left[\mathbf{v}_\perp(0) - \langle\mathbf{v}_\perp\rangle_0\right]^2 + (\mathbf{R} - \mathbf{R}_0)^2$$

$$+ \frac{1}{\omega}\Big(\left[\mathbf{r}_\perp(0) - \langle\mathbf{r}_\perp\rangle_0\right] \cdot \left[\mathbf{v}_\perp(0) - \langle\mathbf{v}_\perp\rangle_0\right]\sin\omega t$$

$$+ \left[\mathbf{v}_\perp(0) - \langle\mathbf{v}_\perp\rangle_0\right] \cdot \left[\mathbf{r}_\perp(0) - \langle\mathbf{r}_\perp\rangle_0\right]\sin\omega t$$

$$+ \left[\mathbf{r}_\perp(0) - \langle\mathbf{r}_\perp\rangle_0\right] \cdot \left\{\left[\mathbf{v}_\perp(0) - \langle\mathbf{v}_\perp\rangle_0\right] \times \hat{\mathbf{B}}\right\}\cos\omega t$$

$$+ \left\{\left[\mathbf{v}_\perp(0) - \langle\mathbf{v}_\perp\rangle_0\right] \times \hat{\mathbf{B}}\right\} \cdot \left[\mathbf{r}_\perp(0) - \langle\mathbf{r}_\perp\rangle_0\right]\cos\omega t\Big)$$

The corresponding expectation value is

$$\left\langle\left[\mathbf{r}_\perp - \langle\mathbf{r}_\perp\rangle_t\right]^2\right\rangle_t = \frac{1}{\omega^2}\left\langle\left[\mathbf{v}_\perp(0) - \langle\mathbf{v}_\perp\rangle_0\right]^2\right\rangle_0 + \left\langle(\mathbf{R} - \mathbf{R}_0)^2\right\rangle$$

$$+ \frac{C}{\omega}\cos\omega t + \frac{D}{\omega}\sin\omega t$$

with

$$D \equiv \left\langle\left\{\left[\mathbf{r}_\perp - \langle\mathbf{r}_\perp\rangle_0\right] \cdot \left[\mathbf{v}_\perp - \langle\mathbf{v}_\perp\rangle_0\right] + \left[\mathbf{v}_\perp - \langle\mathbf{v}_\perp\rangle_0\right] \cdot \left[\mathbf{r}_\perp - \langle\mathbf{r}_\perp\rangle_0\right]\right\}\right\rangle_0$$

$$= \langle\mathbf{r}_\perp \cdot \mathbf{v}_\perp + \mathbf{v}_\perp \cdot \mathbf{r}_\perp\rangle_0 - 2\langle\mathbf{r}_\perp\rangle_0 \cdot \langle\mathbf{v}_\perp\rangle_0$$

and

$$C \equiv \left\langle\left\{\left[\mathbf{r}_\perp - \langle\mathbf{r}_\perp\rangle_0\right] \cdot \left\{\left[\mathbf{v}_\perp - \langle\mathbf{v}_\perp\rangle_0\right] \times \hat{\mathbf{B}}\right\}\right.\right.$$

$$\left.\left.+ \left\{\left[\mathbf{v}_\perp - \langle\mathbf{v}_\perp\rangle_0\right] \times \hat{\mathbf{B}}\right\} \cdot \left[\mathbf{r}_\perp - \langle\mathbf{r}_\perp\rangle_0\right]\right\}\right\rangle_0$$

$$= D - \hat{\mathbf{B}} \cdot \langle\mathbf{r}_\perp \times \mathbf{v}_\perp - \mathbf{v}_\perp \times \mathbf{r}_\perp\rangle_0 + 2\hat{\mathbf{B}} \cdot (\langle\mathbf{r}_\perp\rangle_0 \times \langle\mathbf{v}_\perp\rangle_0)$$

(f) Consider the angular momentum operator $\mathbf{L} = \mathbf{r} \times \mathbf{p}$ and its time derivative

$$\frac{d\mathbf{L}}{dt} = \mathbf{v} \times \mathbf{p} + \mathbf{r} \times \frac{d\mathbf{p}}{dt}$$

The time derivative of the momentum can be obtained from the corresponding Heisenberg equation,

$$\frac{dp_i}{dt} = \frac{i}{\hbar}[H, p_i] = \frac{i\mu}{2\hbar}\left(v_j[v_j, p_i] + [v_j, p_i]v_j\right) = \frac{q}{2c}\left(v_j\frac{\partial A_j}{\partial x_i} + \frac{\partial A_j}{\partial x_i}v_j\right)$$

We can choose the vector potential to be

$$\mathbf{A} = \frac{1}{2}\mathbf{B} \times \mathbf{r}$$

Then, we have[15]

$$\frac{d\mathbf{p}}{dt} = \frac{q}{2c}(\mathbf{v} \times \mathbf{B})$$

Going back to the angular momentum, we have

$$\frac{d\mathbf{L}}{dt} = \mu(\mathbf{v} \times \mathbf{v}) + \frac{q}{2c}\mathbf{v} \times \mathbf{A}(\mathbf{r}) + \frac{q}{2c}\mathbf{r} \times (\mathbf{v} \times \mathbf{B})$$

$$= i\left(\frac{\hbar q}{\mu c}\right)\mathbf{B} - \frac{q}{2c}[\mathbf{v} \times (\mathbf{r} \times \mathbf{B}) - \mathbf{r} \times (\mathbf{v} \times \mathbf{B})]$$

After some algebra, we get

$$[\mathbf{v} \times (\mathbf{r} \times \mathbf{B}) - \mathbf{r} \times (\mathbf{v} \times \mathbf{B})]_i = \frac{2i\hbar}{\mu}B_i + \epsilon_{ijk}B_j(\mathbf{r} \times \mathbf{v})_k$$

Thus, we finally obtain

$$\frac{d\mathbf{L}}{dt} = -\frac{q}{2c}\mathbf{B} \times (\mathbf{r} \times \mathbf{v})$$

It is clear that the parallel component $\hat{\mathbf{B}} \cdot \mathbf{L}$ is conserved.

Problem 7.25 An electron moves under the influence of a uniform magnetic field. The Hamiltonian of the system is

$$H = \frac{1}{2\mu}\left[\mathbf{p} - \frac{e}{c}\mathbf{A}(\mathbf{r})\right]^2 - \frac{e}{\mu c}\mathbf{S} \cdot \mathbf{B}$$

where $\mathbf{S}$ is the spin operator for the electron.

(a) Write down the velocity operator $\mathbf{v}$, derive the commutator of its components and show that the Hamiltonian can be put in the form

$$H = \frac{\mu}{2}(\boldsymbol{\sigma} \cdot \mathbf{v})^2$$

where $\boldsymbol{\sigma}$ represents the Pauli matrices. Show that the Hamiltonian is the sum of three mutually commuting terms

$$H = H_\perp + H_\parallel + H_S$$

depending on independent sets of variables.

---

[15] Similarly,

$$\frac{d\mathbf{v}}{dt} = \frac{q}{\mu c}(\mathbf{v} \times \mathbf{B})$$

(b) Consider the transverse Hamiltonian $H_\perp$ and show that it can be cast in the form of a one-dimensional harmonic oscillator,

$$H_1 = \frac{\varpi^2}{2\mu} + \frac{\mu\omega^2}{2} q^2$$

in terms of new variables q and ϖ that satisfy canonical position–momentum commutation relations. Write down the creation and annihilation operators a, $a^\dagger$ associated with these variables. Discuss the energy eigenvalues and eigenstates of the transverse Hamiltonian.

(c) Show that the angular momentum component along the direction of the magnetic field is a constant of the motion. Consider its common eigenstates with the transverse Hamiltonian and determine its eigenvalues. Construct the spatial eigenfunctions corresponding to these states.

(d) Show that the spin-dependent part of the Hamiltonian can be put in the form

$$H_S = \hbar\omega \left(b^\dagger b - \frac{1}{2} \right), \qquad b \equiv \frac{1}{\hbar} \left(S_y - i S_x \right)$$

Show that b and $b^\dagger$ satisfy $\{b^\dagger, b\} = 1$ and $b^2 = (b^\dagger)^2 = 0$. Show that the *number operator* $\overline{N} = b^\dagger b$ has eigenvalues 0 and 1. Consider the operator $Q \equiv \sqrt{\hbar\omega}\, ab^\dagger$, where a was introduced in part (b), and demonstrate the *anticommutation* relation

$$\{Q, Q^\dagger\} = H_\perp + H_S$$

(e) The energy eigenstates are eigenstates of the number operators $N = a^\dagger a$ and $\overline{N} = b^\dagger b$. Show that the pair of states $|n\ 0\rangle$ and $|n - 1, 1\rangle$ is degenerate, corresponding to energy $n\hbar\omega$. Similarly, the pair $|n\ 1\rangle$ and $|n + 1, 0\rangle$ corresponds to energy $(n + 1)\hbar\omega$. Relate this degeneracy to the commutator of the operator Q and the Hamiltonian.

Solution

(a) The velocity commutator is easily obtained to be

$$[v_i,\ v_j] = -\frac{e}{\mu^2 c}[p_i,\ A_j(\mathbf{r})] + \frac{e}{\mu^2 c}[p_j,\ A_i(\mathbf{r})] = \frac{ie\hbar}{\mu^2 c} \left(\partial_i A_j - \partial_j A_i \right)$$

$$= \frac{ie\hbar}{\mu^2 c}\epsilon_{ijk} B_k = i\hbar \left(\frac{eB}{\mu^2 c} \right) \epsilon_{ijz}$$

Starting from the expression to be derived, we have

$$(\boldsymbol{\sigma} \cdot \mathbf{v})^2 = \sigma_i \sigma_j v_i v_j = \frac{1}{2} \left(\{\sigma_i,\ \sigma_j\} + [\sigma_i,\ \sigma_j] \right) v_i v_j = \left(\delta_{ij} + i\epsilon_{ijk}\sigma_k \right) v_i v_j$$

$$= v^2 + \frac{i}{2}\epsilon_{ijk}\sigma_k \left[v_i,\ v_j \right]$$

$$= v^2 - \frac{q\hbar}{\mu^2 c}\boldsymbol{\sigma} \cdot \mathbf{B}$$

or

$$H = \frac{\mu}{2} (\boldsymbol{\sigma} \cdot \mathbf{v})^2 = \frac{\mu}{2} (\mathbf{v})^2 - \frac{q}{\mu c} (\mathbf{S} \cdot \mathbf{B})$$

Let us take the magnetic field $\mathbf{B} = \hat{\mathbf{z}}B$ and choose for a vector potential

$$\mathbf{A} = \tfrac{1}{2}\mathbf{B} \times \mathbf{r} = \tfrac{1}{2}B\left(-y\hat{\mathbf{x}} + x\hat{\mathbf{y}}\right)$$

We can introduce the frequency[16]

$$\omega \equiv \frac{eB}{\mu c}$$

Then, we obtain

$$v_x = \frac{p_x}{\mu} + \frac{\omega}{2}y, \qquad v_y = \frac{p_y}{\mu} - \frac{\omega}{2}x, \qquad v_z = \frac{p_z}{\mu}$$

Note that

$$[v_x,\ v_y] = i\frac{\hbar\omega}{\mu}, \qquad [v_x,\ v_z] = [v_y,\ v_z] = 0$$

The Hamiltonian is the sum of three commuting terms,

$$H = H_\perp + H_{||} + H_S$$

with

$$H_\perp = \frac{\mu}{2}\left(v_x^2 + v_y^2\right), \qquad H_{||} = \frac{p_z^2}{2\mu}, \qquad H_S = \omega S_z$$

(b) Note that the commutation relation satisfied by the transverse velocity components is of the position–momentum type, namely

$$\left(\frac{\mu}{\omega}\right)[v_x,\ v_y] = i\hbar$$

Thus, we can introduce the operators

$$q \equiv \omega^{-1} v_x, \qquad \varpi \equiv \mu\, v_y$$

which satisfy

$$[q,\ \varpi] = i\hbar$$

It is straightforward to see that

$$v_x^2 + v_y^2 = \omega^2 q^2 + \mu^{-2}\varpi^2$$

and so

$$H_\perp = \frac{\varpi^2}{2\mu} + \frac{\mu\omega^2}{2}q^2$$

---

[16] Since it is the combination eB that appears, for a negatively charged particle we can always take the magnetic field to point towards the negative z-axis.

We can introduce creation–annihilation operators by

$$a \equiv \sqrt{\frac{\mu\omega}{2\hbar}}\, q + \frac{i}{\sqrt{2\hbar\mu\omega}}\,\varpi = \sqrt{\frac{\mu}{2\hbar\omega}}\left(v_x + iv_y\right)$$

which satisfy the standard harmonic-oscillator commutation relation

$$[a,\, a^\dagger] = 1$$

In terms of them, the transverse Hamiltonian is

$$H_\perp = \hbar\omega\left(a^\dagger a + \tfrac{1}{2}\right)$$

with transverse energy eigenvalues

$$E_n^{(\perp)} = \hbar\omega\left(n + \tfrac{1}{2}\right) \qquad (n = 0, 1, \dots)$$

and transverse energy eigenstates generated by the multiple action of $a^\dagger$ on the vacuum $|\,0\,\rangle^{(\perp)}$, defined as a normalized state that is annihilated by a: $a|\,0\,\rangle^{(\perp)} = 0$. The transverse energy eigenstates are

$$|n \cdots\rangle = \frac{\left(a^\dagger\right)^n}{\sqrt{n!}}|\,0\,\rangle^{(\perp)}$$

where the dots signify any additional quantum numbers corresponding to observables that commute with the transverse Hamiltonian.

(c) Let us first calculate the commutators of L_z with the transverse velocity components. They are

$$[L_z,\, v_x] = i\hbar\mathbf{v}_y, \qquad [L_z,\, v_y] = -i\hbar v_x$$

These commutators simply state the fact that the transverse velocity operator behaves as a vector rotating under the action of L_z, the generator of rotations in the transverse plane. These rotations should conserve the square of the velocity operator. It is straightforward to check explicitly that

$$[L_z,\, v_x^2 + v_y^2] = [L_z,\, H_\perp] = [L_z,\, H] = 0$$

This means that the transverse Hamiltonian eigenstates are also eigenstates of L_z.

Using the velocity–angular momentum commutators derived above, it is easy to obtain also the commutators[17]

$$[L_z,\, a] = \hbar a, \qquad [L_z,\, a^\dagger] = -\hbar a^\dagger$$

[17]

$$a = \sqrt{\frac{\mu}{2\hbar\omega}}\left(v_x + iv_y\right)$$

as well as the commutators

$$\left[L_z,\, a^n\right] = n\hbar a^n, \qquad \left[L_z,\, \left(a^\dagger\right)^n\right] = -n\hbar \left(a^\dagger\right)^n$$

From the spatial representation of the angular momentum in terms of the polar coordinates ρ, ϕ it is clear that

$$\langle \rho\,\phi | \cdots \nu \rangle \propto e^{i\nu\phi} \qquad \Longrightarrow \qquad \nu = 0,\ \pm 1,\ \pm 2, \ldots$$

where ν specifies the angular momentum. Consider now the $n = 0$ eigenstate:

$$L_z | 0\,\nu \rangle = \hbar \nu | 0\,\nu \rangle$$

Acting on this state with one of the commutators derived above we have

$$\left[L_z,\, \left(a^\dagger\right)^n\right] | 0\,\nu \rangle = -n\hbar \left(a^\dagger\right)^n | 0\,\nu \rangle$$

giving

$$L_z | n\,\nu \rangle = \hbar(\nu - n) | n\,\nu \rangle$$

The spatial representation of the annihilation operator is

$$a = -i\sqrt{\frac{\hbar}{2\mu\omega}} \left[\frac{\partial}{\partial x} + i\frac{\partial}{\partial y} + \frac{\mu\omega}{2\hbar}(x + iy) \right]$$

This becomes in polar coordinates

$$a = -i\sqrt{\frac{\hbar}{2\mu\omega}}\, e^{i\phi} \left[\frac{\partial}{\partial \rho} + \frac{i}{\rho}\frac{\partial}{\partial \phi} + \left(\frac{\mu\omega}{2\hbar}\right)\rho \right]$$

We require that the ground-state wave functions are annihilated by a:

$$\left[\frac{\partial}{\partial \rho} + \frac{i}{\rho}\frac{\partial}{\partial \phi} + \left(\frac{\mu\omega}{2\hbar}\right)\rho \right] \Psi_{0,\nu}(\rho, \phi) = 0$$

Using the fact that

$$\Psi_{0\nu}(\rho, \phi) = e^{i\nu\phi}\Psi_{0\nu}(\rho, 0)$$

we get

$$\left[\frac{d}{d\rho} - \frac{\nu}{\rho} + \left(\frac{\mu\omega}{2\hbar}\right)\rho \right] \Psi_{0\nu}(\rho, 0) = 0$$

with solution

$$\Psi_{0\nu}(\rho, 0) \propto \rho^\nu e^{-\mu\omega\rho^2/4\hbar}$$

Thus, we have finally

$$\Psi_{0\nu}(\rho, \phi) = N_\nu e^{i\nu\phi} \rho^\nu e^{-\mu\omega\rho^2/4\hbar}$$

The normalization constant is calculated to be

$$N_\nu = (\pi\nu!)^{-1/2} \left(\frac{\mu\omega}{2\hbar}\right)^{(1+\nu)/2}$$

All eigenstates $|n, \nu\rangle$ of higher energy can now be generated by acting on the ground state with the creation operator:[18]

$$a^\dagger = i\sqrt{\frac{\hbar}{2\mu\omega}} e^{-i\phi} \left(-\frac{\partial}{\partial\rho} + \frac{i}{\rho}\frac{\partial}{\partial\rho} + \frac{\mu\omega}{2\hbar}\rho\right)$$

These states will be eigenstates with angular momentum $\hbar(\nu - n)$. Their wave functions are

$$\Psi_{n\nu}(\rho, \phi) = \frac{N_\nu i^n}{\sqrt{n!}} \left[e^{-i\phi}\left(-\frac{\partial}{\partial\rho} + \frac{i}{\rho}\frac{\partial}{\partial\phi} + \frac{\mu\omega}{2\hbar}\rho\right)\right]^n e^{i\nu\phi} \rho^\nu e^{-\mu\omega\rho^2/4\hbar}$$

This can be put in the equivalent form

$$\Psi_{n\nu}(\rho, \phi) = \frac{N_\nu i^n}{\sqrt{n!}} e^{i(\nu-n)\phi} \left(-\frac{\partial}{\partial\rho} - \frac{\nu - n + 1}{\rho} + \frac{\mu\omega}{2\hbar}\rho\right)$$

$$\times \cdots \times \left(-\frac{\partial}{\partial\rho} - \frac{\nu - 1}{\rho} + \frac{\mu\omega}{2\hbar}\rho\right)$$

$$\times \left(-\frac{\partial}{\partial\rho} - \frac{\nu}{\rho} + \frac{\mu\omega}{2\hbar}\rho\right) \rho^\nu e^{-\mu\omega\rho^2/4\hbar}$$

It is clear that near the origin

$$\Psi_{n,\nu}(\rho) \propto \rho^{\nu-n} e^{-\mu\omega\rho^2/4\hbar} + \cdots$$

A singularity at the origin is present unless

$$\nu \geq n$$

This implies that the allowed angular momentum quantum numbers are the positive integers $\ell = n, \, n+1, \, n+2, \ldots$.

[18]
$$\frac{(a^\dagger)^n}{\sqrt{n!}} |0\,\nu\rangle = |n\,\nu\rangle$$

(d) We can use the Pauli representation, i.e. spin algebra, to derive all these relations in a straightforward way. We have

$$b = \begin{pmatrix} 0 & -i \\ 0 & 0 \end{pmatrix}, \qquad b^\dagger = \begin{pmatrix} 0 & 0 \\ i & 0 \end{pmatrix} \qquad \Longrightarrow \qquad b^2 = \left(b^\dagger\right)^2 = 0$$

and

$$b b^\dagger = \begin{pmatrix} 1 & 0 \\ 0 & 0 \end{pmatrix}, \qquad b^\dagger b = \begin{pmatrix} 0 & 0 \\ 0 & 1 \end{pmatrix} \qquad \Longrightarrow \qquad \left\{b, b^\dagger\right\} = 1$$

From the relation

$$b^\dagger b = \begin{pmatrix} 0 & 0 \\ 0 & 1 \end{pmatrix} = \frac{1}{2}(1 - \sigma_3) = \frac{1}{2}\left(1 - \frac{2}{\hbar} S_z\right)$$

we obtain

$$H_S = -\omega S_z = \hbar\omega \left(b^\dagger b - \tfrac{1}{2}\right)$$

Note that the sum of the spin-dependent and the transverse Hamiltonian is

$$H_\perp + H_S = \hbar\omega \left(a^\dagger a + b^\dagger b\right)$$

Introducing the operator

$$Q \equiv \sqrt{\hbar\omega}\, a b^\dagger$$

we note first that its main ingredients, namely a and $b^\dagger$, commute with each other. It is, then, straightforward to show that

$$\left\{Q, Q^\dagger\right\} = \hbar\omega \left(ab^\dagger a^\dagger b + a^\dagger bab^\dagger\right) = \hbar\omega \left(aa^\dagger b^\dagger b + a^\dagger abb^\dagger\right)$$
$$= \hbar\omega \left(b^\dagger b + a^\dagger a \left(b^\dagger b + bb^\dagger\right)\right) = \hbar\omega \left(b^\dagger b + a^\dagger a\right) = H_\perp + H_S$$

(e) From the expression for the Hamiltonian

$$H_\perp + H_S = \omega(N + \overline{N})$$

it is clear that its eigenstates have eigenvalues $\hbar\omega(n + \overline{n})$. Thus, both states of each pair correspond to the same energy eigenvalue. These states are related through the action of the operator Q, namely

$$Q|n\, 0\rangle = \sqrt{\hbar\omega}\, |n - 1\, 1\rangle$$

For any two energy eigenstates related by Q, we have

$$Q|\ldots\rangle = |\ldots\rangle' \qquad \Longrightarrow \qquad HQ|\ldots\rangle = E'|\ldots\rangle'$$
$$\Longrightarrow \qquad [H, Q]|\ldots\rangle = (E' - E)|\ldots\rangle'$$

Thus, they are degenerate if

$$[H, \, Q] = 0$$

This is certainly true here.

Problem 7.26 A particle with electric charge q and mass μ moves under the influence of a uniform magnetic field $\mathbf{B} = \hat{\mathbf{z}}B$. The Hamiltonian is

$$H = \frac{1}{2\mu} \left(\mathbf{p} - \frac{q}{c} \mathbf{A} \right)^2$$

Consider the following three distinct choices of vector potential that lead to the given uniform magnetic field:

$$\mathbf{A}_1 = x B \hat{\mathbf{y}}, \qquad \mathbf{A}_2 = -y B \hat{\mathbf{x}}, \qquad \mathbf{A}_3 = \tfrac{1}{2} B \left(-y \hat{\mathbf{x}} + x \hat{\mathbf{y}} \right)$$

Find the energy eigenvalues and show that the energy eigenfunctions corresponding to the different cases are related by

$$\psi_2 = e^{-iq Bxy/\hbar c} \psi_1, \qquad \psi_3 = e^{-iq Bxy/2\hbar c} \psi_1$$

Solution

The Hamiltonians in all three cases have the familiar form

$$H = \frac{p_z^2}{2\mu} + H^{(\perp)},$$

where

$$H_1^{(\perp)} = \frac{p_x^2}{2\mu} + \frac{1}{2\mu} \left(p_y - \frac{q}{c} Bx \right)^2, \qquad H_2^{(\perp)} = \frac{1}{2\mu} \left(p_x + \frac{q}{c} By \right)^2 + \frac{p_y^2}{2\mu}$$

$$H_3^{(\perp)} = \frac{1}{2\mu} \left(p_x + \frac{q}{2c} By \right)^2 + \frac{1}{2\mu} \left(p_y - \frac{q}{2c} Bx \right)^2$$

In each case the transverse Hamiltonian can be put into a harmonic-oscillator form. Introducing $\varpi = p_x$ and $q = x - c p_y / q B$ in the first case, we have

$$H_1^{(\perp)} = \frac{\varpi^2}{2\mu} + \frac{\mu\omega^2}{2} q^2, \qquad [q, \, \varpi] = i\hbar$$

where $\omega = q B / \mu c$. In an analogous fashion, in the second case, introducing $\varpi' = p_y$ and $q' = y + c p_x / q B$, we get

$$H_2^{(\perp)} = \frac{\varpi'^2}{2\mu} + \frac{\mu\omega^2}{2} q'^2, \qquad [q', \, \varpi'] = i\hbar$$

Finally, introducing $\varpi'' = p_y - qBx/2c$ and $q'' = y/2 + cp_x/qB$ in the third case, we get

$$H_3^{(\perp)} = \frac{\varpi''^2}{2\mu} + \frac{\mu\omega^2}{2}q''^2, \qquad [\varpi'', q''] = i\hbar$$

Thus, in all three cases the eigenvalues are given by

$$E_{kn} = \frac{\hbar^2 k^2}{2\mu} + E_n^{(\perp)}, \qquad E_n^{(\perp)} = \hbar\omega\left(n + \tfrac{1}{2}\right) \qquad (n = 0, 1, \ldots)$$

where k is the wave number in the direction parallel to the magnetic field along which the particle propagates freely.

The Hamiltonians corresponding to the three vector potential choices can be related to each other through a *momentum-translation operator*

$$e^{-i\boldsymbol{\beta}\cdot\mathbf{x}/\hbar}\, \mathbf{p}\, e^{i\boldsymbol{\beta}\cdot\mathbf{x}/\hbar} = \mathbf{p} + \boldsymbol{\beta}$$

We have

$$H_1^{(\perp)} = \frac{p_x^2}{2\mu} + e^{iqBxy/\hbar c}\frac{p_y^2}{2\mu}e^{-iqBxy/\hbar c}$$

$$H_2^{(\perp)} = e^{-iqBxy/\hbar c}\frac{p_x^2}{2\mu}e^{iqBxy/\hbar c} + \frac{p_y^2}{2\mu}$$

Thus, introducing $\zeta = qBxy/\hbar c$, we may write

$$H_2^{(\perp)} = e^{-i\zeta}\, H_1^{(\perp)}\, e^{i\zeta}$$

In an analogous fashion we have

$$H_3^{(\perp)} = e^{-i\zeta/2}\frac{p_x^2}{2\mu}e^{i\zeta/2} + e^{i\zeta/2}\frac{p_y^2}{2\mu}e^{-i\zeta/2}$$

$$= e^{i\zeta/2}\left\{ e^{-i\zeta}\frac{p_x^2}{2\mu}e^{i\zeta} + \frac{p_y^2}{2\mu}\right\}e^{-i\zeta/2}$$

or

$$H_3^{(\perp)} = e^{i\zeta/2}\, H_2^{(\perp)}\, e^{-i\zeta/2} = e^{-i\zeta/2}\, H_1^{(\perp)}\, e^{i\zeta/2}$$

From the Schroedinger equations, since the eigenvalues are the same, we get

$$H_2^{(\perp)}\psi_2 = E^{(\perp)}\psi_2 \qquad \Longrightarrow \qquad e^{-i\zeta}\, H_1^{(\perp)}\, e^{i\zeta}\,\psi_2 = E^{(\perp)}\psi_2$$

or $\psi_2 = e^{-iqBxy/\hbar c}\,\psi_1$. Similarly, we get $\psi_3 = e^{-iqBxy/2\hbar c}\,\psi_1$. Note that the function $\Lambda = \tfrac{1}{2}Bxy$ appearing in the exponential is just the gauge function involved in the

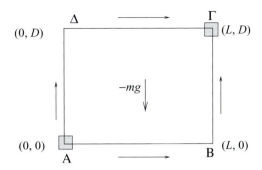

Fig. 36 Particle trajectories in the earth's gravitational field.

gauge transformation among the vector potential choices, namely $\mathbf{A}_1 = \mathbf{A}_3 + \nabla \Lambda$ and $\mathbf{A}_2 = \mathbf{A}_3 - \nabla \Lambda$.

Problem 7.27 Consider a particle of mass m moving in a uniform gravitational field $mg\hat{\mathbf{z}}$.

(a) Starting from the Heisenberg equations of motion, calculate the probability amplitude for finding the particle at a position $\mathbf{r}'$ at time t, if initially (at $t = 0$) the particle is at the position $\mathbf{r}$ (i.e. calculate the propagator).

(b) Consider a classical particle of the same mass moving from the origin to $\mathbf{r}$ in time t and calculate the *classical action* $S = \int_0^t dt'\, (mv^2/2 - mgz)$. Show that the propagator obtained in (a) is equal to the exponential of the classical action.

(c) Consider now an approximately monoenergetic beam of neutrons, which is split into two parts that travel along two different paths $AB\Gamma$ and $A\Delta\Gamma$ and finally meet; see Fig. 36. The size of the wave packet is much smaller than the length of the path travelled and so it is meaningful to speak of a particle trajectory. Calculate the phase difference in the beams induced by the fact that the neutrons travel through regions with different values of the gravitational potential. Consider the de Broglie wavelength of the neutrons given.

Solution

(a) From the Hamiltonian

$$H = \frac{p^2}{2m} + mgz,$$

we obtain

$$\frac{d\mathbf{r}}{dt} = \frac{\mathbf{p}}{m}, \qquad \frac{d\mathbf{p}}{dt} = -mg\hat{\mathbf{z}}$$

and thus

$$\mathbf{p} = -mgt\,\hat{\mathbf{z}} + \mathbf{p}(0), \qquad \mathbf{r} = \mathbf{r}(0) + \frac{\mathbf{p}(0)}{m}t - \tfrac{1}{2}gt^2\hat{\mathbf{z}}$$

The probability amplitude (propagator) for finding the particle at position $\mathbf{r}'$ at time t if initially it is at position $\mathbf{r}$ is

$$\mathcal{K}(\mathbf{r}', \mathbf{r}; t) = \langle \mathbf{r}' | e^{-iHt/\hbar} | \mathbf{r} \rangle$$

Multiplying the propagator by $\mathbf{r}'$, we get

$$\mathbf{r}' \, \mathcal{K}(\mathbf{r}', \mathbf{r}; t) = \langle \mathbf{r}' | \mathbf{r}(0) \, e^{-iHt/\hbar} | \mathbf{r} \rangle = \langle \mathbf{r}' | e^{-iHt/\hbar} \, \mathbf{r}(t) \, | \mathbf{r} \rangle$$

$$= \left\langle \mathbf{r}' \left| e^{-iHt/\hbar} \left[\mathbf{r}(0) + t \frac{\mathbf{p}(0)}{m} - \frac{gt^2}{2} \hat{\mathbf{z}} \right] \right| \mathbf{r} \right\rangle$$

$$= \langle \mathbf{r}' | e^{-iHt/\hbar} \left[\mathbf{r}(0) - \tfrac{1}{2} gt^2 \hat{\mathbf{z}} \right] | \mathbf{r} \rangle + \frac{t}{m} \langle \mathbf{r}' | e^{-iHt/\hbar} \mathbf{p}(0) | \mathbf{r} \rangle$$

From this we obtain[19]

$$\frac{i\hbar t}{m} \nabla \mathcal{K}(\mathbf{r}', \mathbf{r}; t) = \left(\mathbf{r}' - \mathbf{r} + \tfrac{1}{2} gt^2 \hat{\mathbf{z}} \right) \mathcal{K}(\mathbf{r}', \mathbf{r}; t)$$

Integrating as in problem 7.18, we arrive at the expression

$$\mathcal{K}(\mathbf{r}', \mathbf{r}; t) = \mathcal{K}_0(\mathbf{r}', \mathbf{r}; t) \exp\left[-i \frac{mg}{2\hbar}(z + z')t \right] \exp\left(-i \frac{mg^2}{24\hbar} t^3 \right)$$

where $\mathcal{K}_0$ is the free propagator.[20]

(b) A classical particle of mass m obeys the classical equation of motion

$$m\ddot{\mathbf{r}} = -mg\hat{\mathbf{z}} \qquad \Longrightarrow \qquad \mathbf{r} = \mathbf{v}(0)t - \tfrac{1}{2}gt^2\hat{\mathbf{z}}$$

assuming that it is at the origin at time $t = 0$. Its velocity is $\mathbf{v} = \mathbf{v}(0) - gt\hat{\mathbf{z}}$. The classical action will be the time integral of the Lagrangian:

$$\mathcal{S}_c[\mathbf{r}; t] = \int_0^t dt' \left\{ \tfrac{1}{2} m [\mathbf{v}(t)]^2 - mgz(t') \right\}$$

Substituting the classical trajectory obtained above and integrating, we get

$$\mathcal{S}_c[\mathbf{r}; t] = \frac{mr^2}{2t} - \frac{1}{2} mgtz - \frac{1}{24} mg^2 t^3$$

Thus, we can write

$$\mathcal{K}(\mathbf{r}; t) = \left(\frac{m}{2\pi i\hbar t} \right)^{3/2} \exp\left(\frac{i}{\hbar} \mathcal{S}_c[\mathbf{r}; t] \right)$$

[19]

$$\langle x | p = -i\hbar \frac{\partial}{\partial x} \langle x | \qquad \Longrightarrow \qquad p | x \rangle = +i\hbar \frac{\partial}{\partial x} | x \rangle$$

[20]

$$\mathcal{K}_0(\mathbf{r}', \mathbf{r}; t) = \left(\frac{m}{2\pi i\hbar t} \right)^{3/2} \exp\left[i \frac{m}{2\hbar t} (\mathbf{r}' - \mathbf{r})^2 \right]$$

This formula is exact. Note however that, even for a general Lagrangian, in the limit $\hbar \to 0$ we have approximately

$$\mathcal{K} \propto \exp\left(\frac{i}{\hbar}\mathcal{S}_c[x_c]\right)$$

(c) If the size of the neutron wave packet is much smaller than the size of the distance travelled, it makes sense to assume that the integral involved in the probability amplitude is dominated by one path. Then, we can use as an approximate formula the expression

$$\mathcal{K}_{AB\Gamma}(\mathbf{r};t) \propto \left(\frac{m}{2\pi i \hbar t}\right)^{3/2} \exp\left(\frac{i}{\hbar}\mathcal{S}_{AB\Gamma}\right)$$

Alternatively, we have

$$\mathcal{K}_{A\Delta\Gamma}(\mathbf{r};t) \propto \left(\frac{m}{2\pi i \hbar t}\right)^{3/2} \exp\left(\frac{i}{\hbar}\mathcal{S}_{A\Delta\Gamma}\right)$$

The classical action calculated for each path is

$$\mathcal{S}_{A\Delta} = \frac{m}{2t_1}D^2 - mgDt_1 - \frac{mg^2t_1^3}{24}$$

$$\mathcal{S}_{\Delta\Gamma} = \frac{m}{2(t-t_1)}L^2 - mgD(t-t_1)$$

$$\mathcal{S}_{AB} = \frac{m}{2(t-t_1)}L^2, \qquad \mathcal{S}_{B\Gamma} = \frac{m}{2t_1}D^2 - mDt_1 - \frac{mg^2}{24}t_1^3$$

Thus

$$\mathcal{S}_{A\Delta\Gamma} = \frac{m}{2}\left(\frac{D^2}{t_1} + \frac{L^2}{t-t_1}\right) - mgDt - \frac{mg^2}{24}t_1^3$$

$$\mathcal{S}_{AB\Gamma} = \frac{m}{2}\left(\frac{D^2}{t_1} + \frac{L^2}{t-t_1}\right) - mgDt_1 - \frac{mg^2}{24}t_1^3$$

It is clear that between the two amplitudes there is a phase difference

$$\Delta\phi = \frac{1}{\hbar}(\mathcal{S}_{A\Delta\Gamma} - \mathcal{S}_{AB\Gamma}) = -\frac{mg}{\hbar}D(t-t_1) = -2\pi\lambda g\left(\frac{m}{\hbar}\right)^2 LD$$

where we have replaced the horizontal travel time in terms of the de Broglie wavelength λ:

$$t - t_1 = \frac{L}{v} = L\frac{m\lambda}{h}$$

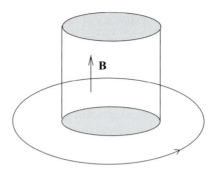

Fig. 37 Magnetic flux quantization. The outer ring represents a possible particle trajectory.

Problem 7.28 A particle of electric charge q and mass μ moves in a region of vanishing magnetic field but non-vanishing vector potential. Consider the simple case where a uniform magnetic field has a constant non-zero value $B\hat{z}$ in a cylindrical region of fixed radius R and infinite height inaccessible to the particle; see Fig. 37. What are the allowed values of B?

Solution

For vanishing magnetic field we have $\mathbf{A} = \nabla\Lambda(\mathbf{r})$. The only constraint on Λ is continuity at the border surface surrounding the region of non-vanishing magnetic field. The Schroedinger equation for the particle in the external region will be

$$\frac{1}{2\mu}\left(\mathbf{p} - \frac{q}{c}\nabla\Lambda\right)^2 \Psi(\mathbf{r}, t) = i\hbar\frac{\partial}{\partial t}\Psi(\mathbf{r}, t)$$

The wave function Ψ vanishes on the surface of the cylinder $\rho = R$. Note however that

$$\nabla - i\frac{q}{\hbar c}\nabla\Lambda = e^{iq\Lambda/\hbar c}\nabla e^{-iq\Lambda/\hbar c}$$

Thus, we may write

$$e^{iq\Lambda/\hbar c}\left(-\frac{\hbar^2}{2\mu}\nabla^2\right)e^{-iq\Lambda/\hbar c}\Psi(\mathbf{r}, t) = i\hbar\frac{\partial}{\partial t}\Psi(\mathbf{r}, t)$$

or

$$\Psi(\mathbf{r}, t) = e^{iq\Lambda/\hbar c}\Psi_0(\mathbf{r}, t)$$

where Ψ_0 satisfies the free Schroedinger equation. Both Ψ and Ψ_0 vanish at the cylinder boundary. We can write

$$\Psi(\mathbf{r}', t) = e^{iq\Lambda(\mathbf{r}')/\hbar c}\int d^3r\, \mathcal{K}_0(\mathbf{r}', \mathbf{r}; t)\Psi_0(\mathbf{r}, 0)$$

$$= e^{iq\Lambda(\mathbf{r}')/\hbar c}\int d^3r\, \mathcal{K}_0(\mathbf{r}', \mathbf{r}; t)e^{-iq\Lambda(\mathbf{r})/\hbar c}\Psi(\mathbf{r}, 0)$$

in terms of the free propagator $\mathcal{K}_0$. Note that $\mathcal{K}_0$ satisfies the appropriate boundary condition since it vanishes on the boundary. The full propagator is

$$\mathcal{K}(\mathbf{r}', \mathbf{r}; t) = \exp\left\{i\frac{q}{\hbar c}\left[\Lambda(\mathbf{r}') - \Lambda(\mathbf{r})\right]\right\}\mathcal{K}_0(\mathbf{r}', \mathbf{r}; t)$$

The appropriate continuous choice of vector potential is

$$\mathbf{A}_< = \frac{B}{2}(-y\hat{\mathbf{x}} + x\hat{\mathbf{y}}) = \frac{B\rho}{2}\hat{\phi}$$

$$\mathbf{A}_> = \frac{BR^2}{2\rho^2}(-y\hat{\mathbf{x}} + x\hat{\mathbf{y}}) = \frac{BR^2}{2\rho}\hat{\phi} = \frac{BR^2}{2}\nabla\phi$$

where ϕ and ρ are the standard cylindrical coordinates. Thus,

$$\Lambda(\phi) = \frac{BR^2}{2}\phi$$

Since

$$\mathcal{K}_0(\mathbf{r}', \mathbf{r}; t) = \mathcal{K}_0(|\mathbf{r}' - \mathbf{r}|; t)$$

with

$$|\mathbf{r}' - \mathbf{r}| = \sqrt{(z - z')^2 + \rho'^2 + \rho^2 - 2\rho\rho'\cos(\phi - \phi')}$$

the propagator takes the form

$$\mathcal{K} = \exp\left[i\frac{qBR^2}{2\hbar c}(\phi' - \phi)\right]\mathcal{K}_0(|\mathbf{r}' - \mathbf{r}|; t)$$

Single-valuedness implies that we require

$$\frac{|q|BR^2}{2\hbar c} = 2\pi n \qquad (n = 1, 2, \ldots)$$

or

$$B = \frac{4\pi n\hbar c}{|q|R^2}$$

8

Many-particle systems

Problem 8.1 Consider a pair of free identical particles of mass m. For simplicity, suppose that they are moving in one dimension and neglect their spin variables. Each particle is described in terms of a real wave function, well-localized around points $+a$ and $-a$ respectively; see Fig. 38. For definiteness, take $\psi_\pm(x) = (\beta/\pi)^{1/4} \exp[-\frac{\beta}{2}(x \mp a)^2]$. A well-localized state corresponds to $\beta \gg 1/a^2$. Write down the wave function of the system and calculate the expectation value of the energy. Show that if the two particles are fermions then there is an *effective repulsion* between them. Compare with the case of two identical bosons.

   Solution   The given one-particle wave functions $\psi_\pm(x)$ are normalized. The two-particle wave function is

$$\psi(x_1, x_2) = N[\psi_+(x_1)\psi_-(x_2) \pm \psi_-(x_1)\psi_+(x_2)]$$

The symmetric case (plus sign) corresponds to bosons and the antisymmetric case (minus sign) to fermions. The normalization constant N is, up to a phase, given by

$$N = \frac{1}{\sqrt{2}}\left(1 \pm I^2\right)^{-1/2} = \frac{1}{\sqrt{2}}\left(1 \pm e^{-2\beta a^2}\right)^{-1/2}$$

where I is the *overlap function*,

$$I = \int_{-\infty}^{\infty} dx\, \psi_+(x)\psi_-(x) = e^{-\beta a^2}$$

Note that a well-localized particle is characterized by a small value of $(\Delta x)^2$. In order for the system to consist of two relatively well-localized particles, we should have

$$(\Delta x)_\pm^2 \ll a^2 \quad \Longrightarrow \quad \frac{1}{2\beta} \ll a^2 \quad \Longrightarrow \quad 2\beta a^2 \gg 1$$

$$-a \qquad a$$

Fig. 38 Wave functions of two-particle system.

The Hamiltonian of the system of the two non-interacting identical particles is just

$$H = \frac{p_1^2}{2m} + \frac{p_2^2}{2m}$$

The expectation value of the energy in the above state will be

$$\mathcal{E} = 4|N|^2(E \pm \Gamma I)$$

where E is the energy expectation value in either of the localized one-particle states,

$$E = -\frac{\hbar^2}{2m} \int_{-\infty}^{\infty} dx \, \psi_\pm(x)\psi_\pm''(x) = \frac{\hbar^2 \beta}{4m}$$

and Γ is the mixed matrix element

$$\Gamma = -\frac{\hbar^2}{2m} \int_{-\infty}^{\infty} dx \, \psi_\pm(x)\psi_\mp''(x) = \frac{\hbar^2 \beta}{4m}(1 - 2\beta a^2)e^{-\beta a^2}$$

The final expression is

$$\mathcal{E} = \left(\frac{\hbar^2 \beta}{2m}\right) \frac{1 \pm e^{-2\beta a^2}(1 - 2\beta a^2)}{1 \pm e^{-2\beta a^2}}$$

A variation in the distance of the two particles will result in a change in energy of the system. This defines an *effective force* between the two particles. If the two particles are fermions, we have

$$\mathcal{F} \equiv -\frac{\partial \mathcal{E}}{\partial a} = \frac{2\hbar^2 \beta^2 a}{m} e^{-2\beta a^2} \frac{-1 + 2\beta a^2 + e^{-2\beta a^2}}{\left(1 - e^{-2\beta a^2}\right)^2}$$

This is always positive, since $e^{-2\beta a^2} + 2\beta a^2 - 1 > 0$. Thus, we always have an *effective repulsion* between the two fermions. This is a reflection of the antisymmetry of the two-particle wave function. In the case of two bosons, we have

$$\mathcal{F} = \frac{2\hbar^2 \beta^2 a}{m} e^{-2\beta a^2} \frac{e^{-2\beta a^2} + 1 - 2a^2\beta}{\left(1 + e^{-2\beta a^2}\right)^2}$$

For well-separated bosons ($2\beta a^2 \gg 1$), this force is attractive. It is repulsive only at very short separations $a \le 0.8\beta^{-1/2}$.

Problem 8.2 Consider two particles, each with orbital angular momentum quantum numbers $\ell = 1$, $m_\ell = 0$. What are the possible values of the total orbital angular momentum? What is the probability[1] that a measurement will find each of these values? Consider the case where the two particles are spin-1/2 fermions. Neglect their interaction and assume that they both have the same radial wave function. What are the total spin and the total angular momentum of the system?

Solution

Going from the eigenstates of L_1^2, L_{1z}, L_2^2, L_{2z} to eigenstates of the total orbital angular momentum, we obtain

$$|\ldots; \ell_1 = 1, 0; \ell_2 = 1, 0\rangle = \sum_\ell \langle \ell\, 0|\ell_1 = 1, 0; \ell_2 = 1, 0\rangle \, |\ldots; \ell\, 0\rangle$$

$$= \langle \ell = 0, 0|\ell_1 = 1, 0; \ell_2 = 1, 0\rangle \, |\ldots; \ell = 0, 0\rangle$$
$$+ \langle \ell = 2, 0|\ell_1 = 1, 0; \ell_2 = 1, 0\rangle \, |\ldots; \ell = 2, 0\rangle$$

The probability of finding vanishing total orbital angular momentum is

$$|\langle 1\, 0;\, 1\, 0|0\, 0\rangle|^2 = \left|\frac{-1}{\sqrt{3}}\right|^2 = \frac{1}{3}$$

This immediately implies that the probability of finding total angular momentum quantum number $\ell = 2$ will be $1 - \frac{1}{3} = \frac{2}{3}$.

In the case where the two particles are spin-1/2 fermions, their total eigenfunction must be antisymmetric. Since they have the same radial wave function, their spatial wave function

$$|n_1 = n, \ell = 1, m_1 = 0; n_2 = n, \ell_2 = 1, m_2 = 0\rangle$$

will be symmetric or, equivalently, $|n_1 = n; n_2 = n; \ell, m = 0\rangle$ with $\ell = 0, 2$ will be symmetric. Thus, the spinor wave function

$$|s_1 = \tfrac{1}{2}, m_{s_1}; s_2 = \tfrac{1}{2}, m_{s_2}\rangle$$

must be antisymmetric. This implies that the total spin vanishes ($s = m_s = 0$), since only this choice (singlet) corresponds to an antisymmetric combination:

$$|s = 0\, m_s = 0\rangle = \tfrac{1}{\sqrt{2}} \left(|\uparrow\rangle_1|\downarrow\rangle_2 - |\downarrow\rangle_1|\uparrow\rangle_2\right)$$

The total angular momentum will have values equal to the total orbital angular momentum: thus $j = 0, 2$ and $m_j = 0$. The probabilities of these values are $\frac{1}{3}$ and $\frac{2}{3}$ respectively.

---

[1] You may use the following Clebsch–Gordan coefficient:

$$\langle \ell_1, m_1 = \ell_1 - 1, \ell_2, m_2 = \ell_2 - 1|\ell_1, \ell_2; \ell = \ell_1 + \ell_2 - 2, m = m_1 + m_2\rangle$$
$$= -\left[\frac{(2\ell_1 - 1)(2\ell_2 - 1)}{(\ell_1 + \ell_2 - 1)(2\ell_1 + 2\ell_2 - 1)}\right]^{1/2}$$

Problem 8.3 Consider a pair of electrons constrained to move in one dimension in a total spin $S = 1$ state. The electrons interact through an attractive potential

$$V(x_1, x_2) = \begin{cases} 0, & |x_1 - x_2| > a \\ -V_0, & |x_1 - x_2| \le a \end{cases}$$

Find the lowest-energy eigenvalue in the case where the total momentum vanishes.

Solution

Since the spin state of the electrons is the triplet[2] $|S = 1, M_S\rangle$ which is symmetric in the interchange of the two electrons, their spatial wave function has to be antisymmetric. In terms of the *centre-of-mass variables*

$$X = \tfrac{1}{2}(x_1 + x_2), \qquad P = p_1 + p_2$$
$$x = x_1 - x_2, \qquad p = \tfrac{1}{2}(p_1 - p_2)$$

the Hamiltonian becomes

$$H = \frac{P^2}{4m} + \frac{p^2}{m} + V(x)$$

with eigenfunctions

$$\Psi(x_1, x_2) = \frac{1}{\sqrt{2\pi}} e^{iKX} \psi_\epsilon(x)$$

Here $\psi_\epsilon(x)$ satisfies the Schroedinger equation

$$\psi_\epsilon''(x) = \frac{m}{\hbar^2}[V(x) - \epsilon]\psi(x)$$

The energy eigenvalues are

$$E = \frac{\hbar^2 K^2}{4m} + \epsilon$$

Interchange between the two electrons corresponds to

$$x_1 \to x_2 \qquad \Longrightarrow \qquad X \to X, \quad x \to -x$$

Thus, an antisymmetric spatial wave function corresponds to an odd $\psi_\epsilon(x)$:

$$\Psi(x_1, x_2) = -\Psi(x_2, x_1) \qquad \Longrightarrow \qquad \psi_\epsilon(-x) = -\psi_\epsilon(x)$$

---

[2] The triplet states are

$$|1\,1\rangle = |\uparrow\rangle_1|\uparrow\rangle_2, \qquad |1\,-1\rangle = |\downarrow\rangle_1|\downarrow\rangle_2$$
$$|1\,0\rangle = \tfrac{1}{\sqrt{2}}(|\uparrow\rangle_1|\downarrow\rangle_2 + |\downarrow\rangle_1|\uparrow\rangle_2)$$

Vanishing total momentum corresponds to $K = 0$. Thus, the lowest-energy eigenstate Ψ_ϵ will be the state

$$\Psi_\epsilon(x_1, x_2) = \frac{1}{\sqrt{2\pi}} \psi_\epsilon(x)$$

where $\psi_\epsilon(x)$ is the odd eigenfunction with the lowest-energy eigenvalue. From the corresponding Schroedinger equation, we obtain

$$\psi_\epsilon(x) = \begin{cases} -A\, e^{\kappa x}, & x < -a \\ B \sin qx, & -a < x < a \\ A\, e^{-\kappa x}, & x > a \end{cases}$$

where

$$E = -\frac{\hbar^2 \kappa^2}{m} < 0, \qquad E + V_0 = \frac{\hbar^2 q^2}{m}$$

From the continuity of the above wave function, we obtain the energy eigenvalue condition

$$\tan aq = -\frac{q}{\kappa}$$

This can be written in terms of $\xi = qa$ and $\beta^2 = mV_0 a^2 / \hbar^2$ as

$$\tan \xi = -\frac{\xi}{\sqrt{\beta^2 - \xi^2}}$$

It has a solution only if

$$\beta^2 = \frac{mV_0 a^2}{\hbar^2} > \frac{\pi^2}{4}$$

Problem 8.4 Consider N identical particles. Assume that their interactions can be neglected and that the Hamiltonian of the system is the sum of N identical one-particle Hamiltonians with known eigenvalues ϵ_i:

$$H = \sum_{a=1}^{N} H_a, \qquad H_a |i\rangle_a = \epsilon_i |i\rangle_a$$

(a) What is the energy of the ground state if these particles are spin-0 bosons? What if they are spin-1/2 fermions?[3]
(b) Consider the case of three such particles and write down the corresponding ground-state wave functions.

---

[3] H_a is spin-independent.

Solution

(a) In the ground state of N identical bosons each boson is in the lowest-energy single-particle state. If ϵ_1 is the energy eigenvalue corresponding to it, the ground-state energy of the system will be

$$E = N\epsilon_1$$

For $N = 2M$ identical fermions, the particles will occupy its one-particle energy eigenstate in pairs of opposite spins. Thus, the ground-state energy will be

$$E = 2(\epsilon_1 + \epsilon_2 + \cdots + \epsilon_M)$$

For $N = 2M + 1$ identical fermions, the situation will be the same with the exception of one unpaired fermion, which will occupy the highest-energy single-particle state. Thus

$$E = 2(\epsilon_1 + \epsilon_2 + \cdots + \epsilon_M) + \epsilon_{M+1}$$

(b) The wave function of three identical bosons has to be totally symmetrized with respect to interchanges of the particles. Since the Hamiltonian is a sum of commuting terms, the energy eigenfunctions will be products of single-particle eigenfunctions. If $\psi_1(x)$ is the eigenfunction corresponding to the lowest-energy eigenvalue ϵ_1, the ground-state wave function will be

$$\Psi(1, 2, 3) = \psi_1(1)\psi_1(2)\psi_1(3) \qquad \Longrightarrow \qquad E = 3\epsilon_1$$

Three identical fermions must have an antisymmetric wave function. The ground state corresponds to a pair of these particles occupying the state ψ_1 of lowest eigenvalue ϵ_1 with opposite spins and the third occupying the state ψ_2 of eigenvalue ϵ_2:

$$\begin{aligned}
\Psi(1, 2, 3) = \frac{1}{3!} \Big\{ &\left[\psi_{1,\uparrow}(1)\psi_{1,\downarrow}(2) - \psi_{1,\downarrow}(1)\psi_{1,\uparrow}(2)\right] \psi_{2,\uparrow}(3) \\
&- \left[\psi_{1,\uparrow}(1)\psi_{1,\downarrow}(3) - \psi_{1,\downarrow}(1)\psi_{1,\uparrow}(3)\right] \psi_{2,\uparrow}(2) \\
&+ \psi_{2,\uparrow}(1)\left[\psi_{1,\uparrow}(2)\psi_{1,\downarrow}(3) - \psi_{1,\downarrow}(2)\psi_{1,\uparrow}(3)\right] \Big\}
\end{aligned}$$

The corresponding energy is

$$E = 2\epsilon_1 + \epsilon_2$$

Problem 8.5 The Hamiltonian of the helium atom[4] can be written as

$$H = H_0 + H_{12}$$

---

[4] We neglect spin–orbit forces, hyperfine interactions, nuclear motion etc.

where H_0 is the sum of two parts H_i, each corresponding to the Hamiltonian of a single electron interacting separately with the nucleus:

$$H_i = \frac{p_i^2}{2m} - \frac{2e^2}{r_i}$$

The term

$$H_{12} = \frac{e^2}{|\mathbf{r}_1 - \mathbf{r}_2|}$$

is the electrostatic repulsion between the two electrons.

(a) What is the ground state of the helium atom if we neglect the electron repulsion? What is the first excited state? Consider now the repulsion term H_{12} as a perturbation and calculate the ground-state energy correction.

(b) Calculate the expectation value of the magnetic dipole moment of the helium atom in the ground state found in (a). Consider the helium atom in the presence of a uniform magnetic field $\mathbf{B} = B\hat{z}$. What is the correction to the energy of the approximate ground state found in (a)?

(c) Consider the correction to the degenerate 1s2s eigenstates of H_0 due to electron repulsion and calculate the splitting.

(d) Consider a helium atom that at time $t = 0$ has one electron in a 1s hydrogen-like state with spin up and an electron in a 2s state with spin down. Write down the state of the atom at a later time $t > 0$. Is it possible that at some time T the spins are reversed? Treating the electron repulsion as a perturbation, calculate this time and comment on the validity of such a calculation.

Solution

(a) The eigenfunctions of H_0 are products of hydrogen-like wave functions with a_0 replaced by $a_0/2$ owing to the change in atomic number of the nucleus. The energy eigenvalues of H_0 will be sums of the energy eigenvalues corresponding to these states. The wave function has to be antisymmetric in the exchange of the two electrons. The lowest energy arises for the spatial wave function

$$\psi_{11}(\mathbf{r}_1, \mathbf{r}_2) = \psi_{100}(r_1)\psi_{100}(r_2)$$

and is

$$E_{11} = -\frac{Z^2 m e^4}{2\hbar^2} - \frac{Z^2 m e^4}{2\hbar^2} = 8E_1$$

where $E_1 = -me^4/2\hbar^2$ is the energy of the ground state of the hydrogen atom. This state is necessarily symmetric. Thus, the spin wave function will have to be antisymmetric. Among the two-spinor wave functions only the *singlet*

$$\chi_{00} \rightarrow |0\,0\rangle = \frac{1}{\sqrt{2}}(|\uparrow\rangle_1|\downarrow\rangle_2 - |\downarrow\rangle_1|\uparrow\rangle_2)$$

is antisymmetric. Thus, the ground state of the helium atom (neglecting the electron repulsion) is

$$\Psi_{1s1s;00}(\mathbf{r}_1, \mathbf{r}_2) = \psi_{100}(r_1)\psi_{100}(r_2)\chi_{00}$$

The next energy level of H_0 corresponds to the combination of a 1s hydrogen-like electronic state with a 2s or a 2p state. Then

$$E_{12} = -\frac{Z^2 m e^4}{2\hbar^2} - \frac{Z^2 m e^4}{8\hbar^2} = 5E_1$$

The spatial wave function will be one of the combinations

$$\psi_{12}^{(\pm)}(\mathbf{r}_1, \mathbf{r}_2) = \frac{1}{\sqrt{2}}\left[\psi_{100}(\mathbf{r}_1)\,\psi_{2\ell m_\ell}(\mathbf{r}_2) \pm \psi_{2\ell m_\ell}(\mathbf{r}_1)\,\psi_{100}(\mathbf{r}_2)\right]$$

These have to be combined with a suitable spin wave function in such a way that the total resulting wave function is antisymmetric. There are two combinations:

$$\Psi_{12;1m_s}^{(-)}(\mathbf{r}_1, \mathbf{r}_2) = \psi_{12}^{(-)}(\mathbf{r}_1, \mathbf{r}_2)\chi_{1m_s}$$
$$\Psi_{12;00}^{(+)}(\mathbf{r}_1, \mathbf{r}_2) = \psi_{12}^{(+)}(\mathbf{r}_1, \mathbf{r}_2)\chi_{00}$$

χ_{1m_s} stands for the symmetric *triplet* spin wave functions

$$|1\,1\rangle = |\uparrow\rangle_1|\uparrow\rangle_2, \qquad |1\,-1\rangle = |\downarrow\rangle_1|\downarrow\rangle_2$$
$$|1\,0\rangle = \frac{1}{\sqrt{2}}\left(|\uparrow\rangle_1|\downarrow\rangle_2 + |\downarrow\rangle_1|\uparrow\rangle_2\right)$$

The expectation value of the electron repulsion term in the ground state is

$$(\Delta E_{12})_{11} = e^2 \int d^3r_1 \int d^3r_2\, \psi_{100}^2(r_1)\frac{1}{|\mathbf{r}_1 - \mathbf{r}_2|}\psi_{100}^2(r_2)$$
$$= \frac{64e^2}{\pi^2 a_0^6}\int_0^\infty dr_1\, r_1^2\, e^{-4r_1/a_0} \int_0^\infty dr_2\, r_2^2\, e^{-4r_2/a_0}\, J_{12}$$

where[5]

$$J_{12} = \int d\Omega_2 \int d\Omega_1 \frac{1}{|\mathbf{r}_1 - \mathbf{r}_2|} = 4\pi \int d\Omega_1 \frac{1}{\sqrt{r_1^2 + r_2^2 - 2r_1r_2\cos\theta}}$$
$$= \frac{8\pi^2}{r_1 r_2}(r_1 + r_2 - |r_1 - r_2|)$$

---

[5] For the first angular integration, say Ω_1, we take the z_1-axis to coincide with $\hat{\mathbf{r}}_2$. Then the remaining angular integration is trivial, giving just 4π.

Substituting back into the radial integral, we obtain eventually[6]

$$(\Delta E_{12})_{11} = \frac{5e^2}{4a_0} = \frac{5}{2}|E_1|$$

(b) The magnetic dipole moment operator is

$$\boldsymbol{\mu} \equiv \frac{e}{2mc}(\mathbf{L}_1 + \mathbf{L}_2 + 2\mathbf{S}_1 + 2\mathbf{S}_2) = \frac{e}{2mc}(\mathbf{L} + 2\mathbf{S})$$

Since

$$(\mathbf{L}_1 + \mathbf{L}_2)\,\psi_{100}(r_1)\psi_{100}(r_2) = 0, \qquad \mathbf{S}|0\,0\rangle = 0$$

the magnetic dipole moment of the above ground state vanishes:

$$\langle 1\,1; 0\,0|\boldsymbol{\mu}|1\,1; 0\,0\rangle = 0$$

The fact that not only does the expectation value of the magnetic dipole moment $\boldsymbol{\mu}$ in the ground state vanish but also this state itself gives zero when $\boldsymbol{\mu}$ acts on it implies that higher-order corrections to the energy involving it will vanish as well, since these corrections depend on $\langle \cdots |\boldsymbol{\mu}|1\,1; 0\,0\rangle$.

In the presence of a uniform magnetic field $\mathbf{B} = B\hat{\mathbf{z}}$ the Hamiltonian H_0 becomes

$$\overline{H}_0 = \frac{1}{2m}\left[\mathbf{p}_1 - \frac{e}{c}\mathbf{A}(\mathbf{r}_1)\right]^2 + \frac{1}{2m}\left[\mathbf{p}_2 - \frac{e}{c}\mathbf{A}(\mathbf{r}_2)\right]^2$$
$$- \frac{2e^2}{r_1} - \frac{2e^2}{r_2} - \frac{e}{mc}\mathbf{B}\cdot(\mathbf{S}_1 + \mathbf{S}_2)$$

where for the vector potential we can choose, up to a gauge transformation,

$$\mathbf{A} = \frac{1}{2}\mathbf{B}\times\mathbf{r} = \frac{B}{2}(-y\hat{\mathbf{x}} + x\hat{\mathbf{y}})$$

The Hamiltonian can be written as

$$\overline{H}_0 = H_0 - \frac{eB}{2mc}(L_z + 2S_z) + \frac{e^2B^2}{8mc^2}\left(r_1^2 + r_2^2 - z_1^2 - z_2^2\right)$$

The two terms due to the magnetic field, treated as a perturbation, will give a correction to the ground-state energy equal to their expectation value in that state, namely

$$(\Delta E)_B = -\mathbf{B}\cdot\langle\boldsymbol{\mu}\rangle + \frac{e^2B^2}{8mc^2}\sum_{i=1,2}\langle(r_i^2 - z_i^2)\rangle$$

6

$$(\Delta E_{12})_{11} = \frac{512e^2}{a_0}\int_0^\infty dx\,xe^{-4x}\int_0^\infty dy\,ye^{-4y}(x + y - |x - y|) = \frac{5e^2}{4a_0}$$

The first term, being proportional to the expectation value of the magnetic dipole moment operator, vanishes. Each of the other two terms equals

$$\frac{e^2 B^2}{8mc^2} \int d^3r \, |\psi_{100}(r)|^2 \left(r^2 - z^2\right)$$

$$= \left(\frac{e^2 B^2}{8mc^2}\right) \left(\frac{8}{\pi a_0^3}\right) \int_0^\infty dr \, r^4 \, e^{-4r/a_0} \int_0^{2\pi} d\phi \int_{-1}^1 d\cos\theta \, \sin^2\theta = \frac{e^2 B^2 a_0^2}{16mc^2}$$

The energy correction can be written as $(\Delta E)_B = -\beta B^2/2$ with $\beta \equiv -e^2 a_0^2/4mc^2$. The parameter β represents the *magnetic susceptibility* of the helium atom. The fact that $\beta < 0$ classifies the He atom as *diamagnetic*.

(c) Among the degenerate states corresponding to the first excited level of H_0 we have the pair

$$\Psi_{12}^{(-)} = \tfrac{1}{\sqrt{2}} \left[\psi_{100}(\mathbf{r}_1)\,\psi_{200}(\mathbf{r}_2) - \psi_{200}(\mathbf{r}_1)\,\psi_{100}(\mathbf{r}_2)\right] \chi_{10} = \psi_{12}^{(-)}\chi_{10}$$

$$\Psi_{12}^{(+)} = \tfrac{1}{\sqrt{2}} \left[\psi_{100}(\mathbf{r}_1)\,\psi_{200}(\mathbf{r}_2) + \psi_{200}(\mathbf{r}_1)\,\psi_{100}(\mathbf{r}_2)\right] \chi_{00} = \psi_{12}^{(+)}\chi_{00}$$

Treating the electron repulsion as a perturbation, we shall have, to first order, the following corrected energy eigenvalues

$$E_{12}^{(\pm)} = E_{12} + \Delta_{12}^{(\pm)}$$

with

$$\Delta_{12}^{(\pm)} \equiv \left\langle \Psi_{12}^{(\pm)}\middle| H_{12}\middle| \Psi_{12}^{(\pm)}\right\rangle = \left\langle \psi_{12}^{(\pm)}\middle| H_{12}\middle| \psi_{12}^{(\pm)}\right\rangle$$

that is,

$$\Delta_{12}^{(\pm)} = \int d^3r_1 \int d^3r_2 \, \frac{e^2}{|\mathbf{r}_1 - \mathbf{r}_2|} \left[\psi_1^2(r_1)\psi_2^2(r_2) \pm \psi_1(r_1)\psi_2(r_1)\psi_1(r_2)\psi_2(r_2)\right]$$

The splitting will be

$$\Delta = \Delta_{12}^{(+)} - \Delta_{12}^{(-)} = 2 \int d^3r_1 \int d^3r_2 \, \frac{e^2}{|\mathbf{r}_1 - \mathbf{r}_2|} \psi_1(r_1)\psi_2(r_1)\psi_1(r_2)\psi_2(r_2)$$

or

$$\Delta = 256 \frac{e^2}{a_0} \int_0^\infty dx \, x(1-x)e^{-3x}(J_1 + x J_2)$$

with

$$J_1 = \int_0^x dy \, y^2(1-y)e^{-3y} = \frac{x^3}{3}e^{-3x}$$

$$J_2 = \int_x^\infty dy \, y(1-y)e^{-3y} = \left(\frac{1}{27} + \frac{x}{9} - \frac{x^3}{3}\right)e^{-3x}$$

Thus, finally

$$\Delta = \left(\frac{512}{1944}\right)\frac{e^2}{2a_0} = -0.264\,E_1$$

(d) The initial state $\Psi(\mathbf{r}_1, \mathbf{r}_2; 0)$ will be the antisymmetric combination

$$\frac{1}{\sqrt{2}}\left[\psi_{100}(r_1)\psi_{200}(\mathbf{r}_2)|\uparrow\rangle_1|\downarrow\rangle_2 - \psi_{200}(\mathbf{r}_1)\psi_{100}(r_2)|\downarrow\rangle_1|\uparrow\rangle_2\right]$$

It can be written in terms of the eigenstates of H_0 encountered in (a), namely

$$\Psi(\mathbf{r}_1, \mathbf{r}_2; 0) = \frac{1}{\sqrt{2}}\left[\psi_{12}^{(-)}(\mathbf{r}_1, \mathbf{r}_2)\,\chi_{10} + \psi_{12}^{(+)}(\mathbf{r}_1, \mathbf{r}_2)\,\chi_{00}\right]$$

The time-evolved state of the atom will be $|\Psi(t)\rangle = e^{-itH/\hbar}|\Psi(0)\rangle$, where H is the full Hamiltonian. Assume now that we find the system at some time T in a state $\Psi(\mathbf{r}_1, \mathbf{r}_2; T)$ with reflected spins,

$$\frac{1}{\sqrt{2}}\left[\psi_{100}(r_1)\psi_{200}(\mathbf{r}_2)|\downarrow\rangle_1|\uparrow\rangle_2 - \psi_{200}(\mathbf{r}_1)\psi_{100}(r_2)|\uparrow\rangle_1|\downarrow\rangle_2\right]$$
$$= \frac{1}{\sqrt{2}}\left[\psi_{12}^{(-)}(\mathbf{r}_1, \mathbf{r}_2)\,\chi_{10} - \psi_{12}^{(+)}(\mathbf{r}_1, \mathbf{r}_2)\,\chi_{00}\right]$$

This means that, up to a phase γ, we have

$$\frac{1}{\sqrt{2}}\left(|\psi_{12}^{(-)}\rangle\,\chi_{10} - |\psi_{12}^{(+)}\rangle\,\chi_{00}\right) = e^{i\gamma}\,e^{-iHT/\hbar}\,\frac{1}{\sqrt{2}}\left(|\psi_{12}^{(-)}\rangle\,\chi_{10} + |\psi_{12}^{(+)}\rangle\,\chi_{00}\right)$$

$$\langle\psi_{12}^{(-)}|e^{-iTH/\hbar}|\psi_{12}^{(-)}\rangle = e^{-i\gamma}, \qquad \langle\psi_{12}^{(+)}|e^{-iTH/\hbar}|\psi_{12}^{(+)}\rangle = -e^{-i\gamma}$$

Treating H_{12} as a perturbation, we can replace it in each exponent by its expectation value in the unperturbed state. This would be legitimate within first-order perturbation theory. Thus, we get

$$\langle\psi_{12}^{(-)}|e^{-iTH_0/\hbar}\,e^{-iT\Delta_{12}^{(-)}/\hbar}|\psi_{12}^{(-)}\rangle = e^{-i\gamma}$$
$$\langle\psi_{12}^{(+)}|e^{-iTH_0/\hbar}\,e^{-iT\Delta_{12}^{(+)}/\hbar}|\psi_{12}^{(+)}\rangle = -e^{-i\gamma}$$

or

$$\exp\left[-\frac{i}{\hbar}\left(E_{12} + \Delta_{12}^{(-)}\right)T\right] = -\exp\left[-\frac{i}{\hbar}\left(E_{12} + \Delta_{12}^{(+)}\right)T\right] = e^{i\gamma}$$
$$\implies \quad T = \frac{\hbar\pi}{\left|\Delta_{12}^{(-)} - \Delta_{12}^{(+)}\right|}$$

Since $T = \hbar\pi/\Delta$ is inversely proportional to the perturbation, it should not be trusted as a quantitative estimate.

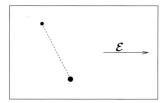

Fig. 39 Charged particles in a uniform electric field.

Problem 8.6 Two ions having equal mass m and electric charges q_1 and q_2 interact through harmonic forces described by the potential

$$V(\mathbf{r}_1, \mathbf{r}_2) = \frac{m\omega^2}{2}(\mathbf{r}_1 - \mathbf{r}_2)^2$$

The system is subject to a uniform electric field $\mathcal{E}$; see Fig. 39. If the system is initially (at $t = 0$) in a state described by a real wave function that is symmetric in the interchange of the two ions, find the expectation value of the total electric dipole moment $\langle \mathbf{D} \rangle_t$ in terms of its initial value. What is the electric polarizability of the system?

Solution

The Hamiltonian of the system is

$$H = \frac{p_1^2}{2m} + \frac{p_2^2}{2m} + \frac{m\omega^2}{2}(\mathbf{r}_1 - \mathbf{r}_2)^2 - q_1\,\mathcal{E}\cdot\mathbf{r}_1 - q_2\,\mathcal{E}\cdot\mathbf{r}_2$$

In terms of the centre-of-mass variables

$$\mathbf{R} = \frac{1}{2}(\mathbf{r}_1 + \mathbf{r}_2), \qquad \mathbf{r} = \mathbf{r}_1 - \mathbf{r}_2$$

$$\mathbf{P} = \mathbf{p}_1 + \mathbf{p}_2, \qquad \mathbf{p} = \frac{1}{2}(\mathbf{p}_1 - \mathbf{p}_2)$$

the Hamiltonian becomes

$$H = \frac{P^2}{4m} - (q_1 + q_2)\,\mathcal{E}\cdot\mathbf{R} + \frac{p^2}{m} + \frac{m\omega^2}{2}r^2 - \frac{1}{2}(q_1 - q_2)\,\mathcal{E}\cdot\mathbf{r}$$

The total electric dipole moment is

$$\mathbf{D} = \mathbf{D}_{\mathrm{CM}} + \mathbf{d} \equiv (q_1 + q_2)\mathbf{R} + \frac{1}{2}(q_1 - q_2)\mathbf{r}$$

The Hamiltonian is a sum of two commuting parts, the centre-of-mass part, describing a particle of mass $2m$ and charge $q_1 + q_2$ under the influence of the electric field, and a *relative* part, describing a particle of mass $m/2$ and charge $(q_1 - q_2)/2$ that, apart from its interaction with the electric field, self-interacts through a harmonic force.

The Heisenberg equations of motion for the centre-of-mass variables are

$$\frac{d\mathbf{R}}{dt} = \frac{\mathbf{P}}{2m}, \qquad \frac{d\mathbf{P}}{dt} = (q_1 + q_2)\mathbf{R}$$

or

$$\mathbf{P}(t) = \mathbf{P}(0) + (q_1 + q_2)\,\mathcal{E}t$$

$$\mathbf{R}(t) = \mathbf{R}(0) + \frac{t}{2m}\mathbf{P}(0) + \frac{t^2}{4m}(q_1 + q_2)\,\mathcal{E}$$

Then we get

$$\langle \mathbf{D}_{\mathrm{CM}} \rangle_t = \langle \mathbf{D}_{\mathrm{CM}} \rangle_0 + \frac{t}{2m}(q_1 + q_2)\langle \mathbf{P} \rangle_0 + \frac{t^2}{4m}(q_1 + q_2)^2\,\mathcal{E}$$

The corresponding equations for the relative variables are

$$\frac{d\mathbf{r}}{dt} = \frac{2\mathbf{p}}{m}, \qquad \frac{d\mathbf{p}}{dt} = -m\omega^2\mathbf{r} + \frac{1}{2}(q_1 - q_2)\,\mathcal{E}$$

These give the second-order equation

$$\frac{d^2\mathbf{r}}{dt^2} = -2\omega^2\mathbf{r} + \frac{q_1 - q_2}{m}\,\mathcal{E}$$

with solution

$$\mathbf{r}(t) = \left[\mathbf{r}(0) - \frac{q_1 - q_2}{2m\omega^2}\mathcal{E} \right] \cos \omega\sqrt{2}t + \frac{\sqrt{2}}{m\omega}\mathbf{p}(0) \sin \omega\sqrt{2}t + \frac{q_1 - q_2}{2m\omega^2}\mathcal{E}$$

From this we get

$$\langle \mathbf{d} \rangle_t = \langle \mathbf{d} \rangle_0 \cos \omega\sqrt{2}t + \frac{q_1 - q_2}{\sqrt{2}m\omega}\langle \mathbf{p} \rangle_0 \sin \omega\sqrt{2}t$$

$$+ \frac{(q_1 - q_2)^2}{2m\omega^2}\,\mathcal{E} \sin^2\frac{\omega t}{\sqrt{2}}$$

If the initial wave function is real, we have for the initial expectation values of $\mathbf{p}_1$ and $\mathbf{p}_2$

$$\langle \mathbf{p}_{1,2} \rangle_0 = -i\hbar \int d^3r_1 \int d^3r_2\, \psi(\mathbf{r}_1, \mathbf{r}_2)\nabla_{1,2}\psi(\mathbf{r}_1, \mathbf{r}_2)$$

$$= -\frac{i\hbar}{2} \int d^3r_1 \int d^3r_2\, \nabla_{1,2}\,[\psi(\mathbf{r}_1, \mathbf{r}_2)]^2 = 0$$

Since $\psi(\mathbf{r}_1, \mathbf{r}_2) = \psi(\mathbf{r}_2, \mathbf{r}_1)$ we have

$$\langle \mathbf{d} \rangle_0 \propto \int d^3r_1 \int d^3r_2\, \psi^2(\mathbf{r}_1, \mathbf{r}_2)(\mathbf{r}_1 - \mathbf{r}_2) = 0$$

Thus

$$\langle \mathbf{D}_{CM} \rangle_t = \langle \mathbf{D}_{CM} \rangle_0 + \frac{t^2}{4m}(q_1 + q_2)^2 \, \mathcal{E}$$

and

$$\langle \mathbf{d} \rangle_t = \frac{(q_1 - q_2)^2}{2m\omega^2} \, \mathcal{E} \sin^2 \frac{\omega t}{\sqrt{2}}$$

The expectation value of the total electric dipole moment will be

$$\langle \mathbf{D} \rangle_t = \langle \mathbf{D} \rangle_0 + \frac{t^2}{4m}(q_1 + q_2)^2 \, \mathcal{E} + \frac{(q_1 - q_2)^2}{2m\omega^2} \, \mathcal{E} \sin^2 \frac{\omega t}{\sqrt{2}}$$

The terms proportional to the electric field correspond to an *induced electric dipole moment*.

The Hamiltonian of the system can be written as

$$H = \frac{P^2}{4m} - (q_1 + q_2)\,\mathcal{E} \cdot \mathbf{R} + \frac{p^2}{m} + \frac{m\omega^2}{2}\left(\mathbf{r} - \frac{q_1 - q_2}{2m\omega^2}\,\mathcal{E}\right)^2 - \frac{(q_1 - q_2)^2}{8m\omega^2}\,\mathcal{E}^2$$

Thus, the energy eigenvalues of the relative part of the system will be the eigenvalues of a shifted oscillator,

$$E_n = \hbar\omega\sqrt{2}\left(n + \frac{3}{2}\right) - \frac{(q_1 - q_2)^2}{8m\omega^2}\,\mathcal{E}^2$$

For any normalizable state, the energy expectation value will have an electric-field-dependent part that is quadratic in the electric field:

$$\Delta E = -\frac{\alpha}{2}\mathcal{E}^2 \quad \Longrightarrow \quad \alpha = \frac{(q_1 - q_2)^2}{4m\omega^2}$$

The parameter α is the *electric polarizability* of the system.

Problem 8.7 Consider a system of N particles with a Hamiltonian that contains *one-body* and *two-body* interaction terms,[7] namely

$$H = \sum_{i=1}^{N} \frac{p_i^2}{2m_i} + \sum_{i=1}^{N} V_i(\mathbf{r}_i) + \frac{1}{2}\sum_{i,j}^{N} V_{ij}(\mathbf{r}_i, \mathbf{r}_j)$$

The probability current density for particle i is defined as

$$\boldsymbol{\mathcal{J}}_i(\mathbf{r}_1, \mathbf{r}_2, \ldots, \mathbf{r}_N; t) = \frac{\hbar}{2m_i i}\left(\Psi^* \nabla_i \Psi - \Psi \nabla_i \Psi\right)$$

where $\Psi(\mathbf{r}_1, \mathbf{r}_2, \ldots, \mathbf{r}_N; t)$ is the wave function of the system.

---

[7] V_{ij} is symmetric.

(a) Show that in the case where only isotropic two-body forces $V_{ij}\left(|\mathbf{r}_i - \mathbf{r}_j|\right)$ are present, the total momentum and total angular momentum are conserved.

(b) Show that the probability density $\rho(\mathbf{r}_1, \mathbf{r}_2, \ldots, \mathbf{r}_N; t) = |\Psi|^2$ satisfies the continuity equation

$$\frac{\partial \rho}{\partial t} + \sum_{j=1}^{N} (\nabla_j \cdot \boldsymbol{\mathcal{J}}_j) = 0$$

(c) Assume that the N particles are identical fermions. How does the probability density $\rho(\mathbf{r}_1, \mathbf{r}_2, \ldots, \mathbf{r}_N)$ behave under interchanges of the particles? Show that the quantity

$$\rho(\mathbf{r}, t) = \left\langle \Psi(t) | \delta(\mathbf{r} - \mathbf{r}_j) | \Psi(t) \right\rangle$$

is independent of j and represents the probability density for finding one particle at a position $\mathbf{r}$ at time t.

(d) Introduce the *particle density operator*

$$\rho(\mathbf{r}) \equiv \frac{1}{N} \sum_{j=1}^{N} \delta(\mathbf{r} - \mathbf{r}_j)$$

Then, show that the probability density for finding a particle at the position $\mathbf{r}$ is given by

$$\rho(\mathbf{r}, t) = \left\langle \Psi(t) | \delta(\mathbf{r} - \mathbf{r}_j) | \Psi(t) \right\rangle$$

Show that we can introduce a *current density operator*

$$\mathbf{J}(\mathbf{r}) \equiv \frac{1}{2mN} \sum_{j=1}^{N} \left[\mathbf{p}_j \, \delta(\mathbf{r} - \mathbf{r}_j) + \delta(\mathbf{r} - \mathbf{r}_j) \, \mathbf{p}_j \right]$$

that satisfies

$$\boldsymbol{\mathcal{J}}(\mathbf{r}, t) = \left\langle \Psi(t) | \mathbf{J}(\mathbf{r}) | \Psi(t) \right\rangle$$

Verify the operator relation

$$\frac{i}{\hbar}[H, \rho(\mathbf{r})] = -\nabla \cdot \mathbf{J}(\mathbf{r})$$

and show that it is equivalent to the continuity equation.

Solution

(a) The equation of motion for the total momentum operator is

$$\dot{\mathbf{P}} = \sum_j \dot{\mathbf{p}}_j = \frac{i}{\hbar} \sum_j [H, \mathbf{p}_j] = -\frac{1}{2} \sum_j \nabla_j \sum_{k,\ell} V_{k\ell}(|\mathbf{r}_k - \mathbf{r}_\ell|)$$

$$= -\frac{1}{2} \sum_{jk} V'_{kj} \frac{\mathbf{r}_k - \mathbf{r}_j}{|\mathbf{r}_k - \mathbf{r}_j|} = 0$$

since V_{ij} and V'_{ij} are symmetric functions; here V'_{kj} abbreviates $V'_{kj}(|\mathbf{r}_k - \mathbf{r}_j|)$.

Similarly,

$$\dot{\mathbf{L}} = \sum_j \dot{\mathbf{L}}_j = \sum_j \mathbf{r}_j \times \dot{\mathbf{p}}_j = -\frac{1}{2} \sum_{jk} V'_{kj} \mathbf{r}_j \times \left(\frac{\mathbf{r}_k - \mathbf{r}_j}{|\mathbf{r}_k - \mathbf{r}_j|} \right) = 0$$

(b) It is straightforward to see that

$$\sum_j \nabla_j \cdot \boldsymbol{\mathcal{J}}_j = \frac{\hbar}{2mi} \sum_j \left[\Psi^* \nabla_j^2 \Psi - (\nabla_j^2 \Psi)^* \Psi \right] = \frac{i}{\hbar} \left[\Psi^* \mathbf{T} \Psi - (\mathbf{T} \Psi)^* \Psi \right]$$

$$= \frac{i}{\hbar} \left[\Psi^* \mathbf{H} \Psi - (\mathbf{H} \Psi)^* \Psi \right] = -\left[\Psi^* \dot{\Psi} + (\dot{\Psi})^* \Psi \right] = -\frac{\partial \rho}{\partial t}$$

(c) For identical fermions the wave function has to be antisymmetric and thus the probability density will be symmetric:

$$\Psi(\dots, \mathbf{r}_i, \dots, \mathbf{r}_j, \dots) = -\Psi(\dots, \mathbf{r}_j, \dots, \mathbf{r}_i, \dots)$$

and

$$\left| \Psi(\dots, \mathbf{r}_i, \dots, \mathbf{r}_j, \dots) \right|^2 = \left| \Psi(\dots, \mathbf{r}_j, \dots, \mathbf{r}_i, \dots) \right|^2$$

We have for the quantity $\rho(\mathbf{r}, t)$

$$\rho(\mathbf{r}, t) = \int d^3 r_1 \cdots \int d^3 r_N \, \rho(\dots, \mathbf{r}_j, \dots) \delta(\mathbf{r} - \mathbf{r}_j)$$

$$= \int d^3 r_1 \cdots \int d^3 r_{j-1} \int d^3 r_{j+1} \cdots \int d^3 r_N \, \rho(\dots, \mathbf{r}_{j-1}, \mathbf{r}, \mathbf{r}_{j+1}, \dots)$$

Since j simply denotes the position of the free variable $\mathbf{r}$ and $\rho(\mathbf{r}_1, \mathbf{r}_2, \dots)$ is symmetric, the right-hand side does not depend on j. It is clear that the interpretation of the quantity $\rho(\mathbf{r}, t)$ must be the probability of finding a particle at $\mathbf{r}$ independently of where the other $N - 1$ particles are.

(d) Starting from the particle density operator definition, we get

$$\langle \Psi(t) | \rho(\mathbf{r}) | \Psi(t) \rangle = \frac{1}{N} \sum_{j=1}^N \langle \Psi(t) | \delta(\mathbf{r} - \mathbf{r}_j) | \Psi(t) \rangle = \rho(\mathbf{r}, t)$$

The left-hand side of the operator relation to be verified gives

$$\frac{i}{\hbar N} \sum_{j=1}^N [H, \delta(\mathbf{r} - \mathbf{r}_j)] = \frac{i}{2m\hbar N} \sum_{j=1}^N [p_j^2, \delta(\mathbf{r} - \mathbf{r}_j)]$$

$$= \frac{1}{2mN} \sum_{j=1}^N \left\{ \mathbf{p}_j \cdot \nabla_j \delta(\mathbf{r} - \mathbf{r}_j) + [\nabla_j \delta(\mathbf{r} - \mathbf{r}_j)] \cdot \mathbf{p}_j \right\}$$

$$= -\frac{1}{2mN} \sum_{j=1}^N \left\{ \mathbf{p}_j \cdot \nabla \delta(\mathbf{r} - \mathbf{r}_j) + [\nabla \delta(\mathbf{r} - \mathbf{r}_j)] \cdot \mathbf{p}_j \right\}$$

$$= -\nabla \cdot \mathbf{J}$$

Taking the expectation value of this operator equation for the state $|\Psi(t)\rangle$ we get

$$\frac{i}{\hbar}\langle\Psi(t)|H\rho|\Psi(t)\rangle - \frac{i}{\hbar}\langle\Psi(t)|\rho H|\Psi(t)\rangle = -\nabla\cdot\mathcal{J}$$

or

$$\left[\frac{d}{dt}\langle\Psi(t)|\right]\rho|\Psi(t)\rangle + \langle\Psi(t)|\rho\frac{d}{dt}|\Psi(t)\rangle = \frac{d}{dt}\langle\Psi(t)|\rho|\Psi(t)$$

$$= \frac{\partial\rho}{\partial t} = -\nabla\cdot\mathcal{J}$$

Problem 8.8 Consider two identical bosons with zero spin subject to a potential $V(|\mathbf{r}_1 - \mathbf{r}_2|)$ that has at least one bound state. The system is in its ground state $|\psi_0\rangle$.

(a) Find the expectation values of the *electric dipole moment* $\mathbf{d}$, the *magnetic dipole moment* $\boldsymbol{\mu}$ and the *electric quadrupole moment* Q_{ij} in the ground state.

(b) Find the matrix elements

$$\langle\ell\, m_\ell;\, \ldots\, |\mu_z|0\, 0;\, \ldots\rangle, \qquad \langle\ell\, m_\ell;\, \ldots\, |d_z|0\, 0;\, \ldots\rangle$$

for any state with ℓ, $m_\ell \neq 0$.

Solution

(a) The two-identical-boson state has to be symmetric in the interchange of the two particles. Thus, the relative wave function will have to be even and so the ground state $|\Psi_0\rangle$ will correspond to an even relative wave function. The operator $\mathbf{d} = q\mathbf{r}$, being a vector operator of odd parity, will have a vanishing expectation value in the parity-even ground state:

$$\mathcal{P}\mathbf{d}\mathcal{P} = -\mathbf{d} \qquad \Longrightarrow \qquad \langle\Psi_0|\mathbf{d}|\Psi_0\rangle = 0$$

In addition, the lowest-energy bound state of the system must be an s-state, with vanishing angular momentum. Thus, since for a spin-0 boson $\boldsymbol{\mu} \propto \mathbf{L}$, we shall have

$$\mathbf{L}|\Psi_0\rangle = 0 \qquad \Longrightarrow \qquad \langle\Psi_0|\boldsymbol{\mu}|\Psi_0\rangle = 0$$

Since the ground state is an s-state, the expectation value of the electric quadrupole moment will be

$$\langle Q_{ij}\rangle \propto \langle x_i x_j - \tfrac{1}{3}r^2\rangle \qquad \Longrightarrow \qquad \langle\Psi_0|Q_{ij}|\Psi_0\rangle = \tfrac{1}{3}\delta_{ij}\langle\Psi_0|Q_{kk}|\Psi_0\rangle = 0$$

(b) We have for the first matrix element

$$\langle\ell\, m_\ell;\, \ldots\, |\mu_z|0\, 0;\, \ldots\rangle \propto \langle\ell\, m_\ell;\, \ldots\, |L_z|0\, 0;\, \ldots\rangle = 0$$

For the other,

$$\langle \ell\, m_\ell;\, \ldots |d_z|0\,0;\, \ldots\rangle = -\langle \ell\, m_\ell;\, \ldots |\mathcal{P}d_z\mathcal{P}|0\,0;\, \ldots\rangle$$
$$= -\varpi\,\langle \ell\, m_\ell;\, \ldots |d_z|0\,0;\, \ldots\rangle \qquad \Longrightarrow \qquad \varpi = -1$$

where ϖ is the parity of the left-hand state. Because a system of two identical bosons cannot be in an odd-parity (relative) state, since odd parity implies antisymmetry, this matrix element has to vanish also.

Problem 8.9 The electromagnetic interaction energy between two magnetic dipole moments is

$$\Delta H = -\frac{1}{4\pi}\left(\boldsymbol{\mu}_1 \cdot \boldsymbol{\mu}_2\right)\nabla^2\frac{1}{r} + \frac{1}{4\pi}\left(\boldsymbol{\mu}_1 \cdot \nabla\right)\left(\boldsymbol{\mu}_2 \cdot \nabla\right)\frac{1}{r}$$

This interaction between the electron and the proton gives rise to the *hyperfine splitting* of electronic energy levels. Find the expectation value of this term for an energy eigenstate of the unperturbed hydrogen atom $|n = 2,\, \ell = m_\ell = 1;\, \ldots\rangle$.

Solution

Neglecting all other perturbations, the Hamiltonian will be

$$H = H_0 + \frac{ge^2}{8\pi m_e m_p c^2}\left[\left(\mathbf{S}_e \cdot \mathbf{S}_p\right)\nabla^2\frac{1}{r} - \left(\mathbf{S}_e \cdot \nabla\right)\left(\mathbf{S}_p \cdot \nabla\right)\frac{1}{r}\right]$$
$$+ \frac{ge^2}{16\pi m_e m_p c^2}\left[\left(\mathbf{L}_e \cdot \mathbf{S}_p\right)\nabla^2\frac{1}{r} - \left(\mathbf{L}_e \cdot \nabla\right)\left(\mathbf{S}_p \cdot \nabla\right)\frac{1}{r}\right]$$

where H_0 is the standard unperturbed hydrogen-atom Hamiltonian. The orbital part vanishes, being proportional to

$$-4\pi\, S_{\mathrm{p}i}\epsilon_{ijk}x_j\nabla_k\delta(\mathbf{r}) - S_{\mathrm{p}\ell}\epsilon_{ijk}x_j\nabla_k\nabla_\ell\frac{1}{r}$$
$$= -4\pi\, S_{\mathrm{p}i}\epsilon_{ijk}\nabla_k\left[x_j\delta(\mathbf{r})\right] + 4\pi\, S_{\mathrm{p}i}\epsilon_{ijk}(\nabla_k x_j)\delta(\mathbf{r}) + 0$$
$$= 0 + 4\pi\, S_{\mathrm{p}i}\epsilon_{ijk}\delta_{kj}\,\delta(\mathbf{r}) = 0$$

where x_j is the jth component of $\mathbf{r}$. The expectation value of the remaining spinor term, for an eigenstate $\psi_{n\ell m_\ell}(\mathbf{r})\chi$ of H_0, the spinor part being arbitrary, is

$$\langle|\Delta H|\rangle = \frac{ge^2}{8\pi m_e m_p c^2}\left[\left(\chi^\dagger\mathbf{S}_e \cdot \mathbf{S}_p\chi\right)I - \left(\chi^\dagger S_{ei} S_{pj}\chi\right)I_{ij}\right]$$

where

$$I = \int d^3r\,|\psi_{n\ell m_\ell}(\mathbf{r})|^2\nabla^2\frac{1}{r} = -4\pi\int d^3r\,|\psi_{n\ell m_\ell}(\mathbf{r})|^2\delta(\mathbf{r})$$
$$= -4\pi\,|\psi_{n\ell m_\ell}(0)|^2$$

and

$$I_{ij} = \int d^3r \, |\psi_{n\ell m_\ell}(\mathbf{r})|^2 \nabla_i \nabla_j r^{-1}$$

For $n = 2$, $\ell = m_\ell = 1$, we get

$$I = 0, \quad I_{ij} = \frac{1}{64\pi a_0^5} \int_0^\infty dr \, r \, e^{-r/a_0} \int d\Omega \, \sin^2\theta \left(-\delta_{ij} + 3\hat{r}_i\hat{r}_j\right)$$

where $\hat{r}_i$, $\hat{r}_j$ are components of the unit vector $\hat{r}$. The non-zero I_{ij} are

$$I_{xx} = I_{yy} = \frac{1}{8 \times 15 a_0^3}, \qquad I_{zz} = -\frac{1}{4 \times 15 a_0^3}$$

Thus, we get

$$\langle \Delta H \rangle_{211} = -\frac{ge^2}{8\pi m_e m_p c^2} \left(\frac{1}{8 \times 15 a_0^3}\right) \chi^\dagger \left[\mathbf{S}_e \cdot \mathbf{S}_e - 3 S_{ez} S_{pz}\right] \chi$$

This can be expressed in terms of the *total spin*

$$\mathbf{S} = \mathbf{S}_e + \mathbf{S}_p$$

as

$$\mathbf{S}_e \cdot \mathbf{S}_p = \tfrac{1}{2}\left(S^2 - \tfrac{3}{2}\hbar^2\right), \qquad S_{p_z} S_{e_z} = \tfrac{1}{2}\left(S_z^2 - \tfrac{1}{2}\hbar^2\right)$$

Then, we can write

$$\mathbf{S}_e \cdot \mathbf{S}_e - 3 S_{ez} S_{pz} = \tfrac{1}{2}\left(\mathbf{S}^2 - 3 S_z^2\right)$$

For total spin eigenstates, this has the expectation value

$$\tfrac{1}{2}\hbar^2 \left[s(s+1) - 3m_s^2\right]$$

Given that the allowed values of s are 0, 1, this gives either 0 or $-\hbar^2/2$, $\hbar^2$ correspondingly.

Problem 8.10 Assume that the Hamiltonian that describes the interaction of electron and proton spins, responsible for the hyperfine splitting, in a hydrogen-like atom has the simplified form

$$H = H_0 + \lambda \left(\mathbf{S}_e \cdot \mathbf{S}_p\right)$$

where

$$H_0 = \frac{p^2}{2m} - \frac{Ze^2}{r}$$

is the standard unperturbed Hamiltonian and λ is a parameter. Ignore all other perturbations.

(a) Express the state with electron quantum numbers $n = 2$, $\ell = m_\ell = 1$, $m_s^{(e)} = \frac{1}{2}$ and proton quantum number $m_s^{(p)} = -\frac{1}{2}$ in terms of the eigenstates of the total spin $\mathbf{S} = \mathbf{S}_e + \mathbf{S}_p$.

(b) The total angular momentum operator is $\mathbf{J} = \mathbf{L} + \mathbf{S}$. What are the allowed values of the quantum number j? Assuming that the state of the system is $|\ell = m_\ell = 1; m_s^{(e)} = \frac{1}{2}, m_s^{(p)} = -\frac{1}{2}\rangle$, what are the probabilities of finding the system with each of these values of j?

(c) Consider a state with $n = 5$, even parity and $j = 3$. What is the total spin of such a state? What is the energy difference between that state and the state with $n = 2$, odd parity and $j = 0$?

(d) If the system starts at time $t = 0$ in the state considered in (a), calculate the probability of finding the system in the same state at a later time t.

Solution

(a) The spin part of the state of the system is $|\uparrow\rangle_e |\downarrow\rangle_p$. The total-spin eigenstates are the *singlet* $|0\,0\rangle$ and the *triplet* $|1\,1\rangle$. In terms of them we have

$$|\uparrow\rangle_e |\downarrow\rangle_p = \frac{1}{\sqrt{2}}(|1\,0\rangle + |0\,0\rangle)$$

(b) The allowed values of the total spin quantum number are $s = 0, 1$. The allowed values of total angular momentum are determined by

$$|\ell - s| \leq j \leq \ell + s$$

which for $\ell = 1$ implies $j = 0, 1, 2$. The relation of the two states $|\ell = m_\ell = 1; s = m_s = 0\rangle$ and $|\ell = m_\ell = 1; s = 1, m_s = 0\rangle$ to the basis $|\ell = 1, s; j, m\rangle$ is found as follows[8]

$$|\ell = m_\ell = 1; s = m_s = 0\rangle = |\ell = 1, s = 0; j = 1, m = 1\rangle$$
$$|\ell = m_\ell = 1; s = 1, m_s = 0\rangle$$
$$= \frac{1}{\sqrt{2}}(|\ell = 1, s = 1; j = 1, m = 1\rangle - |\ell = 1, s = 1; j = 2, m = 1\rangle)$$

Thus, we may write

$$|\ell = m_\ell = 1; m_s^{(e)} = \tfrac{1}{2}, m_s^{(p)} = -\tfrac{1}{2}\rangle$$
$$= \frac{1}{\sqrt{2}}(|\ell = m_\ell = 1; s = m_s = 0\rangle + |\ell = m_\ell = 1; s = 1, m_s = 0\rangle)$$
$$= \frac{1}{\sqrt{2}}|\ell = 1, s = 0; j = 1, m = 1\rangle + \frac{1}{2}|\ell = 1, s = 1; j = 1, m = 1\rangle$$
$$- \frac{1}{2}|\ell = 1, s = 1; j = 2, m = 1\rangle$$

From this expression it is clear that the probability of finding $j = 0$ is zero, the probabilities of finding $j = 1$ are $\frac{1}{2}$ for $s = 0$ and $\frac{1}{4}$ for $s = 1$, and the probability of finding $j = 2$ is $\frac{1}{4}$.

---

[8] See for example problem 5.25.

(c) For such a state, the allowed values of ℓ are 0, 1, 2, 3, 4. Of these, only 0, 2, 4 correspond to even parity. $j = 3$ can be achieved from 2 or 4 with $s = 1$. The state with $n = 2$, $j = 0$ and odd parity corresponds to $\ell = 1$ and $s = 1$. There is no hyperfine splitting difference between these two states and their energy difference is just

$$E_5 - E_2 = -\frac{Z^2 e^2}{2a_0}\left(\frac{1}{25} - \frac{1}{4}\right).$$

(d) The time-evolved state of the system is

$$|\psi(t)\rangle = \exp\left[i\left(\frac{Z^2 e^2}{8a_0\hbar}\right)t\right]\left(\frac{1}{\sqrt{2}}e^{3i\lambda\hbar t/4}|\ell = 1,\, s = 0;\, j = 1,\, m = 1\rangle\right.$$
$$+ \frac{1}{2}e^{-i\lambda\hbar t/4}|\ell = 1,\, s = 1;\, j = 1,\, m = 1\rangle$$
$$\left.- \frac{1}{2}e^{-i\lambda\hbar t/4}|\ell = 1,\, s = 1;\, j = 2,\, m = 1\rangle\right)$$

The probability of finding the system in the same state as initially is

$$\mathcal{P}(t) = |\langle\psi(0)|\psi(t)\rangle|^2 = \cos^2\frac{\lambda\hbar t}{2}$$

Problem 8.11 The interaction between the spin magnetic moment of the electron and the magnetic moment of the proton in a hydrogen atom is

$$\Delta H = -\frac{1}{4\pi}\left(\boldsymbol{\mu}_e \cdot \boldsymbol{\mu}_p\right)\nabla^2\frac{1}{r} + \frac{1}{4\pi}\left(\boldsymbol{\mu}_e \cdot \nabla\right)\left(\boldsymbol{\mu}_p \cdot \nabla\right)\frac{1}{r}$$

Consider a hydrogen atom with its proton replaced by a *deuteron*, a spin-1 bound state of a proton and a neutron. Calculate the *hyperfine splitting* due to this interaction when the atom is in the state

$$|n = 1,\, \ell = m_\ell = 0\rangle\,|\tfrac{1}{2}\,\tfrac{1}{2}\rangle\,|1\,1\rangle$$

Solution
In terms of the spins, the interaction term is

$$\Delta H = \frac{e^2 g_d}{8\pi m_e m_d c^2}\left[(\mathbf{S}_e \cdot \mathbf{S}_d)\nabla^2\frac{1}{r} - (\mathbf{S}_e \cdot \nabla)(\mathbf{S}_d \cdot \nabla)\frac{1}{r}\right]$$

The shift in energy due to this interaction is

$$\langle\Delta H\rangle = \frac{e^2 g_d}{8\pi m_e m_d c^2}\int d^3r\,|\psi_{100}(r)|^2\,\chi^\dagger\left[(\mathbf{S}_e \cdot \mathbf{S}_d)\nabla^2\frac{1}{r} - (\mathbf{S}_e \cdot \nabla)(\mathbf{S}_d \cdot \nabla)\frac{1}{r}\right]\chi$$

that is,

$$\langle\Delta H\rangle = -\frac{e^2 g_d}{3m_e m_d c^2}|\psi_{100}(0)|^2\chi^\dagger(\mathbf{S}_e \cdot \mathbf{S}_d)\chi$$

where $\psi_{100}^2(0) = (\pi a_0^3)^{-1}$. The spin inner product can be written as

$$\mathbf{S_e} \cdot \mathbf{S_d} = \tfrac{1}{2}\left(S^2 - S_e^2 - S_d^2\right) = \tfrac{\hbar^2}{2}\left[s(s+1) - \tfrac{3}{4} - 2\right]$$

where

$$\mathbf{S} = \mathbf{S_e} + \mathbf{S_d}$$

is the total spin of the atom, specified by s. Since $s_d = 1$ and $s_e = \tfrac{1}{2}$, the possible values of the total spin quantum number are

$$s = \tfrac{1}{2}, \ \tfrac{3}{2}$$

The spin state of the system

$$|\uparrow\rangle_e |1\ 1\rangle_d$$

has $m_s = \tfrac{3}{2}$. Therefore, it can only correspond to $s = \tfrac{3}{2}$. Thus, we get

$$\Delta E = -\frac{e^2 g_d \hbar^2}{6\pi m_e m_d c^2 a_0^3} = -\frac{\alpha^4}{6\pi}\left(\frac{m_e}{m_d}\right) m_e c^2$$

Problem 8.12 Consider the HD^+ ion consisting of a proton, a deuteron[9] and an electron. As in the case of the H_2^+ ion, a good approximation is to neglect the nuclear kinetic energies and take as the Hamiltonian

$$H_0 = \frac{p^2}{2m} - \frac{2e^2}{|\mathbf{r} - \mathbf{R}/2|} - \frac{2e^2}{|\mathbf{r} + \mathbf{R}/2|} + \frac{e^2}{R}$$

The system is in the state with spatial wave function[10]

$$\psi(\mathbf{r}) = \frac{N}{\sqrt{\pi a_0^3}}\left[\exp\left(-\frac{|\mathbf{r} - \mathbf{R}/2|}{a_0}\right) + \exp\left(-\frac{|\mathbf{r} + \mathbf{R}/2|}{a_0}\right)\right]$$

Assume that the internuclear distance R is given and that it has its optimal value, determined by the minimization of energy.[11] A simplified version of the *hyperfine splitting* interaction between the electron and nuclear spins has the form $(m_d \approx 2m_p)$

$$\Delta H = \frac{e^2}{4 m_e m_p c^2}\left[2g_p\left(\mathbf{S_e} \cdot \mathbf{S_p}\right)\delta(\mathbf{r} - \mathbf{R}/2) + g_d\left(\mathbf{S_e} \cdot \mathbf{S_d}\right)\delta(\mathbf{r} + \mathbf{R}/2)\right]$$

---

[9] The deuteron is a spin-1 bound state of a proton and a neutron.
[10] Consider the normalization factor N as given.
[11] This is $R \approx 2.45 a_0 \approx 1.3$ Å. The corresponding value of the normalization constant is $N \approx 0.58$.

Assume that the spin part of the wave function is

$$\left|\tfrac{1}{2}\ \tfrac{1}{2}\right\rangle^{(e)} \left|\tfrac{1}{2}\ -\tfrac{1}{2}\right\rangle^{(p)} |1\ 1\rangle^{(d)}$$

and calculate the correction to the energy due to this factor.

Solution

It is straightforward to obtain

$$\langle \Delta H \rangle = \frac{\overline{N}^2 e^2}{4\pi m_e m_p a_0^3 c^2} \left[2g_p \, \chi^\dagger (\mathbf{S}_e \cdot \mathbf{S}_p)\chi + g_d \, \chi^\dagger (\mathbf{S}_e \cdot \mathbf{S}_d)\chi\right]$$

where

$$\overline{N}^2 = N^2 \left(1 + e^{-R/a_0}\right)^2$$

is a dimensionless number.

The spin inner products appearing in the interaction term can be expressed in terms of the square of the total electron–proton spin,

$$\mathbf{S}_{ep} \equiv \mathbf{S}_e + \mathbf{S}_p$$

and the square of the total electron–deuteron spin,

$$\mathbf{S}_{ed} \equiv \mathbf{S}_e + \mathbf{S}_p$$

as

$$\mathbf{S}_e \cdot \mathbf{S}_p = \tfrac{1}{2}\left(S_{ep}^2 - \tfrac{3}{2}\hbar^2\right)$$

and

$$\mathbf{S}_e \cdot \mathbf{S}_d = \tfrac{1}{2}\left(S_{ed}^2 - \tfrac{3}{4}\hbar^2 - 2\hbar^2\right)$$

The allowed values of s_{ep} are 0 and 1, while the allowed values of s_{ed} are $\tfrac{1}{2}$ and $\tfrac{3}{2}$. Since

$$|\uparrow\rangle_e |\downarrow\rangle_p = \tfrac{1}{\sqrt{2}}\left(|0\ 0\rangle_{ep} + |1\ 0\rangle_{ep}\right)$$

we have

$$\chi^\dagger \tfrac{1}{2}\left(S_{ep}^2 - \tfrac{3}{2}\hbar^2\right)\chi = \tfrac{1}{4}\left(\langle 0\ 0| + \langle 1\ 0|\right)\left(S_{ep}^2 - \tfrac{3}{2}\hbar^2\right)\left(|0\ 0\rangle_{ep} + |1\ 0\rangle_{ep}\right)$$

$$= \tfrac{1}{4}\hbar^2 \left(\langle 0\ 0| + \langle 1\ 0|\right)\left(-\tfrac{3}{2}|0\ 0\rangle_{ep} + \tfrac{1}{2}|1\ 0\rangle_{ep}\right) = -\tfrac{1}{4}\hbar^2$$

However, the z-component of $\mathbf{S}_{ed}$ is given by $m_{ed} = \tfrac{3}{2}$ and this can only occur for $s_{ed} = \tfrac{3}{2}$. Therefore,

$$\chi^\dagger \tfrac{1}{2}\left(S_{ed}^2 - \tfrac{11}{4}\hbar^2\right)\chi = \tfrac{1}{2}\hbar^2$$

The energy shift will be

$$\Delta E = \left(\frac{\overline{N}^2 e^2}{4\pi m_e m_p a_0^3 c^2} \right) \frac{\hbar^2}{2} \left(-g_p + g_d \right)$$

or

$$\Delta E = \overline{N}^2 \left(\frac{\alpha^4}{8\pi} \right) \left(\frac{m_e}{m_p} \right) m_e c^2 \left(-g_p + g_d \right)$$

The dimensionless coefficient is $\overline{N}^2 \approx 0.4$.

Problem 8.13 The *deuteron* is a $j = 1$ and even-parity bound state of a proton and a neutron. Consider a toy model of the deuteron in which the dominant centrally symmetric part of the interaction of the proton and the neutron is approximated by a harmonic oscillator potential. Use the fact that the deuteron *quadrupole moment*[12] has a known small but non-zero value and obtain the form of the deuteron state.[13] Using the general form of the *spin–orbit* coupling

$$\frac{1}{2m^2 c^2} \frac{V'(r)}{r} (\mathbf{L} \cdot \mathbf{S}),$$

calculate the corresponding correction to the energy.

Solution

The total angular momentum $\mathbf{J}_d$ of the deuteron arises from the total proton–neutron spin $\mathbf{S}$ and their relative orbital angular momentum, namely

$$\mathbf{J}_d = \mathbf{L} + \mathbf{S}, \qquad \mathbf{S} = \mathbf{S}_p + \mathbf{S}_n$$

The possible values of s are 0 and 1. The first would require $\ell = 1$, which corresponds to negative parity. Therefore, $s = 0$ is excluded. With $s = 1$, the value $j = 1$ can be achieved either with $\ell = 0$ or with $\ell = 2$. This can be expressed, for $m_j = 0$, as

$$|j = 1, \, m_j = 0\rangle = C_0 |0 \, 0\rangle \, |1 \, 0\rangle + C_2 |2 \, 0\rangle \, |1 \, 0\rangle$$

---

[12] Consider the definition $Q_{ij} = x_i x_j - \frac{1}{3} \delta_{ij} r^2$.
[13] You may use the harmonic-oscillator radial wave functions

$$R_{20}(r) = N_0 r^2 \exp\left[-\left(\frac{m\omega}{2\hbar} \right) r^2 \right], \qquad R_{22}(r) = N_2 \left[\frac{3}{2} - \left(\frac{m\omega}{\hbar} \right) r^2 \right] \exp\left[-\left(\frac{m\omega}{2\hbar} \right) r^2 \right]$$

$$N_0 = \frac{4}{\sqrt{15}} \pi^{-1/4} \left(\frac{m\omega}{\hbar} \right)^{7/4}, \qquad N_2 = \sqrt{\frac{8}{3}} \pi^{-1/4} \left(\frac{m\omega}{\hbar} \right)^{3/4}$$

The harmonic oscillator mass $m \approx m_p/2$ is the reduced mass of the neutron–proton system.

The admixture $|C_2|/|C_0|$ of the $\ell = 2$ component is a measure of the departure from spherical symmetry. The lowest value of the radial quantum number n compatible with $\ell = 2$ is $n = 2$.

The expectation value of the *electric quadrupole moment* operator in the deuteron state will be[14]

$$\langle Q_{ij} \rangle = |C_0|^2 \langle 2; 0\ 0|Q_{ij}|2; 0\ 0\rangle + |C_2|^2 \langle 2; 2\ 0|Q_{ij}|2; 2\ 0\rangle$$
$$+ C_0^* C_2 \langle 2; 0\ 0|Q_{ij}|2; 2\ 0\rangle + \text{c.c.}$$

The first expectation value vanishes owing to spherical symmetry. Assuming that the $\ell = 2$ admixture is small, we can take $C_2 \approx \epsilon$ and $C_0 \approx 1 + O(\epsilon^2)$. Then

$$\langle Q_{ij} \rangle \approx \epsilon \langle 2; 0\ 0|Q_{ij}|2; 2\ 0\rangle + \text{c.c.}$$

This matrix element is

$$Q_{ij} = \frac{4}{3}\sqrt{\frac{8}{5\pi}}\left(\frac{m\omega}{\hbar}\right)^{5/2} J I_{ij}$$

where

$$J = \int_0^\infty dr\, r^6 \left[\frac{3}{2} - \left(\frac{m\omega}{\hbar}\right) r^2\right] e^{-m\omega r^2/2\hbar} = -\sqrt{\pi}\,\frac{15}{8}\left(\frac{\hbar}{m\omega}\right)^{7/2}$$

and

$$I_{ij} = \int d\Omega\, Y_{00}(\Omega) Y_{20}(\Omega) \left(\hat{x}_i \hat{x}_j - \frac{1}{3}\delta_{ij}\right) = \frac{1}{3\sqrt{5}}\,\text{Diag}\,(-1,\, -1,\, 2)$$

Finally, we get

$$\langle Q_{ij} \rangle = Q\,\text{Diag}\,(1,\, 1,\, -2)$$

with

$$Q \approx \frac{\sqrt{2}}{3}\left(\frac{\hbar}{m\omega}\right)\epsilon$$

from which we can deduce the admixture ϵ of the $\ell = 2$ component.

The *spin–orbit* correction to the Hamiltonian is

$$\Delta H_{\text{SO}} = \frac{\omega^2}{2mc^2}\mathbf{L}\cdot\mathbf{S} = \frac{\omega^2}{4mc^2}\left(\mathbf{J}^2 - \mathbf{L}^2 - \mathbf{S}^2\right)$$

The corresponding correction to the energy is

$$\Delta E_{\text{SO}} = -\frac{3(\hbar\omega)^2}{2mc^2}\epsilon^2$$

---

[14] The first entry in the states is the radial quantum number $n = 2$. We will drop the common spinor part.

Problem 8.14 Consider a simplified form for the hyperfine splitting interaction between the electron and proton spins:

$$H = \frac{p^2}{2m} - \frac{e^2}{r} + \lambda \left(\mathbf{S}_p \cdot \mathbf{S}_e \right)$$

Assume that the atom is initially (at $t = 0$) in the ground state of the spin-independent part of the Hamiltonian and with the proton spin 'up' and the electron spin 'down', namely the state $\psi_{100}(r)\chi_\uparrow^{(p)}\chi_\downarrow^{(e)}$.

(a) Find the wave function at any later time $t > 0$.
(b) What is the probability of finding the spin of the proton pointing down?
(c) Calculate the expectation value of the magnetic dipole moment of the system at any time.
(d) Consider now the influence of a uniform (weak) magnetic field on the system. What are the new eigenstates and energy eigenvalues?

Solution
(a) The evolved state will be

$$|\psi(t)\rangle = e^{-i H t/\hbar}|1\,0\,0\rangle|\downarrow\rangle^{(e)}|\uparrow\rangle^{(p)}$$
$$= e^{-i H_0 t/\hbar}|1\,0\,0\rangle \exp\left[-\frac{i\lambda}{\hbar} \left(\mathbf{S}_p \cdot \mathbf{S}_e \right) t \right]|\downarrow\rangle^{(e)}|\uparrow\rangle^{(p)}$$

For the spin part, we have

$$\mathbf{S}_p \cdot \mathbf{S}_e = \tfrac{1}{2} \left(\mathbf{S}^2 - S_e^2 - S_p^2 \right) = \tfrac{1}{2} \left(\mathbf{S}^2 - \tfrac{3}{2}\hbar^2 \right)$$

and

$$|s = 1, \, m_s = 0\rangle = \tfrac{1}{\sqrt{2}} \left(|\uparrow\rangle^{(e)}|\downarrow\rangle^{(p)} + |\downarrow\rangle^{(e)}|\uparrow\rangle^{(p)} \right)$$
$$|s = 0, \, m_s = 0\rangle = \tfrac{1}{\sqrt{2}} \left(|\uparrow\rangle^{(e)}|\downarrow\rangle^{(p)} - |\downarrow\rangle^{(e)}|\uparrow\rangle^{(p)} \right)$$

Thus,

$$|\psi(t)\rangle = \tfrac{1}{\sqrt{2}} e^{-i E_1 t/\hbar}|1\,0\,0\rangle \exp\left[-\frac{i\lambda}{2\hbar} \left(\mathbf{S}^2 - \frac{3\hbar^2}{2} \right) t \right] (|1\,0\rangle - |0\,0\rangle)$$

or

$$|\psi(t)\rangle = \tfrac{1}{\sqrt{2}} e^{-i E_1 t/\hbar} e^{3i\lambda\hbar t/4}|1\,0\,0\rangle \left(e^{-i\lambda\hbar t}|1\,0\rangle - |0\,0\rangle \right)$$

with $E_1 = -e^2/2a_0$. This can be transformed back to electron–proton spin eigenstates by writing

$$|\psi(t)\rangle = e^{-i E_1 t/\hbar} e^{i\lambda\hbar t/4}|1\,0\,0\rangle |\chi\rangle$$

where the spin part $|\chi\rangle$ is given by

$$|\chi\rangle = -i \sin \frac{\lambda \hbar t}{2} |\uparrow\rangle^{(e)} |\downarrow\rangle^{(p)} + \cos \frac{\lambda \hbar t}{2} |\downarrow\rangle^{(e)} |\uparrow\rangle^{(p)}$$

(b) The total spin commutes with the Hamiltonian and is a constant of the motion. Thus, proton spin down is at all times accompanied by electron spin up. From the evolved state it is clear that the probability amplitude for finding the proton spin pointing down will be

$$\langle \psi(t)| \uparrow\rangle^{(e)} |\downarrow\rangle^{(p)} = -i e^{-iE_1 t/\hbar} e^{i\lambda\hbar t/4} \sin \frac{\lambda \hbar t}{2}$$

The corresponding probability is

$$\mathcal{P}(t) = \sin^2 \frac{\lambda \hbar t}{2}$$

This probability becomes unity periodically, at times

$$t_n = (2n + 1)\frac{\pi}{\hbar \lambda} \qquad (n = 0, 1, \ldots)$$

(c) The magnetic dipole moment of the system is

$$\boldsymbol{\mu} = \left(\frac{e}{2m_e c\hbar}\right)\mathbf{L} + \left(\frac{g_e e}{2m_e c\hbar}\right)\mathbf{S}_e + \left(\frac{-g_p e}{2m_p c\hbar}\right)\mathbf{S}_p$$

$$= \frac{1}{2\hbar}\left(\mu_e \mathbf{L} + 2\mu_e \mathbf{S}_e + 2\mu_p \mathbf{S}_p\right)$$

with $g_e = 2$. Taking the expectation value in the evolved state we get, after some algebra,

$$\langle \boldsymbol{\mu}\rangle = \frac{\mu_e}{\hbar} \sin^2 \frac{\lambda \hbar t}{2} \langle\uparrow |\mathbf{S}_e| \uparrow\rangle + \frac{\mu_p}{\hbar} \cos^2 \frac{\lambda \hbar t}{2} \langle\uparrow |\mathbf{S}_p| \uparrow\rangle$$
$$+ \frac{\mu_p}{\hbar} \sin^2 \frac{\lambda \hbar t}{2} \langle\downarrow |\mathbf{S}_p| \downarrow\rangle + \frac{\mu_e}{\hbar} \cos^2 \frac{\lambda \hbar t}{2} \langle\downarrow |\mathbf{S}_e| \downarrow\rangle$$

Since $\mathbf{S} = \frac{1}{2}(S_+ + S_-)\hat{\mathbf{x}} + \frac{1}{2i}(S_+ - S_-)\hat{\mathbf{y}} + S_z\hat{\mathbf{z}}$, we get

$$\langle \boldsymbol{\mu}\rangle = \frac{1}{2}(\mu_p - \mu_e)\cos \lambda \hbar t \, \hat{\mathbf{z}}$$

(d) The magnetic field will induce on the Hamiltonian the additional term

$$\Delta H = -\boldsymbol{\mu} \cdot \mathbf{B}$$

The total Hamiltonian is a sum of an orbital and a spin part that commute, namely

$$H_{tot} = \frac{p^2}{2m} - \frac{e^2}{r} - \frac{\mu_e B}{\hbar} L_z + \frac{\lambda}{2}\left(\mathbf{S}^2 - \frac{3\hbar^2}{2}\right) - \frac{B}{\hbar}\left(\mu_e S_{ez} + \mu_p S_{pz}\right)$$

The eigenfunctions will be a product of an orbital and a spinor part,

$$|E\rangle = |n\ \ell\ m_\ell\rangle\,|\ldots\rangle_s$$

The eigenvalues will be

$$E = E_n - \mu_e B m_\ell + \Delta E_s$$

The spinor parts $|\ldots\rangle_s$ and ΔE_s can be determined as follows. Observe that

$$[S_z,\ H_{tot}] = -\frac{B}{\hbar}\left[S_z,\ \mu_e S_{ez} + \mu_p S_{pz}\right] = 0$$

Acting with $\mu_e S_{ez} + \mu_p S_{pz}$ on an arbitrary combination of the states $|s = 1,\ m_s = 0\rangle$ and $|s = 0,\ m_s = 0\rangle$, we obtain

$$\left(\mu_e S_{ez} + \mu_p S_{pz}\right)(a|1\ 0\rangle + b|0\ 0\rangle) = (\mu_e - \mu_p)\frac{\hbar}{2}\,(a|0\ 0\rangle + b|1\ 0\rangle)$$

From that, we obtain for the spin part of the Hamiltonian

$$H_s = \frac{\lambda}{2}\left(\mathbf{S}^2 - \frac{3\hbar^2}{2}\right) - \frac{B}{\hbar}\mu_{s,z}$$

which, acting on the above state, gives

$$H_s(a|1\ 0\rangle + b|0\ 0\rangle)$$
$$= \left[\frac{\lambda\hbar^2 a}{4} - \frac{Bb(\mu_e - \mu_p)}{2}\right]|1\ 0\rangle - \left[\frac{3\hbar^2\lambda b}{4} + \frac{B(\mu_e - \mu_p)a}{2}\right]|0\ 0\rangle$$

Demanding that this is equal to

$$H_s\left(a|1\ 0\rangle + b|0\ 0\rangle\right) = (\Delta E)_s\left(a|1\ 0\rangle + b|0\ 0\rangle\right)$$

leads to

$$\left(\Delta E_s - \frac{\lambda\hbar^2}{4}\right)\left(\Delta E_s + \frac{3\lambda\hbar^2}{4}\right) = \frac{B^2(\mu_e - \mu_p)^2}{4}$$

and

$$\frac{a}{b} = -\frac{B(\mu_e - \mu_p)}{2\Delta E_s - \lambda\hbar^2/2}$$

The energy eigenvalues are

$$\Delta E_s = -\frac{\lambda\hbar^2}{4} \mp \frac{\lambda\hbar^2}{2}\sqrt{1 - \left[B(\mu_e - \mu_p)/\lambda\hbar^2\right]^2}$$

The corresponding spin eigenstates will be the above $|1\ 0\rangle$ and $|0\ 0\rangle$ combinations with coefficient values given by

$$\frac{a}{b} = \frac{B(\mu_e - \mu_p)/\lambda\hbar^2}{-1 \pm \sqrt{1 - \left[B(\mu_e - \mu_p)/\lambda\hbar^2\right]^2}}$$

9

Approximation methods

Problem 9.1 Consider a particle of mass μ and charge e in the central potential

$$
V(r) = \begin{cases} -\dfrac{e^2}{r}, & 0 < r < R \\[2mm] -\dfrac{e^2}{r} e^{-\lambda(r-R)}, & R < r < \infty \end{cases}
$$

This potential differs from the Coulomb potential only in the region $r > R$, where the Coulomb force is *screened*. The difference becomes negligible if the parameter $\lambda \to 0$. Consider this difference as a perturbation and calculate the first-order correction to the energy of the ground state.

Solution

The Hamiltonian can be written as

$$
H = H_0 + \Delta V
$$

with

$$
H_0 = \frac{p^2}{2\mu} - \frac{e^2}{r}
$$

and

$$
\Delta V = \Theta(r - R) \frac{e^2}{r} \left[1 - e^{-\lambda(r-R)} \right]
$$

Treating ΔV as a perturbation, we can compute the first-order energy shift of the ground state. It will be

$$
\Delta E_1^{(1)} = \langle 1\,0\,0 | \Delta V | 1\,0\,0 \rangle = \frac{4e^2}{a_0^3} \int_R^\infty dr\, r\, e^{-2r/a_0} \left[1 - e^{-\lambda(r-R)} \right]
$$

where $|1\,0\,0\rangle \to \psi_{100} = (\pi a_0^3)^{-1/2} e^{-r/a_0}$ is the unperturbed ground state.

Performing the integral, we get

$$\Delta E_1^{(1)} = \lambda e^2 e^{-2R/a_0} \left[1 + \frac{R}{a_0} + \frac{\lambda a_0}{4} \left(1 + \frac{2R}{a_0} \right) \right] \left(1 + \frac{\lambda a_0}{2} \right)^{-2}$$

This vanishes in the limit $\lambda \to 0$. It also vanishes, as it should, in the limit $R \to \infty$. Expanding in λ, we obtain to first order

$$\Delta E_1^{(1)} \approx \lambda e^2 e^{-2R/a_0} \left(1 + \frac{R}{a_0} \right)$$

Problem 9.2 A particle of mass m moves in one dimension subject to a harmonic oscillator potential $\frac{1}{2} m\omega^2 x^2$. The particle oscillation is perturbed by an additional weak anharmonic force described by the potential $\Delta V = \lambda \sin \kappa x$. Find the corrected ground state and calculate the expectation value of the position operator in that state.

Solution

The corrected ground state will be

$$\overline{|0\rangle} = |0\rangle - \frac{\lambda}{\hbar \omega} \sum_{n=1}^{\infty} \frac{1}{n} \langle n | \sin \kappa x | 0 \rangle |n\rangle$$

The relevant matrix element can be written as the imaginary part of

$$\langle n | e^{i\kappa x} | 0 \rangle = \left\langle n \left| \exp \left[i\kappa \sqrt{\frac{\hbar}{2m\omega}} (a + a^\dagger) \right] \right| 0 \right\rangle$$

Since the commutator of the operators appearing in the exponent is a c-number, we can make use of the operator identity

$$e^{\mathbf{A} + \mathbf{B}} = e^{\mathbf{A}} e^{\mathbf{B}} e^{-[\mathbf{A}, \mathbf{B}]/2}$$

and obtain

$$\langle n | e^{i\kappa x} | 0 \rangle = \left\langle n \left| \exp \left(i\kappa \sqrt{\frac{\hbar}{2m\omega}} a^\dagger \right) \exp \left(i\kappa \sqrt{\frac{\hbar}{2m\omega}} a \right) \right| 0 \right\rangle e^{-\hbar \kappa^2 / 4m\omega}$$

$$= \left\langle n \left| \exp \left(i\kappa \sqrt{\frac{\hbar}{2m\omega}} a^\dagger \right) \right| 0 \right\rangle e^{-\hbar \kappa^2 / 4m\omega}$$

$$= \left\langle n \left| \sum_{n'=0}^{\infty} \frac{1}{n'!} \left(i\kappa \sqrt{\frac{\hbar}{2m\omega}} \right)^{n'} (a^\dagger)^{n'} \right| 0 \right\rangle e^{-\hbar \kappa^2 / 4m\omega}$$

$$= \frac{1}{\sqrt{n!}} \left(i\kappa \sqrt{\frac{\hbar}{2m\omega}} \right)^n e^{-\hbar \kappa^2 / 4m\omega}$$

Thus, the corrected state is

$$\overline{|0\rangle} = |0\rangle - \frac{\lambda}{2i\hbar\omega} e^{-\hbar\kappa^2/4m\omega} \sum_{n=1}^{\infty} \frac{1}{n} \frac{1}{\sqrt{n!}} \left(i\kappa\sqrt{\frac{\hbar}{2m\omega}} \right)^n |n\rangle - \text{c.c.}$$

$$= |0\rangle - \frac{\lambda}{\hbar\omega} e^{-\hbar\kappa^2/4m\omega} \sum_{\nu=0}^{\infty} \frac{(-1)^\nu}{(2\nu+1)} \frac{1}{\sqrt{(2\nu+1)!}} \left(\kappa\sqrt{\frac{\hbar}{2m\omega}} \right)^{2\nu+1} |2\nu+1\rangle$$

The expectation value of the position operator in this state will be, to first order,

$$\overline{\langle 0|x|0\rangle}$$

$$= -\frac{2\lambda}{\hbar\omega} e^{-\hbar\kappa^2/4m\omega} \sum_{\nu=0}^{\infty} \frac{(-1)^\nu}{(2\nu+1)} \frac{1}{\sqrt{(2\nu+1)!}} \left(\kappa\sqrt{\frac{\hbar}{2m\omega}} \right)^{2\nu+1} \langle 0|x|2\nu+1\rangle$$

$$= -\frac{2\lambda}{\hbar\omega} e^{-\hbar\kappa^2/4m\omega} \sum_{\nu=0}^{\infty} \frac{(-1)^\nu}{(2\nu+1)} \frac{\kappa^{2\nu+1}}{\sqrt{(2\nu+1)!}} \left(\frac{\hbar}{2m\omega} \right)^{\nu+1} \langle 0|(a+a^\dagger)|2\nu+1\rangle$$

$$= -\frac{2\lambda}{\hbar\omega} e^{-\hbar\kappa^2/4m\omega} \left(\kappa\sqrt{\frac{\hbar}{2m\omega}} \right) \sqrt{\frac{\hbar}{2m\omega}} = -\frac{\lambda\kappa}{m\omega^2} e^{-\hbar\kappa^2/4m\omega}$$

Problem 9.3 A particle of mass m moves in one dimension subject to an anharmonic potential that is close to but not exactly a harmonic oscillator potential, namely

$$V(x) = \frac{m\omega^2 x^2}{2} \left(\frac{x}{a} \right)^{2\lambda}$$

where a is a parameter with the dimensions of length and $\lambda \ll 1$ is a dimensionless exponent. We can write this potential as

$$V(x) = \frac{m\omega^2 x^2}{2} + \Delta V$$

with

$$\Delta V(x) = \frac{m\omega^2 x^2}{2} \left[\left(\frac{x}{a} \right)^{2\lambda} - 1 \right]$$

(a) Treating ΔV as a small perturbation, calculate the first-order correction to the ground-state energy.

(b) Consider now a trial wave function for the ground state,

$$\psi(x, \beta) = \left(\frac{\beta}{\pi} \right)^{1/4} e^{-\beta x^2/2}$$

and calculate the expectation value of the energy without making any use of the smallness of λ. Find the value of the parameter β for which the ground-state energy has a minimum and write down an expression for it as a function of λ.

(c) Now take into consideration the smallness of λ and see whether the value of the ground-state energy coincides with the one obtained in part (a) from perturbation theory.

Solution

(a) We have

$$E_0^{(1)} = \langle 0|\Delta V|0\rangle = \frac{m\omega^2}{2}\sqrt{\frac{m\omega}{\hbar\pi}}\int_{-\infty}^{\infty} dx\, e^{-m\omega x^2/\hbar}x^2\left[\left(\frac{x}{a}\right)^{2\lambda} - 1\right]$$

$$= -\frac{\hbar\omega}{4} + \frac{\hbar\omega}{2\sqrt{\pi}}\left(\frac{\hbar}{m\omega a^2}\right)^{\lambda} J(\lambda)$$

where

$$J(\lambda) \equiv \int_{-\infty}^{\infty} dy\, e^{-y^2}(y^2)^{1+\lambda} = \Gamma(\lambda + 3/2) \approx \Gamma(3/2)\left[1 + \lambda\,\psi(3/2)\right]$$

Substituting the values of the above functions,[1,2] we obtain[3]

$$E_0^{(1)} = \lambda\frac{\hbar\omega}{4}\left[\psi\left(\frac{3}{2}\right) + \ln\left(\frac{\hbar}{m\omega a^2}\right)\right] = \lambda\frac{\hbar\omega}{2}\left[1 - \frac{\gamma}{2} + \frac{1}{2}\ln\left(\frac{\hbar}{4m\omega a^2}\right)\right]$$

(b) The expectation value of the Hamiltonian with respect to the trial state $\psi(x, \beta)$ is

$$\mathcal{E}(\beta) = \langle\psi|H|\psi\rangle = \frac{\hbar^2\beta}{4m} + \frac{m\omega^2}{2\sqrt{\pi}}a^{-2\lambda}\beta^{-\lambda-1}\Gamma\left(\lambda + \frac{3}{2}\right)$$

where the gamma function arises from the integral over the potential energy. Minimizing with respect to β, we obtain a minimum at

$$\beta_0 = \left[\frac{\hbar^2\sqrt{\pi}a^{\lambda}}{2m^2\omega^2(\lambda + 1)\Gamma(\lambda + 3/2)}\right]^{-1/(\lambda+2)}$$

The minimal ground-state energy value corresponding to β_0 is

$$\mathcal{E}_0 = (\hbar\omega)^{\lambda/(\lambda+2)}\left(\frac{\hbar^2}{ma^2}\right)^{\lambda/(\lambda+2)} I(\lambda)$$

where

$$I(\lambda) \equiv [\Gamma(\lambda + 3/2)]^{1/(\lambda+2)}\, 2^{-2\lambda+3/(\lambda+2)}\pi^{-1/2(\lambda+2)}$$
$$\times \left[(\lambda + 1)^{1/(\lambda+2)} + (\lambda + 1)^{-\lambda+1/(\lambda+2)}\right]$$

---

[1] $\Gamma(3/2) = \int_0^{\infty} dx\,\sqrt{x}\,e^{-x} = \sqrt{\pi}/2$ and $\psi(z) \equiv \Gamma'(z)/\Gamma(z)$.
[2] $\psi(3/2) = -\gamma + 2 - 2\ln 2$, where $\gamma = 0.57721\ldots$ is the Euler constant.
[3] We have expanded $(\hbar/m\omega a^2)^{\lambda} = \exp\left[\lambda\ln(\hbar/m\omega a^2)\right]$ as approximately $1 + \lambda\,\ln(\hbar/m\omega a^2)$.

(c) Expanding around $\lambda \to 0$, we obtain[4]

$$\mathcal{E}_0 \approx \frac{\hbar\omega}{2} + \lambda\frac{\hbar\omega}{4}\left[\psi\left(\frac{3}{2}\right) + \ln\left(\frac{\hbar}{m\omega a^2}\right)\right] + O(\lambda^2)$$

which coincides with the expression obtained perturbatively in (a).

Problem 9.4 Replace the nucleus of a hydrogen-like atom with a uniform electric charge distribution of radius $R \ll a_0$. What is the resulting electrostatic potential $V_R(r)$? The difference

$$\Delta V(r) = V_R(r) - \left(-\frac{e^2}{r}\right)$$

will be proportional to the assumed extension R of the nucleus.

(a) Considering ΔV as a perturbation, calculate the correction to the ground-state energy to first order.[5]
(b) Do the same for the 2s and 2p states.[6]

Solution

(a) Assuming a uniform electric charge density

$$\rho = \frac{|e|}{4\pi R^3/3}$$

and using the standard Coulomb law expression for the electrostatic potential,

---

[4] The expansions of the factors involved are

$$(\hbar\omega)^{\lambda/(\lambda+2)} \approx \hbar\omega\left(1 - \frac{\lambda}{2}\ln\hbar\omega + \cdots\right), \qquad \left(\frac{\hbar^2}{ma^2}\right)^{\lambda/(\lambda+2)} \approx 1 + \frac{\lambda}{2}\ln\left(\frac{\hbar^2}{ma^2}\right) + \cdots$$

$$[\Gamma(\lambda + 3/2)]^{1/(\lambda+2)}2^{-2\lambda+3/(\lambda+2)} \approx \frac{\pi^{1/4}}{\sqrt{2}}\left[1 - \frac{\lambda}{4}\ln\left(\frac{\sqrt{\pi}}{2}\right) + \frac{\lambda}{2}\psi\left(\frac{3}{2}\right) + \cdots\right]$$

$$2^{-2\lambda+3/(\lambda+2)} \approx \frac{\sqrt{2}}{4}\left(1 - \frac{\lambda}{4}\ln 2 + \cdots\right), \qquad \pi^{-1/2(\lambda+2)} \approx \pi^{-1/4}\left(1 + \frac{\lambda}{8}\ln\pi + \cdots\right)$$

$$(1 + \lambda)^{1/(\lambda+2)} + (\lambda + 1)^{-\lambda+1/(\lambda+2)} \approx 2 + O(\lambda^2)$$

[5] You will need the integrals of the type

$$I_n(\lambda) = \int_0^\lambda dx\, x^n e^{-x} = n!\left(1 - e^{-\lambda}\sum_{\nu=0}^n \frac{\lambda^\nu}{\nu!}\right)$$

[6] The corresponding wave functions are

$$\psi_{100}(r) = (\pi a_0^3)^{-1/2}e^{-r/a_0}, \qquad \psi_{200}(r) = (32\pi a_0^3)^{-1/2}(2 - r/a_0)e^{-r/2a_0}$$

$$\psi_{210}(\mathbf{r}) = (32\pi a_0^3)^{-1/2}(r/a_0)e^{-r/2a_0}$$

we obtain

$$V_R(r) = -\frac{3e^2}{4\pi R^3} \int_{r' \leq R} d^3 r' \frac{1}{|\mathbf{r}' - \mathbf{r}|}$$

$$= -\frac{3e^2}{2R^3} \int_r^R dr' r'^2 \int_{-1}^1 (d\cos\theta) \frac{1}{\sqrt{r^2 + r'^2 - 2rr'\cos\theta}}$$

or

$$V_R(r) = \begin{cases} -\dfrac{e^2}{r}, & r > R \\ -\dfrac{e^2}{R}\left[\dfrac{3}{2} - \dfrac{1}{2}\left(\dfrac{r}{R}\right)^2\right], & r \leq R \end{cases}$$

The difference from the standard point-like Coulomb potential is

$$\Delta V(r) = \begin{cases} 0, & r > R \\ e^2\left(\dfrac{1}{r} + \dfrac{1}{2}\dfrac{r^2}{R^3} - \dfrac{3}{2R}\right), & r \leq R \end{cases}$$

The first-order correction to the ground-state energy will be

$$E_{10}^{(1)} = \langle 1\,0\,0|\Delta V|1\,0\,0\rangle = \frac{1}{4\pi a_0^3} \int d^3 r\, e^{-2r/a_0}\, \Delta V(r)$$

$$= \frac{e^2}{a_0} \int_0^\lambda dx\, e^{-2x}\left(x + \frac{x^4}{2\lambda^3} - \frac{3x^2}{2\lambda}\right)$$

where we have introduced the parameter

$$\lambda \equiv \frac{R}{a_0} \ll 1$$

Performing the integral, we obtain

$$E_{10}^{(1)} = \frac{e^2}{8a_0\lambda^3}\left[3 - 3\lambda^2 + 2\lambda^3 - e^{-2\lambda}\left(3 + 6\lambda + 3\lambda^2\right)\right]$$

Expanding the exponential around $\lambda = 0$ and keeping terms up to fifth order, we get

$$E_{10}^{(1)} = -\frac{e^2}{2a_0}\left[\frac{4\lambda^2}{5} + O(\lambda^3)\right]$$

(b) The energy correction for the 2s state is

$$E_{20}^{(1)} = \frac{1}{32\pi a_0^3} \int d^3 r\left(2 - \frac{r}{a_0}\right)^2 e^{-r/a_0}\, \Delta V(r)$$

$$= \frac{e^2}{8a_0} \int_0^\lambda dx\,(2 - x)^2 e^{-x}\left(x + \frac{x^4}{2\lambda^3} - \frac{3x^2}{2\lambda}\right)$$

Performing the integral, we obtain

$$E_{20}^{(1)} = \frac{e^2}{2a_0} \left(\frac{1}{8\lambda^3} \right) I(\lambda)$$

where

$$I(\lambda) \equiv 336 - 24\lambda^2 + 4\lambda^3 - e^{-\lambda} \left(336 + 336\lambda + 144\lambda^2 + 36\lambda^3 + 6\lambda^4 \right)$$

In the limit of small λ this gives

$$E_{20}^{(1)} \approx \frac{e^2}{2a_0} \left(\frac{\lambda^2}{10} \right)$$

The corresponding correction for the 2p state is

$$E_{21}^{(1)} = \frac{1}{32\pi a_0^3} \int d^3r \left(\frac{r}{a_0} \right)^2 e^{-r/a_0} \cos^2\theta \, \Delta V(r)$$

Thus

$$E_{21}^{(1)} = \frac{e^2}{24a_0} \int_0^\lambda dx \, x^2 \left(x + \frac{x^4}{2\lambda^3} - \frac{3x^2}{2\lambda} \right)$$

$$= \frac{e^2}{48a_0\lambda^3} \left[720 - 72\lambda^2 + 12\lambda^3 - e^{-\lambda} \left(720 + 720\lambda + 288\lambda^2 + 60\lambda^3 + 6\lambda^4 \right) \right]$$

Expanding, we are led to

$$E_{21}^{(1)} \approx -\frac{e^2}{2a_0} \left(\frac{\lambda^4}{240} \right)$$

Note that the 2p-state correction is strongly suppressed in comparison to that from the s-states:

$$\left| \frac{E_{20}^{(1)}}{E_{21}^{(1)}} \right| \approx \frac{\lambda^2}{24}$$

Problem 9.5 A hydrogen atom is subject to a uniform electric field $\mathcal{E}$. The electric field is sufficiently weak to be treated as a perturbation.

(a) Calculate the energy eigenvalue corrections for the first excited level ($n = 2$).
(b) Write down an expression for the *induced electric dipole moment* of the ground state to the lowest non-trivial order. Obtain an upper estimate for the *polarizability* of the atom.

Solution

(a) In the absence of an electric field, the first excited level consists of the four degenerate states $|2\,0\,0\rangle$, $|2\,1\,0\rangle$, $|2\,1\,1\rangle$ and $|2\,1\,-1\rangle$. The matrix elements of the perturbing Hamiltonian in these states are $-e\mathcal{E}\langle 2\ \ell\ m|z|2\ \ell'\ m'\rangle$. It is not difficult to see that on the one hand

$$\langle \ell'\ m'|z|\ell\ m\rangle = \delta_{mm'}\langle \ell'\ m|z|\ell\ m\rangle$$

On the other hand, parity considerations imply that

$$\langle \ell'\ m'|z|\ell\ m\rangle = -\langle \ell'\ m'|\mathcal{P}z\mathcal{P}|\ell\ m\rangle = -(-1)^{\ell+\ell'}\langle \ell'\ m'|z|\ell\ m\rangle$$

which shows that these matrix elements vanish unless $\ell' + \ell = 2\nu + 1$, where ν is an integer. Thus, all these matrix elements vanish except

$$\langle 2\ 0\ 0|z|2\ 1\ 0\rangle = \frac{1}{32\pi a_0^3}\int d^3r\, r\cos^2\theta\, \frac{r}{a_0}\left(2 - \frac{r}{a_0}\right)e^{-r/a_0} = -3a_0$$

Therefore, the states $|2\ 1\ \pm 1\rangle$ do not receive any correction to first order and both correspond to the eigenvalue $E_2 = -e^2/8a_0$. In contrast, the states $|2\ 0\ 0\rangle$ and $|2\ 1\ 0\rangle$ mix and split. The new eigenvalues are the eigenvalues $E_2 \mp 3e\mathcal{E}a_0$ of the matrix

$$\begin{pmatrix} E_2 & 3e\mathcal{E}a_0 \\ 3e\mathcal{E}a_0 & E_2 \end{pmatrix}$$

The corresponding eigenvectors are

$$\frac{1}{\sqrt{2}}\left(|2\ 0\ 0\rangle \mp |2\ 1\ 0\rangle\right)$$

(b) The first-order correction to the ground state reads

$$|1\ 0\ 0\rangle^{(1)} = -e\mathcal{E}\sum_{n'}{}'\frac{\langle n'\ \ell'\ m'|z|1\ 0\ 0\rangle}{E_1 - E_{n'}}|n'\ \ell'\ m'\rangle$$

$$= -e\mathcal{E}\sum_{n'=2,\ \ell'=1,\ 3,\ldots}\frac{\langle n'\ \ell'\ 0|z|1\ 0\ 0\rangle}{E_1 - E_{n'}}|n'\ \ell'\ 0\rangle$$

Using this expression, we can write down to first order the expectation value of the electric dipole moment in the ground state:

$$\langle \mathbf{d}\rangle = e\langle 1\ 0\ 0|\mathbf{r}|1\ 0\ 0\rangle$$

$$- 2e^2\mathcal{E}\sum_{n'=2}^{\infty}\sum_{\ell'=1,\ 3,\ldots}\frac{\langle n'\ \ell'\ 0|z|1\ 0\ 0\rangle}{E_1 - E_{n'}}\langle 1\ 0\ 0|\mathbf{r}|n'\ \ell'\ 0\rangle$$

The first term (*the permanent electric dipole moment*) vanishes owing to parity. The second term, linear in the electric field, represents the *induced electric dipole moment.* It is non-vanishing only in the direction of the electric field.[7] Thus

$$\langle d_{\text{ind}} \rangle = -2e^2 \mathcal{E} \sum_{n'=2}^{\infty} \sum_{\ell'=1,3,\ldots} \frac{|\langle n' \ \ell' \ 0|z|1 \ 0 \ 0\rangle|^2}{E_1 - E_{n'}}$$

Note that this sum should be extended to the continuous part of the spectrum, including the scattering states in the form of an integral. The proportionality coefficient relating the induced electric dipole moment to the electric field is the *polarizability*,

$$\alpha = 2e^2 \sum_{n=2}^{\infty} \sum_{\ell=1,3,\ldots} \frac{|\langle n \ \ell \ 0|z|1 \ 0 \ 0\rangle|^2}{E_n - E_1}$$

Again, to the sum we should add an integral that includes the continuum states. The squared matrix element in the sum is multiplied by the positive number[8]

$$\frac{1}{E_n - E_1} = \frac{1}{|E_1|} \frac{n^2}{n^2 - 1} \leq \frac{4}{3|E_1|}$$

This bound comes from the maximal value[9] of $n^2/(n^2 - 1)$ obtained for the lowest value $n = 2$. Thus, we may write

$$\alpha \leq \frac{8e^2}{3|E_1|} \sum_{n=2}^{\infty} \sum_{\ell=1,3,\ldots} |\langle n \ \ell \ 0|z|1 \ 0 \ 0\rangle|^2$$

$$= \frac{8e^2}{3|E_1|} \sum_{n=1}^{\infty} \sum_{\ell=0}^{\infty} \sum_{m=-\ell}^{\ell} \langle 1 \ 0 \ 0|z|n \ \ell \ m\rangle \langle n \ \ell \ m|z|1 \ 0 \ 0\rangle$$

$$= \frac{8e^2}{3|E_1|} \langle 1 \ 0 \ 0|z^2|1 \ 0 \ 0\rangle = \frac{16a_0^3}{3}$$

Problem 9.6 An ion of charge q and mass μ is bound in a molecule with forces that can be approximated by the isotropic oscillator potential[10] $\mu\omega^2 r^2/2$.

---

[7] The matrix elements $\langle n \ 1 \ 0|x|1 \ 0 \ 0\rangle$ and $\langle n \ 1 \ 0|y|1 \ 0 \ 0\rangle$ vanish.
[8] The corresponding coefficient in the continuum sum is

$$\frac{1}{|E_1|} \frac{1}{1 + (ka_0)^2} \leq \frac{1}{|E_1|}$$

[9] Actually, $n^2/(n^2 - 1)$ is a slowly varying quantity throughout the sum, starting from its maximal value $4/3$ at $n = 2$ and approching 1 for $n \to \infty$.
[10] You can make use of the radial eigenfunctions of the isotropic harmonic oscillator ($\gamma \equiv \mu\omega/\hbar$),

$$R_{00}(r) = \frac{2\gamma^{3/4}}{\pi^{1/4}} e^{-\gamma r^2/2}, \qquad R_{11}(r) = \frac{2\sqrt{2}\gamma^{5/4}}{\sqrt{3}\pi^{1/4}} r e^{-\gamma r^2/2}$$

$$R_{20}(r) = \frac{\sqrt{6}\gamma^{3/4}}{\pi^{1/4}} \left(1 - \frac{2\gamma r^2}{3}\right) e^{-\gamma r^2/2}, \qquad R_{22}(r) = \frac{4\gamma^{7/4}}{\pi^{1/4}\sqrt{15}} r^2 e^{-\gamma r^2/2}$$

(a) Show that the electric dipole moment of the system in a given direction $d = qz$ satisfies the sum rule

$$\sum_{n', \ell', m'} (E_n - E_{n'}) \left| \langle n\, \ell\, m|\, d\, |n'\, \ell'\, m' \rangle \right|^2 = -\frac{\hbar^2 q^2}{2\mu}$$

Calculate the matrix elements $\langle n'\, \ell'\, m'|d|0\, 0\, 0\rangle$ and verify that they satisfy the above sum rule for $n = \ell = m = 0$.

(b) The system is perturbed by a weak electric field $\mathcal{E} = \mathcal{E}\hat{z}$. Calculate the first- and second-order corrections to the ground-state energy. Determine the ground-state correction to first order and compute the expectation value of the potential energy to this order.

(c) Calculate the matrix element $\langle n'\, \ell'\, m'|z|1\, 1\, 0\rangle$ and verify the above sum rule.

(d) Calculate the second-order correction to the ground state and compute the mean square radius of the system in the ground state to this order.

(e) Write down the second-order correction to the ground-state energy and compare it with the exact ground-state energy eigenvalue.

Solution

(a) It is straightforward to show that

$$[[H, z], z] = -\hbar^2/\mu$$

Taking the expectation value with respect to the state $|n\, \ell\, m\rangle$ and inserting a complete set of states $\{|n'\, \ell'\, m'\rangle\}$, we arrive at the desired sum rule.

The matrix element $\langle n'\, \ell'\, m'|z|0\, 0\, 0\rangle$ is equal to

$$\int d\Omega\, Y_{\ell'm'}^*(\theta,\, \phi)\, Y_{00}(\theta,\, \phi)\, \cos\theta \int_0^\infty dr\, r^3 R_{n'\ell'}(r)R_{00}(r)$$

$$= \frac{1}{\sqrt{3}}\delta_{m'0} \int d\Omega\, Y_{\ell'm'}^*(\Omega)\, Y_{10}(\Omega) \int_0^\infty dr\, r^3 R_{n'\ell'}(r)R_{00}(r)$$

$$= \frac{1}{\sqrt{3}}\delta_{m'0}\delta_{\ell'1} \int_0^\infty dr\, r^2 R_{n'\ell'}(r)r R_{00}(r)$$

$$= \frac{1}{\sqrt{3}}\delta_{m'0}\delta_{\ell'1} \sqrt{\frac{3\hbar}{2\mu\omega}} \int_0^\infty dr\, r^2 R_{n'\ell'}(r)R_{11}(r) = \sqrt{\frac{\hbar}{2\mu\omega}}\, \delta_{m'0}\delta_{\ell'1}\delta_{n'1}$$

This value can be immediately inserted into the sum rule and reduces it to a trivial identity.

(b) The first-order correction to the ground-state energy vanishes, since

$$E_0^{(1)} = -q\mathcal{E}\langle 0\, 0\, 0|z|0\, 0\, 0\rangle = 0$$

The next correction is

$$
E_0^{(2)} = q^2 \mathcal{E}^2 \sum_{n' \neq 0, \, \ell', \, m'} \frac{|\langle n' \, \ell' \, m' | z | 0 \, 0 \, 0 \rangle|^2}{E_0 - E_{n'}}
$$

$$
= -\frac{q^2 \mathcal{E}^2}{\hbar \omega} \sum_{n' \neq 0, \, \ell', \, m'} \frac{1}{n'} \left(\frac{\hbar}{2 \mu \omega} \right) \delta_{m' 0} \delta_{\ell' 1} \delta_{n' 1} = -\frac{q^2 \mathcal{E}^2}{2 \mu \omega^2}
$$

The first-order correction to the ground state is

$$
|0 \, 0 \, 0 \rangle^{(1)} = -q \mathcal{E} \sum_{n' \neq 0, \, \ell', \, m'} \frac{\langle n' \, \ell' \, m' | z | 0 \, 0 \, 0 \rangle}{E_0 - E_{n'}} |n' \, \ell' \, m' \rangle
$$

$$
= \frac{q \mathcal{E}}{\hbar \omega} \sqrt{\frac{\hbar}{2 \mu \omega}} |1 \, 1 \, 0 \rangle
$$

Thus, the corrected ground state will be

$$
\overline{|0 \, 0 \, 0 \rangle} = |0 \, 0 \, 0 \rangle + \frac{q \mathcal{E}}{\hbar \omega} \sqrt{\frac{\hbar}{2 \mu \omega}} |1 \, 1 \, 0 \rangle
$$

The expectation value of the potential energy in this state, to linear order in the electric field, will be

$$
\langle V(r) \rangle_0 = \frac{\mu \omega^2}{2} \langle 0 \, 0 \, 0 | r^2 | 0 \, 0 \, 0 \rangle + \frac{\mu q \mathcal{E} \omega}{\hbar} \sqrt{\frac{\hbar}{2 \mu \omega}} \, \mathrm{Re} \left(\langle 0 \, 0 \, 0 | r^2 | 1 \, 1 \, 0 \rangle \right)
$$

Note however that the matrix element $\langle 1 \, 1 \, 0 | r^2 | 0 \, 0 \, 0 \rangle \propto \int d\Omega \, Y_{10}^* Y_{00}$ vanishes. Thus the expectation value will be the same as in the unperturbed case, namely

$$
\langle V(r) \rangle_0 = \frac{1}{2} \mu \omega^2 \langle 0 \, 0 \, 0 | r^2 | 0 \, 0 \, 0 \rangle = \frac{3}{4} \hbar \omega
$$

(c) The above matrix element is

$$
\langle n' \, \ell' \, m' | z | 1 \, 1 \, 0 \rangle = \delta_{m' 0} \int d\Omega \, Y_{\ell' 1}^* Y_{10} \cos \theta \int_0^\infty dr \, r^3 R_{n' \ell'}(r) R_{11}(r)
$$

Note that, since $Y_{20} = \sqrt{5/16\pi} \, (3 \cos^2 \theta - 1)$, we can write

$$
\cos \theta \, Y_{10}(\Omega) = \sqrt{\frac{3}{4\pi}} \cos^2 \theta = \frac{1}{\sqrt{3}} Y_{00} + \frac{2}{\sqrt{15}} Y_{20}
$$

Thus, we obtain that the matrix element $\langle n' \, \ell' \, m' | z | 1 \, 1 \, 0 \rangle$ is equal to

$$
\frac{1}{\sqrt{3}} \delta_{m' 0} \delta_{\ell' 0} \int_0^\infty dr \, r^3 R_{n' 0} R_{11} + \frac{2}{\sqrt{15}} \delta_{m' 0} \delta_{\ell' 2} \int_0^\infty dr \, r^3 R_{n' 2} R_{11}
$$

Note now that

$$r R_{11} = \sqrt{\frac{\hbar}{\mu\omega}} \left(\sqrt{\frac{3}{2}} R_{00} - R_{20} \right)$$

and also

$$r R_{11} = \sqrt{\frac{\hbar}{\mu\omega}} \sqrt{\frac{5}{2}} R_{22}$$

Therefore, we have

$$\langle n'\, \ell'\, m'|z|1\, 1\, 0 \rangle = \sqrt{\frac{\hbar}{\mu\omega}} \left(\frac{1}{\sqrt{2}} \delta_{m'0} \delta_{\ell'0} \delta_{n'0} - \frac{1}{\sqrt{3}} \delta_{m'0} \delta_{\ell'0} \delta_{n'2} + \sqrt{\frac{2}{3}} \delta_{m'0} \delta_{\ell'2} \delta_{n'2} \right)$$

The sum rule has the form

$$\sum_{n',\ell',m'} (n'-1)|\langle n'\, \ell'\, m'|z|1\, 1\, 0 \rangle|^2 = \frac{\hbar}{2\mu\omega}$$

and it is immediately verified.

(d) The second-order correction to the ground state is

$$|0\, 0\, 0\rangle^{(2)} = \left(\frac{q\mathcal{E}}{\hbar\omega}\right)^2 \sum_{n'\neq 0,\, \ell',\, m'} \sum_{n''\neq 0,\, \ell'',\, m''} \frac{1}{n'n''}$$

$$\times \langle n'\, \ell'\, m'|z|n''\, \ell''\, m'' \rangle \langle n''\, \ell''\, m''|z|0\, 0\, 0 \rangle \, |n'\, \ell'\, m'\rangle$$

$$= \left(\frac{q\mathcal{E}}{\hbar\omega}\right)^2 \sqrt{\frac{\hbar}{2\mu\omega}} \sum_{n'\neq 0,\, \ell',\, m'} \frac{1}{n'} \langle n'\, \ell'\, m'|z|1\, 1\, 0 \rangle \, |n'\, \ell'\, m'\rangle$$

Substituting the matrix element determined in (c), we get

$$|0\, 0\, 0\rangle^{(2)} = \left(\frac{q\mathcal{E}}{\hbar\omega}\right)^2 \frac{\hbar}{2\mu\omega} \left(-\frac{1}{\sqrt{6}}|2\, 0\, 0\rangle + \frac{1}{\sqrt{3}}|2\, 2\, 0\rangle \right)$$

The perturbed ground state complete to second order is

$$\overline{|0\, 0\, 0\rangle} = |0\, 0\, 0\rangle + \frac{q\mathcal{E}}{\hbar\omega}\sqrt{\frac{\hbar}{2\mu\omega}}|1\, 1\, 0\rangle$$

$$+ \left(\frac{q\mathcal{E}}{\hbar\omega}\right)^2 \frac{\hbar}{2\mu\omega} \left(-\frac{1}{\sqrt{6}}|2\, 0\, 0\rangle + \frac{1}{\sqrt{3}}|2\, 2\, 0\rangle \right)$$

From this state we obtain the mean square radius,

$$\overline{\langle r^2\rangle}_0 = \langle r^2\rangle_0 + \left(\frac{q\mathcal{E}}{\hbar\omega}\right)^2 \frac{\hbar}{2\mu\omega} \left(\frac{1}{2}\langle 1\, 1\, 0|r^2|1\, 1\, 0\rangle - \frac{2}{\sqrt{6}}\langle 0\, 0\, 0|r^2|2\, 0\, 0\rangle \right)$$

$$= \frac{3}{2}\frac{\hbar}{\mu\omega} + \frac{7}{4}\left(\frac{q\mathcal{E}}{\hbar\omega}\right)^2 \left(\frac{\hbar}{\mu\omega}\right)^2$$

We thus have

$$\langle 0\,0\,0|r^2|2\,0\,0\rangle = -\frac{\sqrt{6}}{2}\left(\frac{\hbar}{\mu\omega}\right), \qquad \langle 1\,1\,0|r^2|1\,1\,0\rangle = \frac{5}{2}\left(\frac{\hbar}{\mu\omega}\right)$$

(e) The second-order correction to the ground-state energy is

$$E_0^{(2)} = -\frac{q^2\mathcal{E}^2}{\hbar\omega}\sum_{n'\neq 0}\sum_{\ell',m'}\frac{1}{n'}\,\big|\,\langle 0\,0\,0|z|n'\,\ell'\,m'\rangle\,\big|^2$$

$$= -\frac{q^2\mathcal{E}^2}{2\mu\omega^2}\sum_{n'\neq 0}\sum_{\ell',m'}\frac{1}{n'}\delta_{n'1}\delta_{\ell'1}\delta_{m'0} = -\frac{q^2\mathcal{E}^2}{2\mu\omega^2}$$

The system can be solved exactly by writing the potential as

$$V(\mathbf{r}) = \frac{\mu\omega^2}{2}\left(\mathbf{r} - \frac{q}{\mu\omega^2}\mathcal{E}\right)^2 - \frac{q^2\mathcal{E}^2}{2\mu\omega^2}$$

$$= \exp\left(-\frac{iq}{\hbar\mu\omega^2}\mathbf{p}\cdot\mathcal{E}\right)\left(\frac{\mu\omega^2 r^2}{2} - \frac{q^2\mathcal{E}^2}{2\mu\omega^2}\right)\exp\left(\frac{iq}{\hbar\mu\omega^2}\mathbf{p}\cdot\mathcal{E}\right)$$

From this it follows that the eigenvalues of the system are just

$$E_n = \hbar\omega\left(n + \frac{3}{2}\right) - \frac{q^2\mathcal{E}^2}{2\mu\omega^2}$$

while the eigenstates are the translated states

$$\exp\left(-\frac{iq}{\hbar\mu\omega^2}\mathbf{p}\cdot\mathcal{E}\right)|n\,\ell\,m\rangle.$$

Notice that the second-order perturbation theory gives the exact result.

Problem 9.7 A particle of charge q and mass μ is bound in the ground state of an isotropic harmonic oscillator potential. Consider a perturbation in the form of a weak time-dependent spatially uniform electric field $\mathcal{E}(t) = \mathcal{E}_0\Theta(t)\cos\overline{\omega}t\,e^{-t/\tau}$. Calculate the probability of finding the system in an excited state at time $t \gg \tau$, up to first order.

Solution

The perturbing potential is $H'(t) = -q\mathcal{E}_0 z\Theta(t)\cos\overline{\omega}t\,e^{-t/\tau}$. The probability of finding the system in an excited state $|n\,\ell\,m\rangle \neq |0\,0\,0\rangle$ will be

$$\mathcal{P} = \frac{1}{\hbar^2}\left|\int_0^t dt'\,e^{i(E_n - E_0)t'/\hbar}\,\langle n\,\ell\,m|H'(t')|0\,0\,0\rangle\right|^2$$

The matrix element of the perturbation $\langle n \, \ell \, m | H'(t') | 0 \, 0 \, 0 \rangle$ is

$$-q\mathcal{E}_0 \, e^{-t'/\tau} \, \cos \overline{\omega} t' \, \langle n \, \ell \, m | z | 0 \, 0 \, 0 \rangle = -q\mathcal{E}_0 \sqrt{\frac{\hbar}{2\mu\omega}} e^{-t'/\tau} \cos \overline{\omega} t' \, \delta_{n1}\delta_{\ell 1}\delta_{m0}$$

For the last step, $\langle n \, \ell \, m | z | 0 \, 0 \, 0 \rangle = \sqrt{\hbar/2\mu\omega} \, \delta_{n1}\delta_{\ell 1}\delta_{m0}$, see the previous problem.

Thus the probability, to first order, is non-zero only for a transition to the first excited level. It is

$$\mathcal{P}(t) = \delta_{n1}\delta_{\ell 1}\delta_{m0} \frac{q^2\mathcal{E}_0^2}{2\mu\hbar\omega} \left| \int_0^t dt' \, e^{i(E_n-E_0)t'/\hbar} \, e^{-t'/\tau} \, \cos \overline{\omega} t' \right|^2$$

Integrating, we obtain

$$\int_0^t dt' \, e^{i(\omega+\overline{\omega})t'-t'/\tau} = \frac{e^{i(\omega+\overline{\omega})t-t/\tau} - 1}{i(\omega+\overline{\omega}) - \tau^{-1}} \approx \frac{1}{-i(\omega+\overline{\omega}) + \tau^{-1}}$$

Thus

$$\left| \int_0^t \cdots \right|^2 \approx \frac{1}{4} \left| \frac{1}{-i(\omega+\overline{\omega}) + \tau^{-1}} + \frac{1}{-i(\omega-\overline{\omega}) + \tau^{-1}} \right|^2$$

$$= \tau^2 \frac{\left[1 + (\tau\omega)^2 + (\tau\overline{\omega})^2\right]^2 + (\tau\omega)^2 \left[1 + (\tau\omega)^2 - (\tau\overline{\omega})^2\right]^2}{\left[1 + \tau^2(\omega+\overline{\omega})^2\right]^2 \left[1 + \tau^2(\omega-\overline{\omega})^2\right]^2}$$

Problem 9.8 Consider a hydrogen atom in its ground state, which, beyond time $t = 0$, is subject to a spatially uniform time-dependent electric field $\mathcal{E}_0 e^{-t/\tau}$. Treating the electric field as a perturbation, calculate to first order the probability of finding the atom in the first excited state ($n = 2, \ell = 1, m$).

Solution

The perturbing Hamiltonian is $H' = -e\mathcal{E}_0 z e^{-t'/\tau}$. The first-order expression for the probability of the transition $1s \rightarrow 2p$ is

$$\mathcal{P}_{1s \rightarrow 2p}(t) = \frac{e^2\mathcal{E}_0^2}{\hbar^2} \left| \int_0^t dt' \, e^{i(E_2-E_1)t'/\hbar} \, e^{-t'/\tau} \langle 2 \, 1 \, m | z | 1 \, 0 \, 0 \rangle \right|^2$$

We obtain for the matrix element

$$\langle 2 \, 1 \, m | z | 1 \, 0 \, 0 \rangle = \delta_{m0} \frac{1}{\pi a_0^4 \sqrt{32}} \int_0^\infty dr \, r^4 e^{-3r/2a_0} \int d\Omega \, \cos^2 \theta$$

$$= a_0 \delta_{m0} \frac{1}{3\sqrt{2}} \left(\frac{2}{3}\right)^5 4! = a_0 \delta_{m0} \frac{256}{\sqrt{2} \times 243}$$

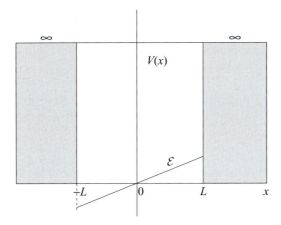

Fig. 40 Linear perturbation in an infinite square well.

Inserting this into the expression for the probability and performing the integration, we obtain

$$\mathcal{P}_{1s \to 2p}(t) = \delta_{m0} \frac{e^2 \mathcal{E}_0^2 a_0^2}{\hbar^2} \left(\frac{256}{\sqrt{2}(243)} \right)^2 \frac{1 - 2e^{-t/\tau} \cos \omega t + e^{-2t/\tau}}{1/\tau^2 + \omega^2}$$

where

$$\omega \equiv \frac{E_2 - E_1}{\hbar} = \frac{3e^2}{8\hbar a_0}$$

At very late times $t \gg \tau$, the probability is

$$\mathcal{P}_{1s \to 2p} \approx \delta_{m0} \frac{e^2 \mathcal{E}_0^2 a_0^2}{\hbar^2} \left(\frac{256}{\sqrt{2}(243)} \right)^2 \frac{\tau^2}{1 + (\omega\tau)^2}$$

Problem 9.9 A particle of mass m and charge q moves in one dimension between the impenetrable walls of an infinite square-well potential

$$V(x) = \begin{cases} 0, & |x| < L \\ \infty, & |x| > L \end{cases}$$

(a) Consider a weak uniform electric field of strength $\mathcal{E}$ that acts on the particle; see Fig. 40. Calculate the first non-trivial correction to the particle's ground-state energy.[11] What is the probability of finding the particle in the first excited state?

---

[11] You may use the sum

$$\sum_{n=1}^{\infty} \left[\frac{1}{(4n^2 - 1)^3} + \frac{4}{(4n^2 - 1)^4} + \frac{4}{(4n^2 - 1)^5} \right] = \frac{1}{2} - \frac{\pi^2}{64} \left(\frac{7}{4} + \frac{\pi^2}{12} \right)$$

(b) Consider now the case of a time-dependent electric field of the form $\mathcal{E}(t) = \mathcal{E}_0 \Theta(t) e^{-t/\tau}$. Calculate the transition probability from the ground state of the system to the first excited state in first-order time-dependent perturbation theory for times $t \gg \tau$.

Solution

(a) The perturbing potential is $H' = -q\mathcal{E}x$. The first-order correction to the energy eigenvalues vanishes owing to parity, as will now be seen. Each of the energy eigenfunctions

$$\psi_n(x) = \frac{1}{2\sqrt{L}} \left[e^{in\pi x/2L} + (-1)^{n+1} e^{-in\pi x/2L} \right]$$

with $n = 1, 2, \ldots$, has parity $(-1)^{n+1}$. Thus

$$E_n^{(1)} = -q\mathcal{E}\langle \psi_n | x | \psi_n \rangle = -(-1)q\mathcal{E}\langle \psi_n | \mathcal{P}x\mathcal{P} | \psi_n \rangle = q\mathcal{E}\langle \psi_n | x | \psi_n \rangle = 0$$

The second-order perturbation theory correction reads[12]

$$E_n^{(2)} = -\frac{8mL^2q^2\mathcal{E}^2}{\hbar^2\pi^2} \sum_{n' \neq n} \frac{|\langle \psi_n | x | \psi_{n'} \rangle|^2}{n'^2 - n^2}$$

For $n = 1$, we need the matrix elements $\langle \psi_1 | x | \psi_{n'} \rangle$. Note that, owing to parity, only the matrix elements $\langle \psi_1 | x | \psi_{2v} \rangle$ with $v = 1, 2, \ldots$, are non-zero. After performing an integral, we obtain

$$\langle \psi_1 | x | \psi_{2v} \rangle = \frac{8iL}{\pi^2}(-1)^v \left[\frac{1}{4v^2 - 1} + \frac{2}{(4v^2 - 1)^2} \right]$$

and thus

$$|\langle \psi_1 | x | \psi_{2v} \rangle|^2 = \frac{64L^2}{\pi^4} \left[\frac{1}{(4v^2 - 1)^2} + \frac{4}{(4v^2 - 1)^3} + \frac{4}{(4v^2 - 1)^4} \right]$$

Thus, the energy eigenvalue correction is

$$\begin{aligned}
E_1^{(2)} &= -\frac{8^3mL^4q^2\mathcal{E}^2}{\hbar^2\pi^6} \sum_{v=1}^{\infty} \left[\frac{1}{(4v^2 - 1)^3} + \frac{4}{(4v^2 - 1)^4} + \frac{4}{(4v^2 - 1)^5} \right] \\
&= -\frac{8^3mL^4q^2\mathcal{E}^2}{\hbar^2\pi^6} \left(\frac{1}{2} - \frac{21\pi^2}{768} - \frac{\pi^4}{768} \right) \\
&= -\frac{8mL^4q^2\mathcal{E}^2}{\hbar^2\pi^2} \left(\frac{32}{\pi^4} - \frac{7}{4\pi^2} - \frac{1}{12} \right)
\end{aligned}$$

---

[12] The unperturbed energy eigenvalues are $E_n = \hbar^2\pi^2n^2/8mL^2$.

The first-order correction to the ground state is

$$|\psi_1\rangle^{(1)} = \frac{8mL^2q\mathcal{E}}{\hbar^2\pi^2} \sum_{v=1}^{\infty} \frac{1}{4v^2-1} \langle\psi_{2v}|x|\psi_1\rangle \, |\psi_{2v}\rangle$$

$$= \frac{8^2 imL^3q\mathcal{E}}{\hbar^2\pi^4} \sum_{v=1}^{\infty}(-1)^v \left[\frac{1}{(4v^2-1)^2} + \frac{2}{(4v^2-1)^4}\right]|\psi_{2v}\rangle$$

The probability of finding the particle in the first excited state ($n = 2$) is therefore

$$\mathcal{P}_{1\to2} = \left|\langle\psi_2\overline{|\psi_1\rangle}\right|^2 = \left|\langle\psi_2|\psi_1\rangle^{(1)}\right|^2 = (11)^2 \left(\frac{8}{9}\right)^4 \left(\frac{mL^3q\mathcal{E}}{\hbar^2\pi^4}\right)^2$$

(b) The probability for a transition from the ground state to a different eigenstate ψ_n is given by

$$\mathcal{P}_{1\to n}(t) = \frac{q^2\mathcal{E}_0^2}{\hbar^2} \left|\int_0^t dt' \, e^{i(E_n-E_1)t'/\hbar - t'/\tau} \langle\psi_n|x|\psi_1\rangle\right|^2$$

From parity, it is clear again that only transitions $1 \to 2v$ can occur. The corresponding matrix element squared is

$$|\langle\psi_1|x|\psi_{2v}\rangle|^2 = \frac{64L^2}{\pi^4} \left[\frac{1}{(4v^2-1)^2} + \frac{4}{(4v^2-1)^3} + \frac{4}{(4v^2-1)^4}\right]$$

Thus, we may write

$$\mathcal{P}_{1\to2}(t) = \frac{64L^2q^2\mathcal{E}_0^2}{\hbar^2\pi^4} \left(\frac{5}{9}\right)^2 \left|\frac{e^{i(E_2-E_1)t/\hbar - t/\tau} - 1}{i(E_2-E_1)/\hbar - \tau^{-1}}\right|^2$$

For times $t \gg \tau$, the probability becomes independent of time:

$$\mathcal{P}_{1\to2} \approx \frac{64L^2q^2\mathcal{E}_0^2}{\hbar^2\pi^4} \left(\frac{5}{9}\right)^2 \frac{\tau^2}{1 + \left(3\hbar\pi^2/8mL^2\right)^2\tau^2}$$

Note that the characteristic time of the unperturbed system is $\tau_0 = \hbar/E_1 = 8mL^2/\hbar\pi^2$. Thus we may write

$$\mathcal{P}_{1\to2} \approx \frac{64L^2q^2\mathcal{E}_0^2}{\hbar^2\pi^4} \left(\frac{5}{9}\right)^2 \left(\frac{8mL^2}{\hbar\pi^2}\right)^2 \frac{(\tau/\tau_0)^2}{1 + 9(\tau/\tau_0)^2}$$

This can also be written as

$$P = P_{\text{static}} \left(\frac{15}{11}\right)^2 \frac{9(\tau/\tau_0)^2}{1 + 9(\tau/\tau_0)^2}$$

in terms of the static probability obtained in (a).

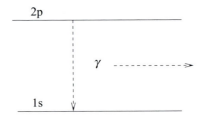

Fig. 41 Photon emission through $2p \to 1s + \gamma$.

Problem 9.10 Consider a hydrogen atom in a 2p state that is perturbed by a plane electromagnetic wave[13] of wave number $\mathbf{k}$ and frequency $\omega = ck$:

$$\mathbf{A}(\mathbf{r},\, t) = 2\mathbf{A}_0 \, \cos(\omega t - \mathbf{k} \cdot \mathbf{r})$$

The positive-frequency part[14] of this vector potential gives a semi-classical description of the emission of a photon. Assume that the wavelength $\lambda = 2\pi / k$ is much larger than the effective dimension of the atom a_0.

(a) Calculate the probability per unit time of finding the atom in its ground state ($n = 1$, $\ell = m = 0$) in first-order perturbation theory (Fig. 41). Assume that we are interested in times t that satisfy $t \gg \omega^{-1}$ and $t \gg \hbar/|E_1|$. Express your result in terms of the matrix element of the electric dipole moment of the atom.

(b) Integrate over all possible wave numbers $\mathbf{k}$ of the electromagnetic wave and obtain the $2p \to 1s$ transition rate per unit solid angle corresponding to the direction of the emitted photon $\hat{\mathbf{k}}$. Determine the amplitude A_0 from the condition that the energy density calculated from the above electromagnetic wave must coincide with the energy density corresponding to one photon per unit volume.

(c) Obtain the total transition rate assuming that we do not measure the direction and polarization of the emitted photon.

Solution

(a) The perturbing Hamiltonian is

$$H' = -\frac{e}{2\mu c} \, (\mathbf{p} \cdot \mathbf{A} + \mathbf{A} \cdot \mathbf{p}) = \overline{H}(0) \, e^{i\omega t} + \overline{H}^{\dagger}(0) \, e^{-i\omega t}$$

$$\overline{H}(0) = -\frac{e}{2\mu c} \, (\mathbf{p} \cdot \mathbf{A}_0 + \mathbf{A}_0 \cdot \mathbf{p}) \, e^{-i\mathbf{k}\cdot\mathbf{r}}$$

Note that

$$p_i \, e^{-i\mathbf{k}\cdot\mathbf{r}} = e^{-i\mathbf{k}\cdot\mathbf{r}} p_i - \hbar k_i e^{-i\mathbf{k}\cdot\mathbf{r}} \qquad \Longrightarrow \qquad \mathbf{A}_0 \cdot \mathbf{p}\, e^{-i\mathbf{k}\cdot\mathbf{r}} = e^{-i\mathbf{k}\cdot\mathbf{r}} \mathbf{A}_0 \cdot \mathbf{p}$$

since $\mathbf{A}_0 \cdot \mathbf{k} = 0$.

---

[13] The vector potential is *transverse*, i.e. $\nabla \cdot \mathbf{A} = \hat{\mathbf{k}} \cdot \mathbf{A}_0 = 0$.

[14] The vector potential can be written as a sum of a *positive-frequency* and a *negative-frequency* term, $\mathbf{A} = \mathbf{A}_0 e^{i\mathbf{k}\cdot\mathbf{x}} e^{-i\omega t} + \mathbf{A}_0 e^{-i\mathbf{k}\cdot\mathbf{x}} e^{i\omega t}$.

The transition probability is ($\alpha = (E_2 - E_1)/\hbar = 3e^2/8a_0\hbar$)

$$\mathcal{P}_{2p \to 1s} = \frac{1}{\hbar^2} \left| \int_{-\infty}^{t} dt' \left[e^{i(-\alpha+\omega)t'} + e^{i(-\alpha-\omega)t'} \right] \langle 1\ 0\ 0 | \overline{H}(0) | 2\ 1\ m \rangle \right|^2$$

Since the extent of the atom is roughly a_0 and we have $ka_0 \ll 1$, we make the expansion $e^{-i\mathbf{k}\cdot\mathbf{r}} \approx 1 - i\mathbf{k} \cdot \mathbf{r} + \cdots$ and keep the first term (the dipole approximation). Then we get, for the matrix element,

$$\langle 1\ 0\ 0 | H'(0) | 2\ 1\ m \rangle \approx -\frac{e}{\mu c} \mathbf{A}_0 \cdot \langle 1\ 0\ 0 | \mathbf{p} | 2\ 1\ m \rangle$$

We can proceed by noting that

$$[H_0, \mathbf{r}] = -\frac{i\hbar}{\mu} \mathbf{p},$$

where H_0 is the unperturbed hydrogen-atom Hamiltonian. Then

$$\langle 2\ 1\ m | \mathbf{p} | 1\ 0\ 0 \rangle = \frac{i\mu}{\hbar} (E_2 - E_1) \langle 2\ 1\ m | \mathbf{r} | 1\ 0\ 0 \rangle = \frac{i\mu\alpha}{e} \langle 2\ 1\ m | \mathbf{d} | 1\ 0\ 0 \rangle$$

where $\mathbf{d}$ is the electric dipole operator. We have

$$\langle 2\ 1\ m | \overline{H}(0) | 1\ 0\ 0 \rangle = \frac{i\alpha}{c} \mathbf{A}_0 \cdot \langle 2\ 1\ m | \mathbf{d} | 1\ 0\ 0 \rangle = \frac{i\alpha}{c} A_0 \langle 2\ 1\ m | \hat{\mathbf{e}} \cdot \mathbf{d} | 1\ 0\ 0 \rangle$$

in terms of the *polarization (unit) vector* $\hat{\mathbf{e}}$ for the electromagnetic field, introduced as $\mathbf{A}_0 = A_0 \hat{\mathbf{e}}$. Note that $\hat{\mathbf{k}} \cdot \hat{\mathbf{e}} = 0$.

The time integration gives for large times $t \gg \alpha^{-1}$, ω^{-1}

$$\left| \int_{-\infty}^{t} dt'\, e^{i(\omega-\alpha)t'} + \int_{-\infty}^{t} dt'\, e^{i(-\omega-\alpha)t'} \right|^2$$

$$= 2\pi [\delta(\alpha - \omega) + \delta(\alpha + \omega)] \left[\int_{-\infty}^{t} dt'\, e^{-i(\omega-\alpha)t'} + \int_{-\infty}^{t} dt'\, e^{i(\omega+\alpha)t'} \right]$$

$$= 2\pi\delta(\alpha - \omega) \left(t + \int_{-\infty}^{\infty} dt'\, e^{-2i\omega t'} \right) = 2\pi t\delta(\alpha - \omega)$$

Thus, the probability per unit time is

$$\frac{\mathcal{P}_{2p \to 1s}}{t} = \frac{2\pi A_0^2 \alpha^2}{c^2 \hbar} \delta(E_2 - E_1 - \hbar\omega) \left| \langle 2\ 1\ m | (\hat{\mathbf{e}} \cdot \mathbf{d}) | 1\ 0\ 0 \rangle \right|^2$$

The delta function enforces the conservation of energy between the energy of the atom and the energy of the emitted photon.

(b) In order to get the transition rate we multiply the probability per unit time by the number of available electromagnetic field modes. If the whole system is put in a box of volume $V \to \infty$, the allowed wave numbers will be labelled by integers

n_x, n_y, n_z that are related to the wave number through

$$k_x = \frac{2\pi n_x}{V^{1/3}}, \qquad k_y = \frac{2\pi n_y}{V^{1/3}}, \qquad k_z = \frac{2\pi n_z}{V^{1/3}}$$

The number of available states will be

$$\sum_{n_x, n_y, n_z} \approx \int dn_x \int dn_y \int dn_z = V \int \frac{d^3 k}{(2\pi)^3} = \frac{V}{8\pi^3 c^3} \int_0^\infty d\omega\, \omega^2\, d\Omega_{\hat{\mathbf{k}}}$$

Thus, the rate for the transition $2p \rightarrow 1s$ per unit solid angle will be

$$\frac{d\mathcal{W}}{d\Omega} = \frac{V}{8\pi^3 c^3} \int_0^\infty d\omega\, \omega^2\, \frac{\mathcal{P}_{2p\rightarrow 1s}}{t} = \frac{V A_0^2 \omega^4}{4\hbar^2 \pi^2 c^5} \left| \langle 2\, 1\, m | \, \hat{\mathbf{e}} \cdot \mathbf{d} \, | 1\, 0\, 0 \rangle \right|^2$$

The time-averaged energy density of the electromagnetic field is[15]

$$\langle U \rangle = \frac{1}{2}\left\langle \frac{1}{c^2}\dot{\mathbf{A}}^2 + (\nabla \times \mathbf{A})^2 \right\rangle = \frac{2\omega^2}{c^2} A_0^2$$

Since the transition occurs by the emission of one photon of energy $\hbar\omega$, we can determine the amplitude of the corresponding classical electromagnetic field by equating its energy density with the energy density corresponding to the emitted photon,

$$\frac{\omega^2}{c^2} A_0^2 = \frac{\hbar\omega}{V}$$

and the rate becomes

$$\left.\frac{d\mathcal{W}}{d\Omega}\right|_{2p\rightarrow 1s} = \frac{\omega^3}{4\hbar\pi^2 c^3} \left| \langle 2\, 1\, m | \, (\hat{\mathbf{e}} \cdot \mathbf{d}) \, | 1\, 0\, 0 \rangle \right|^2$$

(c) The dipole-moment matrix element can be calculated in terms of the given hydrogen-atom states and the polarization vector of the electromagnetic wave. It is

$$\langle 2\, 1\, m | \mathbf{d} | 1\, 0\, 0 \rangle$$

$$= \frac{e}{32\pi a_0^4} \int_0^\infty dr\, r^2 r r e^{-3r/2a_0} \int d\Omega\, Y_{1m}^* Y_{00} \hat{\mathbf{r}}$$

$$= \frac{e 4! a_0}{\pi\sqrt{32}} \left(\frac{2}{3}\right)^5 \int d\Omega\, Y_{1m}^* \left[-\frac{1}{2}\sqrt{\frac{2}{3}}(\hat{\mathbf{x}} - i\hat{\mathbf{y}})Y_{1,-1} + \frac{1}{2}\sqrt{\frac{2}{3}}(\hat{\mathbf{x}} + i\hat{\mathbf{y}})Y_{11} + \frac{1}{\sqrt{3}}\hat{\mathbf{z}}Y_{10} \right]$$

$$= \frac{e 4! a_0}{\pi\sqrt{32}} \left(\frac{2}{3}\right)^5 \left[-\frac{1}{2}\sqrt{\frac{2}{3}}(\hat{\mathbf{x}} - i\hat{\mathbf{y}})\delta_{m,-1} + \frac{1}{2}\sqrt{\frac{2}{3}}(\hat{\mathbf{x}} + i\hat{\mathbf{y}})\delta_{m,1} + \frac{1}{\sqrt{3}}\hat{\mathbf{z}}\delta_{m,0} \right]$$

---

[15] The time average required is $\langle \sin^2(\mathbf{k}\cdot\mathbf{r} - \omega t)\rangle = T^{-1}\int_0^T dt\, \sin^2(\mathbf{k}\cdot\mathbf{r} - \omega t) = 1/2$.

The square of this matrix element, which appears in the 2p → 1s transition rate, is

$$| \langle 2\ 1\ m| \ (\hat{\mathbf{e}} \cdot \mathbf{d})\ |1\ 0\ 0\rangle |^2 = \frac{3e^2 a_0^2}{\pi^2} \left(\frac{2}{3}\right)^{10} \left[\left(\hat{e}_x^2 + \hat{e}_y^2\right)\left(\delta_{m,-1} + \delta_{m,1}\right) + 2\hat{e}_z^2\delta_{m,0}\right]$$

Note also that $| \langle 2\ 1\ m| \mathbf{d}\ |1\ 0\ 0\rangle |^2 = 6 \left(\frac{2}{3}\right)^{10} e^2 a_0^2/\pi^2$. Therefore, we can write

$$| \langle 2\ 1\ m| \ \hat{\mathbf{e}} \cdot \mathbf{d}\ |1\ 0\ 0\rangle |^2 = \tfrac{1}{2} | \langle \mathbf{d}\rangle |^2 \left[\left(\hat{e}_x^2 + \hat{e}_y^2\right)\left(\delta_{m,-1} + \delta_{m,1}\right) + 2\hat{e}_z^2\delta_{m,0}\right]$$

and

$$\left.\frac{dW}{d\Omega}\right|_{2p \to 1s} = \frac{\omega^3}{8\hbar\pi^2 c^3} | \langle \mathbf{d}\rangle |^2 \left[\left(\hat{e}_x^2 + \hat{e}_y^2\right)\left(\delta_{m,-1} + \delta_{m,1}\right) + 2\hat{e}_z^2\delta_{m,0}\right]$$

If we take $\hat{\mathbf{k}} = \hat{\mathbf{z}}$ then the polarization vector must have $\hat{e}_z = 0$. We can consider the two linearly independent polarizations $\hat{\mathbf{e}}^{(1)} = \hat{\mathbf{x}}$ and $\hat{\mathbf{e}}^{(2)} = \hat{\mathbf{y}}$. For a general polarization, the differential rate is

$$\frac{e^2}{4\pi\hbar c} \left(\frac{\omega^3}{2\pi c^2}\right) | \langle \mathbf{r}\rangle |^2 \left[\sin^2\vartheta \left(\delta_{m,-1} + \delta_{m,1}\right) + 2\cos^2\vartheta\, \delta_{m,0}\right]$$

Integrating over angles, we obtain

$$\mathcal{W}_{2p \to 1s+\gamma} = \frac{e^2}{4\pi\hbar c} \left(\frac{4\omega^3}{3c^2}\right) | \langle \mathbf{r}\rangle |^2$$

Problem 9.11 A hydrogen atom is subject to a perturbing electromagnetic wave with vector potential[16]

$$\mathbf{A}(\mathbf{r}, t) = 2\mathbf{A}_0 \cos(\mathbf{k} \cdot \mathbf{r} - \omega t)$$

The wavelength of the electromagnetic wave is much larger than the effective size of the atom, i.e. $\lambda \gg a_0$. The atom makes a quantum transition from a state $|A\rangle = |n\ \ell\ m\ m_s\rangle$ to another state $|B\rangle = |n'\ \ell'\ m'\ m_s'\rangle$. Within the framework of first-order perturbation theory, find the requirements on the parity and angular momentum of the initial and final states for such a transition to occur (*the selection rules*).

Solution
The perturbing potential is

$$H'(t) = -\frac{e}{2\mu c} \left(\mathbf{A} \cdot \mathbf{p} + \mathbf{p} \cdot \mathbf{A}\right) - g\frac{e}{2\mu c}\mathbf{S} \cdot (\nabla \times \mathbf{A})$$

---

[16] The vector potential is transverse, i.e. $\nabla \cdot \mathbf{A} = 0$, or, equivalently, $\mathbf{k} \cdot \mathbf{A}_0 = 0$. The frequency is $\omega(k) = ck$.

The probability for a transition $A \to B$ to occur is, to first order,

$$\mathcal{P}_{A \to B} = \frac{1}{\hbar^2} \left| \int_{-\infty}^{t} dt' \, e^{i(E_B - E_A)t'/\hbar} \langle B; \ldots | H'(t') | A; \ldots \rangle \right|^2$$

The ellipses (dots) in the initial and final states refer to the photon that is absorbed or emitted. In our approximation, these states are just products of an atomic state and a photonic one. Thus, the photonic part is all lumped into a factor $\langle \ldots \mathbf{A}_0 \ldots \rangle = \tilde{\alpha}_0 \, \hat{\mathbf{e}}(k)$, where $\hat{\mathbf{e}}$ is a polarization vector ($\hat{\mathbf{k}} \cdot \hat{\mathbf{e}} = 0$). Finally, we get

$$\mathcal{P}_{A \to B} = \frac{e^2 |\tilde{\alpha}_0|^2}{4\mu^2 c^2 \hbar^2} \left| \int_{-\infty}^{t} dt' \, e^{i(E_B - E_A)t'/\hbar} \langle B | X | A \rangle \right|^2$$

where

$$X = \cos(\mathbf{k} \cdot \mathbf{r} - \omega t') \, \hat{\mathbf{e}} \cdot \mathbf{p} + \mathbf{p} \cdot \hat{\mathbf{e}} \, \cos(\mathbf{k} \cdot \mathbf{r} - \omega t')$$
$$- g \, \mathbf{S} \cdot (\hat{\mathbf{e}} \times \mathbf{k}) \, \sin(\mathbf{k} \cdot \mathbf{r} - \omega t')$$

At this point, we note that, considering the jth component of $\mathbf{p}$,

$$p_j \cos(\mathbf{k} \cdot \mathbf{r} - \omega t) = \cos(\mathbf{k} \cdot \mathbf{r} - \omega t) p_j + i \hbar k_j \sin(\mathbf{k} \cdot \mathbf{r} - \omega t)$$

Thus

$$\hat{\mathbf{e}} \cdot \mathbf{p} \cos(\mathbf{k} \cdot \mathbf{r} - \omega t) = \cos(\mathbf{k} \cdot \mathbf{r} - \omega t) \, \hat{\mathbf{e}} \cdot \mathbf{p}$$

since $\hat{\mathbf{e}} \cdot \mathbf{k} = 0$.

Since the atomic wave functions fall off rapidly beyond a_0 and $k a_0 \ll 1$, we can safely expand the cosine and keep at most linear terms:

$$\cos(\mathbf{k} \cdot \mathbf{r} - \omega t) \approx \cos \omega t + \mathbf{k} \cdot \mathbf{r} \, \sin \omega t + \cdots$$

Returning now to the atomic matrix element, we get

$$\langle B | X | A \rangle = 2 \hat{\mathbf{e}} \cdot \langle B | \mathbf{p} | A \rangle \cos \omega t' + 2 \langle B | (\mathbf{k} \cdot \mathbf{r})(\hat{\mathbf{e}} \cdot \mathbf{p}) | A \rangle \sin \omega t'$$
$$+ g (\hat{\mathbf{e}} \times \mathbf{k}) \cdot \langle B | \mathbf{S} | A \rangle \sin \omega t' + O(k^2)$$

The spin term does not play any role since the unperturbed Hamiltonian does not depend on spin and the spatial states are orthogonal to the spin states.

The dominant term will be proportional to the matrix element of the momentum operator. This transition is called *electric dipole transition* (E1) because of the relation of the momentum matrix element to the electric dipole operator:

$$\langle B | \mathbf{p} | A \rangle = \frac{i \mu}{\hbar} \langle B | [H_0 \, \mathbf{p}] | A \rangle$$
$$= \frac{i \mu}{\hbar} (E_B - E_A) \langle B | \mathbf{r} | A \rangle = \frac{i \mu}{e \hbar} (E_B - E_A) \langle B | \mathbf{d} | A \rangle$$

Since the momentum or the dipole moment operators are of odd parity, the parities of the initial and final states must be different, otherwise the matrix element vanishes and this transition cannot occur. Thus

$$\mathcal{P}\mathbf{d}\mathcal{P} = -\mathbf{d} \qquad \Longrightarrow \qquad \varpi_B \varpi_A = -1$$

However, since these operators are *vector operators* they can be written as

$$\mathbf{V} = \tfrac{1}{\sqrt{2}}\left[(V_{-1} - V_{+1})\hat{\mathbf{x}} - \tfrac{i}{\sqrt{2}}(V_1 + V_{+1})\hat{\mathbf{y}} + V_0\hat{\mathbf{z}}\right]$$

For any component V_q of such an operator, we may deduce, through the *Wigner–Eckart theorem*,

$$\langle j_B\, m_B | V_q | j_A\, m_A \rangle = \frac{1}{\sqrt{2j_A + 1}}\, \langle j_A\, 1\,;\, m_A\, q \mid j_B\, m_B \rangle\, \langle B || V_q || A \rangle$$

that the angular momentum quantum numbers must differ by unity. Otherwise, this transition vanishes. Thus,

$$\text{E1} \qquad \Longrightarrow \qquad |j_B - j_A| = 1$$

If the states are such that the electric-dipole matrix element vanishes, the dominant remaining term contains the product $x_i p_j$, which can be written as

$$\tfrac{1}{2}\left(x_i p_j + p_i x_j\right) + \tfrac{1}{2}\left(x_i p_j - p_i x_j\right)$$

The first term gives rise to the *electric quadrupole transition* (E2); the corresponding matrix element is

$$\langle B | \left(x_i p_j + p_i x_j\right) | A \rangle\, k_i \hat{\epsilon}_j$$

Note, however, the identity

$$[H_0, x_i x_j] = -\frac{i\hbar}{2\mu}\left(p_i x_j + p_j x_i + x_j p_i + x_i p_j\right)$$

$$= -\frac{\hbar^2}{\mu}\delta_{ij} - \frac{i\hbar}{\mu}\left(x_j p_i + x_i p_j\right)$$

Thus, we may write

$$\langle B | \left(x_i p_j + p_i x_j\right) | A \rangle k_i\, \hat{\epsilon}_j = \langle B |[H_0, x_i x_j]| A \rangle\, k_i \hat{\epsilon}_j + \frac{\hbar^2}{\mu}\mathbf{k} \cdot \hat{\mathbf{e}}$$

The last term is, of course, zero. Because of transversality, though, we can add to the commutator any term proportional to δ_{ij}. Thus, we can introduce the *quadrupole moment operator*

$$Q_{ij} \equiv x_i x_j - \tfrac{1}{3}r^2 \delta_{ij}$$

and then the matrix element becomes

$$\langle B|Q_{ij}|A\rangle \, k_i \, \hat{\epsilon}_j$$

Since the quadrupole moment is a parity-even operator, the corresponding selection rule will be

$$\text{E2} \quad \Longrightarrow \quad \varpi_A = \varpi_B$$

Applying the Wigner–Eckart theorem to the second-rank tensor operator, we obtain

$$\langle j_B \, m_B \, |Q_q| j_A \, m_A\rangle \propto \langle j_A \, 2 \, ; \, m_A \, q \, | \, j_B \, m_B\rangle$$

and so

$$\text{E2} \quad \Longrightarrow \quad |j_B - j_A| \le 2 \le j_A + j_B$$

The remaining term gives rise to the *magnetic dipole transition* (M1) and corresponds to the matrix element

$$\langle B| \left(x_i p_j - p_i x_j\right) |A\rangle \, k_i \, \hat{\epsilon}_j = (\mathbf{k} \times \hat{\mathbf{e}}) \cdot (\mathbf{r} \times \mathbf{p})$$
$$= (\mathbf{k} \times \hat{\mathbf{e}}) \cdot \langle B|\mathbf{L}|A\rangle$$

Since the angular momentum is a parity-even vector operator, the corresponding selection rules are

$$\text{M1} \quad \Longrightarrow \quad \varpi_A = \varpi_B, \qquad |j_A - j_B| = 1$$

Problem 9.12 A particle of mass m moves in one dimension under the influence of forces given by the *double-well* potential

$$V(x) = \frac{m\omega^2}{4a^2} \left(x^2 - a^2\right)^2$$

(a) Consider the propagator and replace the time variable t by the *Euclidean time* τ, via $t \to -i\tau$. Then show in general that the lowest-energy eigenvalue can be obtained from the limit

$$E_0 = -\lim_{T \to \infty} \frac{\hbar}{T} \ln \mathcal{K}(x', x; -iT)$$

where T is the Euclidean time and $\mathcal{K}$ is the propagator for a transition of the particle from the point x to the point x'.

(b) For the potential given above, obtain a solution of the classical equations of motion, with the time variable replaced via $t \to -i\tau$, in which the particle starts at the local minimum $x = -a$ at $\tau = -T/2 \to -\infty$ and ends up at the other local minimum, $x' = +a$, at $\tau = T/2 \to \infty$. Discuss also the possibility of a solution that is the sum of two such solutions with widely separated centres.

(c) Consider a classical solution that is the sum of N widely separated solutions of the type derived in (b). In the limit $\hbar \to 0$ each of these solutions gives an approximation to the propagator through the formula

$$K(a, -a; -iT) \approx \sum_N A_N \exp\left(-\mathcal{S}_c^{(N)}[x]/\hbar\right)$$

where A_N is an unknown constant and $\mathcal{S}_c^{(N)}[x]$ is the classical action[17] in Euclidean time for each of the solutions. Derive the dependence on N of each of the above quantities and perform the summation.

(d) Calculate the ground-state energy of the system in terms of the given parameters and the unknown factor A_N.

Solution

(a) From the expression for the propagator in terms of the energy eigenstates,

$$K(x', x; t) = \sum_{n=0}^{\infty} e^{-iE_n t/\hbar} \psi_n(x)\psi_n^*(x')$$

by going to imaginary values of the time via $t \to -i\tau$ we get

$$K(x', x; -i\tau) = \sum_{n=0}^{\infty} e^{-E_n\tau/\hbar} \psi_n(x)\psi_n^*(x') = e^{-E_0\tau/\hbar}\psi_0(x)\psi_0^*(x') + \cdots$$

In the limit $\tau \to \infty$ it is clear that the surviving term will be the one corresponding to the smallest eigenvalue, namely E_0. Thus, we can write

$$E_0 = -\lim_{T\to\infty}\left[\frac{\hbar}{T}\ln K(x', x; -iT)\right]$$

(b) The classical equations of motion are

$$m\frac{d^2x(\tau)}{d\tau^2} = V'(x)$$

Multiplying by the 'velocity' $\dot{x}(\tau)$, we get

$$\frac{m}{2}\frac{d}{d\tau}[\dot{x}(\tau)]^2 = \frac{dV(x)}{d\tau} \quad\Longrightarrow\quad \frac{d}{d\tau}\left[\frac{m(\dot{x})^2}{2} - V(x)\right] = 0$$

which gives an integral of the motion

$$E = \frac{m(\dot{x})^2}{2} - V(x)$$

---

[17]
$$\mathcal{S}_c = \int_{-T/2}^{T/2} d\tau\left[\frac{m(\dot{x})^2}{2} + V(x)\right]$$

We are interested in a solution with boundary conditions $x(\pm\infty) = \pm a$. Such a solution is very easily obtained if we make the additional assumption that the velocity vanishes at the endpoints, i.e. $\dot{x}(\pm\infty) = 0$. Then, since the potential vanishes at these points also, the conserved quantity E must be zero. Thus

$$\frac{dx}{\sqrt{V(x)}} = d\tau \sqrt{\frac{2}{m}} \quad \Longrightarrow \quad \tau\omega = 2a \int^{x(\tau)} \frac{dx}{a^2 - x^2} = -\ln\left[\frac{a - x(\tau)}{a + x(\tau)}\right]$$

or

$$x(\tau) = a \tanh \frac{\omega}{2}(\tau - \tau_0)$$

Note the arbitrary integration constant τ_0.

It is not difficult to see that if two such solutions are each localized at two widely separated points ($|\tau_1 - \tau_2| \gg \omega^{-1}$), the sum $x_1(\tau) + x_2(\tau)$ will be an approximate solution since any products $x_1 x_2$ will have small overlap. The same is true for a sum of an arbitrary number of solutions.

Substituting the classical trajectory into the classical action, we obtain

$$\mathcal{S}_c = \frac{3m\omega a^2}{4} \int_{-\infty}^{\infty} dz \frac{1}{\cosh^4 z} = m\omega a^2$$

This is independent of the centre point. Thus, for a sum of N such solutions the action will be $N\mathcal{S}_c = Nm\omega a^2$.

(c) Each of the centres τ_i (with $i = 1, 2, \ldots, N$) can have any value between $-T/2$ and $T/2$. Thus, we need to make N integrations:

$$\frac{1}{N!} \int_{-T/2}^{T/2} d\tau_1 \int_{-T/2}^{T/2} d\tau_2 \cdots \int_{-T/2}^{T/2} d\tau_N = \frac{T^N}{N!}$$

The factor $N!$ is present since any two of these centres are interchangeable. We may, then, write

$$A_N = A_0^N \frac{T^N}{N!}$$

where A_0 is an unknown factor corresponding to a simple amplitude. Thus we get

$$\mathcal{K}(a, -a; -iT) \approx \sum_N \frac{1}{N!} T^N A_0^N e^{-N\mathcal{S}_c/\hbar}$$

Note that in order to create a string of solutions starting at $-a$ and ending at a, we need an *odd* number of them. Thus, we must have

$$\mathcal{K}(a, -a; -iT) \approx \sum_{n=0}^{\infty} \frac{1}{(2n+1)!} T^{2n+1} A_0^{2n+1} e^{-(2n+1)\mathcal{S}_c/\hbar}$$

$$= \sinh(A_0 T e^{-\mathcal{S}_c/\hbar})$$

This amplitude has to be normalized. We know the corresponding amplitude for the harmonic oscillator,[18]

$$\mathcal{K}_0(0, 0, -iT) = |\psi_0(0)|^2 e^{-\omega T/2} + \cdots = \sqrt{\frac{m\omega}{\pi\hbar}} e^{-\omega T/2}$$

Therefore, we can write

$$\mathcal{K}(a, -a; -iT) \approx \sqrt{\frac{m\omega}{\pi\hbar}} e^{-\omega T/2} \sinh(A_0 T e^{-\mathcal{S}_c/\hbar})$$

(d) The expression for the propagator just obtained can be written as

$$\mathcal{K}(a, -a; -iT) \approx \frac{1}{2}\sqrt{\frac{m\omega}{\pi\hbar}} \left\{ \exp\left[-\left(\frac{\omega}{2} + A_0 e^{-\mathcal{S}_c/\hbar}\right) T\right] \right.$$
$$\left. - \exp\left[-\left(\frac{\omega}{2} - A_0 e^{-\mathcal{S}_c/\hbar}\right) T\right] \right\}$$

From this we can read off the two lowest-lying energy levels:

$$E = \frac{\hbar\omega}{2} \pm \hbar A_0 e^{-\mathcal{S}_c/\hbar} = \frac{\hbar\omega}{2} \pm \hbar A_0 e^{-m\omega a^2/\hbar}$$

Which of the two is the lower depends on the sign of the unknown parameter A_0.

Problem 9.13 Consider the one-dimensional potential

$$V(x) = \frac{\lambda x^4}{4} + \frac{\lambda a x^3}{4} - \frac{\lambda a^2 x^2}{8}$$

(a) Find the points of classical equilibrium for a particle of mass m moving under the influence of this potential.

(b) Using the variational method, consider the trial wave function

$$\psi(x) = \left(\frac{\beta}{\pi}\right)^{1/4} e^{-\beta(x-x_0)^2/2}$$

where x_0 is the global minimum found in (a). Evaluate the expectation value of the energy for this wave function and find the equation defining the optimal values of the parameter β, in order to get an estimate of the ground-state energy. Now take a special, but reasonable, value of the coupling constant, $\lambda = \hbar^2/ma^6$, and obtain the corresponding estimate of the ground-state energy.

(c) Write the potential in terms of the variable $x - x_0$ and, for small values of it, obtain the frequency ω of small oscillations around the global minimum.

---

[18] In the limit $2a \to 0$, we should obtain a simple harmonic oscillator.

Solution

(a) The classical equilibrium points are the minima of the potential, so that

$$m\ddot{x} = -V'(x) \quad \Longrightarrow \quad V'(x_0) = 0, \quad V''(x_0) > 0$$

The extrema of the potential are the solutions of

$$V'(x) = \frac{\lambda}{4}\left(4x^3 + 3ax^2 - a^2x\right) = \frac{\lambda x}{4}(4x^2 + 3ax - a^2) = 0$$

The second derivative of the potential is

$$V''(x) = \frac{\lambda}{4}\left(12x^2 + 6ax - a^2\right)$$

We obtain a maximum at

$$x = 0, \quad V''(0) = -\frac{\lambda a^2}{4} < 0$$

and two minima at

$$x_- = -a, \quad x_+ = \frac{a}{4}$$

with

$$V''(-a) = \frac{5\lambda a^2}{4} > 0, \quad V''\left(\frac{a}{4}\right) = \frac{5\lambda a^2}{16} > 0$$

Of these two minima, $x_- = -a$ is the global minimum since it corresponds to the lower energy:

$$V(-a) = -\frac{\lambda a^4}{8} < V\left(\frac{a}{4}\right) = -\frac{3\lambda a^4}{4^5}$$

(b) Consider the trial wave function with $x_0 = -a$. The expectation value of the kinetic energy, thanks to translational invariance, does not depend on x_0. It is

$$\langle T \rangle = -\frac{\hbar^2}{2m}\sqrt{\frac{\beta}{\pi}} \int_{-\infty}^{\infty} dx\, e^{-\beta x^2/2}\left(e^{-\beta x^2/2}\right)'' = \frac{\hbar^2 \beta}{4m}$$

The expectation value of the potential is

$$\langle V \rangle = \frac{\lambda}{4}\sqrt{\frac{\beta}{\pi}} \int_{-\infty}^{\infty} dx\, e^{-\beta x^2}\left[(x-a)^4 + a(x-a)^3 - \frac{a^2}{2}(x-a)^2\right]$$

$$= \frac{\lambda}{4}\left(\frac{3}{4\beta^2} + \frac{5a^2}{4\beta} - \frac{a^4}{2}\right)$$

Thus we get

$$E(\beta) = \frac{\hbar^2}{4ma^2} \left[\beta a^2 + \bar{\lambda} \left(\frac{3}{4a^4\beta^2} + \frac{5}{4a^2\beta} - \frac{1}{2} \right) \right]$$

where we have introduced the dimensionless coupling $\bar{\lambda} \equiv \lambda ma^6/\hbar^2$. Minimizing the energy with respect to $\xi = \beta a^2$, we get the equation

$$\xi^3 - \frac{3\bar{\lambda}}{2} - \frac{5\bar{\lambda}\,\xi}{4} = 0$$

For $\bar{\lambda} = 1$ there is a special exact solution, namely

$$\bar{\lambda} = 1 \quad \Longrightarrow \quad \xi_0 = \frac{3}{2}, \quad E_0 = \frac{\hbar^2}{2ma^2} \left(\frac{13}{12} \right)$$

(c) In terms of $\delta x = x - x_0 = x + a$, the potential is

$$V(x) = \frac{5\lambda a^2}{8}(\delta x)^2 - \frac{3\lambda a}{4}(\delta x)^3 + \frac{\lambda}{4}(\delta x)^4 - \frac{\lambda a^4}{8}$$

The frequency of small oscillations is determined by the identification

$$\frac{m\omega^2}{2} = \frac{5\lambda a^2}{8} \quad \Longrightarrow \quad \omega = \sqrt{\frac{5}{4} \left(\frac{\lambda a^2}{m} \right)}$$

Alternatively, we may express the characteristic length of the potential in terms of the characteristic oscillation length

$$a = \left(\frac{5\bar{\lambda}}{4} \right)^{1/4} \sqrt{\frac{\hbar}{m\omega}}$$

Problem 9.14 A particle of mass m and positive charge q, moving in one dimension, is subject to a uniform electric field $\mathcal{E}[\Theta(x) - \Theta(-x)]$; see Fig. 42.

(a) Consider a trial wave function $\psi(x) \propto e^{-\alpha|x|}$ and estimate the ground-state energy by minimizing the expectation value of the energy.
(b) Obtain an estimate of the ground-state energy by applying the Bohr–Sommerfeld WKB quantization rule.

Solution
(a) Introducing the normalized trial wave function

$$\psi(x) = \sqrt{\alpha}\, e^{-\alpha|x|}$$

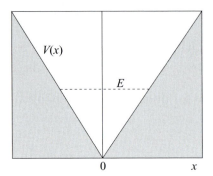

Fig. 42 Particle in an electric field with energy E.

we obtain the expectation value of the kinetic energy,

$$
\begin{aligned}
\langle T \rangle &= -\frac{\hbar^2 \alpha}{2m} \int_{-\infty}^{\infty} dx\, e^{-\alpha|x|} \left(e^{-\alpha|x|} \right)'' \\
&= -\frac{\hbar^2 \alpha}{2m} \int_{-\infty}^{\infty} dx\, e^{-\alpha|x|} \left\{ -\alpha [\Theta(x) - \Theta(-x)] e^{-\alpha|x|} \right\}' \\
&= -\frac{\hbar^2 \alpha}{2m} \int_{-\infty}^{\infty} dx\, \left[-2\alpha\delta(x) + \alpha^2 \right] e^{-2\alpha|x|} = \frac{\hbar^2 \alpha^2}{m} - \frac{\hbar^2 \alpha^2}{2m} = \frac{\hbar^2 \alpha^2}{2m}
\end{aligned}
$$

The potential experienced by the particle is $V(x) = q\mathcal{E} |x|$ and its expectation value is

$$
\langle V \rangle = q\mathcal{E}\alpha \int_{-\infty}^{\infty} dx\, |x| e^{-2\alpha|x|} = 2q\mathcal{E}\alpha \frac{1}{4\alpha^2} = \frac{q\mathcal{E}}{2\alpha}
$$

The expectation value of the energy,

$$
E(\alpha) = \frac{\hbar^2 \alpha^2}{2m} + \frac{q\mathcal{E}}{2\alpha}
$$

has a minimum at

$$
E'(\alpha_0) = 0 \quad \Longrightarrow \quad \frac{\hbar^2 \alpha_0}{m} - \frac{q\mathcal{E}}{2\alpha_0^2} = 0 \quad \Longrightarrow \quad \alpha_0 = \left(\frac{q\mathcal{E}m}{2\hbar^2} \right)^{1/3}
$$

This corresponds to a minimum, since $E''(\alpha_0) = 3\hbar^2/m > 0$. The energy value is

$$
E_0 = \frac{3}{2^{5/3}} \left[\frac{(q\mathcal{E})^2 \hbar^2}{m} \right]^{1/3}
$$

(b) The *turning points* found using the WKB method are $x_0 = \pm E/q\mathcal{E}$. Thus we get

$$\int_{-E/q\mathcal{E}}^{E/q\mathcal{E}} dx \sqrt{2m(E - q\mathcal{E}|x|)} = \left(n + \tfrac{1}{2}\right) \hbar\pi$$

Performing the integral, we obtain for the ground state ($n = 0$)

$$E_0' = \frac{(3\pi)^{2/3}}{2^{5/3}} \left[\frac{(q\mathcal{E})^2\hbar^2}{m}\right]^{1/3}$$

Note that this value is slightly larger than that obtained above, E_0, with the variational method:

$$\frac{E_0'}{E_0} = \frac{(3\pi)^{2/3}}{3} \approx \frac{4.46}{3} \approx 1.49$$

10

Scattering

Problem 10.1 Consider the scattering of particles of a given energy in a central potential of finite range.

(a) Show that the energy eigenfunctions $\psi_{\mathbf{k}}^{(+)}(\mathbf{r})$ depend only on r, an angle and the energy E. The angle can be taken to be $\cos^{-1}(\hat{\mathbf{k}} \cdot \hat{\mathbf{r}})$. On which variables does the *scattering amplitude* $f_{\mathbf{k}}(\mathbf{r})$ depend?

(b) Calculate the asymptotic probability current density corresponding to $\psi_{\mathbf{k}}^{(+)}(\mathbf{r})$ and show that it can always be written as

$$\boldsymbol{\mathcal{J}} = \boldsymbol{\mathcal{J}}_{\mathrm{i}} + \boldsymbol{\mathcal{J}}_{\mathrm{sc}} + \Delta \boldsymbol{\mathcal{J}}$$

where the first term corresponds to the incident particles and the second to the scattered particles.[1] What is the explanation of the third term? Show that

$$\oint_{S_{\infty}} d\mathbf{S} \cdot \boldsymbol{\mathcal{J}}_{\mathrm{sc}} = -\oint_{S_{\infty}} d\mathbf{S} \cdot \Delta \boldsymbol{\mathcal{J}}$$

where S_{∞} is a spherical surface of infinite radius with origin at the centre of the potential.

(c) Use the above to prove the *optical theorem*,

$$\sigma = \frac{4\pi}{k} \, \mathrm{Im}\left[f_{\mathbf{k}}(\hat{\mathbf{k}}) \right]$$

---

[1] You may use the relation

$$\int_{-1}^{1} d\cos\theta \, F(\theta) e^{-ikr\cos\theta} = \frac{i}{kr} \left[F(0)e^{-ikr} - F(\pi)e^{ikr} \right] + O\left(\frac{1}{r^2}\right)$$

Solution

(a) The integral solution to the Schroedinger equation in the asymptotic region is[2]

$$\psi_{\mathbf{k}}(\mathbf{r}) \approx \frac{e^{ikr\cos\theta}}{(2\pi)^{3/2}} - \frac{e^{ikr}}{r} \left(\frac{m}{2\pi\hbar^2}\right) \int d^3r'\, e^{-ik\hat{\mathbf{r}}\cdot\mathbf{r}'}\, V(r')\, \psi_{\mathbf{k}}(\mathbf{r}')$$

$$= \frac{1}{(2\pi)^{3/2}} \left[e^{ikr\cos\theta} + \frac{e^{ikr}}{r} f_k(\theta) \right]$$

It is clear that the second term depends only on $\mathbf{k}$ and $\mathbf{r}$. Thus the right-hand side depends on k, r and θ.

(b) Calculating first the gradient

$$\nabla \psi_k(r, \theta) = \frac{1}{(2\pi)^{3/2}} \left\{ i\mathbf{k} e^{ikr\cos\theta} + \frac{e^{ikr}}{r^2} \left[\mathbf{r} f_k(\theta)(-1 + ikr) + \hat{\boldsymbol{\theta}} \frac{\partial f_k(\theta)}{\partial\theta} \right] \right\}$$

and substituting it into the probability current density, we obtain

$$\boldsymbol{\mathcal{J}} = \boldsymbol{\mathcal{J}}_i + \boldsymbol{\mathcal{J}}_{sc} + \Delta\boldsymbol{\mathcal{J}}$$

where

$$\boldsymbol{\mathcal{J}}_i = \frac{1}{(2\pi)^3}\left(\frac{\hbar\mathbf{k}}{m}\right), \qquad \boldsymbol{\mathcal{J}}_{sc} = \frac{1}{(2\pi)^3}\left(\frac{\hbar k}{m}\right)\frac{|f_k(\theta)|^2}{r^2}\,\hat{\mathbf{r}}$$

and

$$\Delta\boldsymbol{\mathcal{J}} = \frac{\hbar}{2mi(2\pi)^3}\left(\frac{1}{r}\right)\left[i\mathbf{k} f_k^*(\theta)e^{-ikr(1-\cos\theta)} + \hat{\mathbf{r}}\left(ik - \frac{1}{r}\right) f_k(\theta)e^{ikr(1-\cos\theta)} \right.$$

$$\left. + \hat{\boldsymbol{\theta}} \frac{\partial f_k(\theta)}{\partial\theta} e^{ikr(1-\cos\theta)} - \text{c.c.} \right] + O(r^{-3})$$

This expression represents the interference between the incident and the scattered waves.

Integrating on the surface of a sphere centred at the origin and taking its radius to infinity, we must have, from probability conservation,

$$\oint_{S_\infty} d\mathbf{S}\cdot\boldsymbol{\mathcal{J}} = 0 \qquad \Longrightarrow \qquad \oint_{S_\infty} d\mathbf{S}\cdot\boldsymbol{\mathcal{J}}_i = -\oint_{S_\infty} d\mathbf{S}\cdot(\boldsymbol{\mathcal{J}}_{sc} + \Delta\boldsymbol{\mathcal{J}})$$

Note however that

$$\oint_{S_\infty} d\mathbf{S}\cdot\boldsymbol{\mathcal{J}}_i = \frac{\hbar}{m(2\pi)^3}\oint d\mathbf{S}\cdot\mathbf{k} = \frac{\hbar k R^2}{m(2\pi)^3}\int_{-1}^{1}(d\cos\theta)\cos\theta = 0$$

---

[2] The approximation

$$|\mathbf{r} - \mathbf{r}'| \approx r - \hat{\mathbf{r}}\cdot\mathbf{r}'$$

is valid in the asymptotic region.

Therefore

$$\oint_{S_\infty} d\mathbf{S} \cdot \boldsymbol{\mathcal{J}}_{\text{sc}} = -\oint_{S_\infty} d\mathbf{S} \cdot \Delta\boldsymbol{\mathcal{J}}$$

(c) The total cross section σ is given by[3]

$$\sigma = \int d\Omega \, |f_k(\theta)|^2 = \frac{1}{r^2} \oint_{S_r} d\mathbf{S} \, |f_k(\theta)|^2 = \frac{1}{r^2} \oint_{S_R} d\mathbf{S} \cdot \hat{\mathbf{r}} |f_k(\theta)|^2$$

$$= \frac{m(2\pi)^3}{\hbar k} \oint_{S_\infty} d\mathbf{S} \cdot \boldsymbol{\mathcal{J}}_{\text{sc}} = -\frac{m(2\pi)^3}{\hbar k} \oint_{S_\infty} d\mathbf{S} \cdot \Delta\boldsymbol{\mathcal{J}}$$

As is clear from the expression for $\Delta\boldsymbol{\mathcal{J}}$ derived above, the angular ($\propto \hat{\boldsymbol{\theta}}$) component will not contribute. We get

$$\sigma = -\frac{1}{2kr} \oint_{S_\infty} d\mathbf{S} \cdot \left[\mathbf{k} \, f_k^*(\theta) \, e^{-ikr(1-\cos\theta)} + \hat{\mathbf{r}} \, k f_k(\theta) \, e^{ikr(1-\cos\theta)} + \text{c.c.} \right]$$

$$= -\pi r \int_{-1}^{1} d\cos\theta (1 + \cos\theta) \left[f_k(\theta) e^{ikr} e^{-ikr\cos\theta} + \text{c.c.} \right]$$

$$= -\frac{\pi}{k} \left[2i f_k(0) - 2i f_k^*(0) \right] = \frac{4\pi}{k} \text{Im} \left[f_k(0) \right]$$

Problem 10.2 Consider an attractive *delta-shell* potential ($\lambda > 0$)

$$V(r) = -\frac{\hbar^2 \lambda}{2\mu} \delta(r - a)$$

(a) Calculate the *phase shift* $\delta_\ell(k)$, where ℓ is the angular momentum quantum number.
(b) In the case $\ell = 0$, investigate the existence of bound states by examining the analytic properties of the partial scattering amplitude. Are there any resonances?

Solution
(a) The radial energy eigenfunction can be written as

$$R_{k,\ell}(r) = \begin{cases} R_{k,\ell}^{(<)}(r) = A j_\ell(kr), & 0 < r < a \\ R_{k,\ell}^{(>)}(r) = B \left[j_\ell(kr) \cos\delta_\ell - n_\ell(kr) \sin\delta_\ell \right], & a < r < \infty \end{cases}$$

Continuity of the wave function at $r = a$ implies that

$$R_{k,\ell}^{(<)}(a) = R_{k,\ell}^{(>)}(a) \implies \frac{A}{B} = \cos\delta_\ell - \sin\delta_\ell \frac{n_\ell(ka)}{j_\ell(ka)}$$

Discontinuity of the radial derivative at the same point gives

$$R'_{k,\ell}(a + \epsilon) - R'_{k,\ell}(a - \epsilon) = -\lambda R_{k,\ell}(a)$$

---

[3] We take $r \to \infty$.

or

$$B\left[j'_\ell(ka)\,\cos\delta_\ell - n'_\ell(ka)\,\sin\delta_\ell\right] - Aj'_{k,\ell}(ka) = -\lambda A j_{k,\ell}(ka)$$

Finally, using both the equations relating A and B, we obtain the required expression for the phase shift:

$$\tan\delta_\ell = \frac{\lambda j_\ell^2(ka)}{j_\ell(ka)n'_\ell(ka) - n_\ell(ka)j'_\ell(ka) + \lambda n_\ell(ka)j_\ell(ka)}$$

We can also write down an expression for the *partial scattering amplitude*. It is

$$S_\ell(k) = \exp\left[2i\delta_\ell(k)\right] - 1 = -\frac{2\tan\delta_\ell}{i + \tan\delta_\ell}$$

or[4]

$$S_\ell(k) = \frac{2i\lambda j_\ell^2(ka)}{j_\ell(ka)n'_\ell(ka) - n_\ell(ka)j'_\ell(ka) + \lambda j_\ell(ka)h_\ell^{(-)}(ka)}$$

(b) For s-waves ($\ell = 0$), we have

$$S_0(k) = \frac{2i\lambda a\,\sin\,ka}{ka/\sin\,ka - \lambda a e^{ika}}$$

Now introducing[5]

$$ka \equiv i\xi, \qquad \lambda a \equiv g$$

we get

$$S_0 = -\frac{2g\,\sinh\xi}{\xi/\sinh\xi - ge^{-\xi}}$$

The condition for bound states is

$$e^{-2\xi} = 1 - \frac{2\xi}{g}$$

It is clear that for $g > 1$ there is a single solution to this equation. Thus, there is one bound state provided that

$$g > 1$$

The absolute square of the partial scattering amplitude is the appropriate quantity that will appear in the scattering cross section. Reintroducing $\zeta = ka$, we have

$$\tfrac{1}{4}|S_0|^2 = \frac{g^2\sin^2\zeta}{(\zeta/\sin\zeta - g\cos\zeta)^2 + g^2\sin^2\zeta}$$

---

[4] $h_\ell^{(\pm)} = n_\ell \pm ij_\ell$.
[5] Bound states will correspond to imaginary values of the wave number, i.e. negative energies.

It is clear that the function on the right-hand side becomes largest, i.e. unity, when

$$\frac{\zeta}{\sin \zeta} = g \cos \zeta$$

Approximate solutions of this equation are

$$\zeta_n \sim 2\pi n \left(1 + \frac{1}{g}\right) + O\left(\frac{1}{g^2}\right)$$

These points are solutions for values of the coupling much larger than the integer n, namely

$$g \gg n$$

The related values of the energy correspond to *resonances*. Near a resonance, we can write

$$\frac{1}{4}|S_0|^2 \sim \frac{(2\pi n/g)^4}{(2\pi n/g)^4 + (\zeta - \zeta_n)^2}$$

Problem 10.3 Particles of a given energy scatter on an infinitely hard sphere of radius a.

(a) Calculate the *phase shift* $\delta_\ell(k)$.
(b) For s-waves ($\ell = 0$), find the values of the energy for which the partial cross section becomes maximal.
(c) Consider the case of low energies ($ka \ll 1$), write an approximate expression for δ_ℓ and explain why the cross section is dominated by s-waves and is isotropic. Compare the low-energy cross section with the *geometric value* πa^2.

Solution
(a) The radial wave function, written in terms of the phase shift, is

$$R_{k,\ell}(r) = A\left[j_\ell(kr)\cos\delta_\ell - n_\ell(kr)\sin\delta_\ell\right]$$

It has to vanish for $r \leq a$. Thus, we obtain

$$\tan\delta_\ell = \frac{j_\ell(ka)}{n_\ell(ka)}$$

(b) The total cross section can be written as

$$\sigma = \frac{4\pi}{k^2}\sum_{\ell=0}^{\infty}(2\ell+1)\sin^2\delta_\ell(k)$$

The s-wave partial cross section is

$$\sigma_0 = \frac{4\pi}{k^2} \sin^2 \delta_0 = \frac{4\pi}{k^2} \left(\frac{1}{1 + \cot^2 \delta_0} \right)$$

From (a) we have

$$\tan \delta_0 = \frac{j_0(ka)}{n_0(ka)} = -\tan(ka) \qquad \Longrightarrow \qquad \cot^2 \delta_0 = \cot^2 ka$$

Thus

$$\sigma_0 = \frac{4\pi}{k^2} \left(\frac{1}{1 + \cot^2 ka} \right)$$

and its maximal values are achieved for

$$k = \frac{\pi}{2a}(2n + 1) \qquad (n = 0, 1, \ldots)$$

(c) From the behaviour of the spherical Bessel functions near zero, we obtain

$$\tan \delta_\ell \sim -\frac{(ka)^{2\ell+1}}{(2\ell + 1)!!(2\ell - 1)!!} = -(2\ell + 1)\frac{(ka)^{2\ell+1}}{[(2\ell + 1)!!]^2}$$

It is clear that the scattering is dominated by $\ell = 0$, for which both $\tan \delta_\ell$ and the cross section have their largest values.

The low-energy cross section for s-waves is

$$\sigma_0 \approx \frac{4\pi}{k^2} \frac{1}{1 + (ka)^{-2}} \approx \frac{4\pi}{k^2} (ka)^2 = 4\pi a^2$$

This is four times the *geometric cross section*, πa^2. The reason is that the whole surface of the sphere participates in the quantum mechanical scattering process, not just the two-dimensional section of the sphere.

Problem 10.4 Consider the scattering of a particle from a real spherically symmetric potential. If $d\sigma(\theta)/d\Omega$ is the differential cross section and σ is the total cross section, show that

$$\sigma \leq \frac{4\pi}{k} \sqrt{\frac{d\sigma(0)}{d\Omega}}$$

Verify this inequality explicitly for a general central potential using the partial-wave expansion of the scattering amplitude and the cross section.

Solution

The differential cross section is related to the scattering amplitude through

$$\frac{d\sigma(\theta)}{d\Omega} = |f_k(\theta)|^2$$

Since $|f|^2 = (\text{Re } f)^2 + (\text{Im } f)^2 \geq (\text{Im } f)^2$, we can write

$$\frac{d\sigma(\theta)}{d\Omega} \geq [\text{Im } f_k(\theta)]^2$$

On the other hand, from the *optical theorem* we have

$$\sigma = \frac{4\pi}{k} \text{Im } f_k(0) \leq \frac{4\pi}{k} \sqrt{\frac{d\sigma(0)}{d\Omega}}$$

For a central potential the scattering amplitude is

$$f_k(\theta) = \frac{1}{k} \sum_{\ell=0}^{\infty} (2\ell + 1) e^{i\delta_\ell} \sin \delta_\ell \, P_\ell(\cos \theta)$$

and, in terms of this, the differential cross section is

$$\frac{d\sigma(\theta)}{d\Omega} = \frac{1}{k^2} \sum_{\ell=0}^{\infty} \sum_{\ell'=0}^{\infty} (2\ell + 1)(2\ell' + 1) \, e^{i\delta_\ell - i\delta_{\ell'}} \sin \delta_\ell \sin \delta_{\ell'} \, P_\ell(\cos \theta) P_{\ell'}(\cos \theta)$$

The total cross section is

$$\sigma = \frac{4\pi^2}{k^2} \sum_{\ell=0}^{\infty} (2\ell + 1) \sin^2 \delta_\ell$$

Using the fact that $P_\ell(1) = 1$, we obtain

$$\frac{d\sigma(0)}{d\Omega} = \frac{1}{k^2} \left| \sum_{\ell=0}^{\infty} (2\ell + 1) e^{i\delta_\ell} \sin \delta_\ell \right|^2$$

$$= \frac{1}{k^2} \left| \sum_{\ell=0}^{\infty} (2\ell + 1) \left(\sin \delta_\ell \cos \delta_\ell + i \sin^2 \delta_\ell \right) \right|^2$$

$$= \frac{1}{k^2} \left[\sum_{\ell=0}^{\infty} (2\ell + 1) \sin \delta_\ell \cos \delta_\ell \right]^2 + \frac{1}{k^2} \left[\sum_{\ell=0}^{\infty} (2\ell + 1) \sin^2 \delta_\ell \right]^2$$

$$\implies \quad \frac{d\sigma(0)}{d\Omega} \geq \frac{1}{k^2} \left[\sum_{\ell=0}^{\infty} (2\ell + 1) \sin^2 \delta_\ell \right]^2 = \frac{k^2 \sigma^2}{16\pi^2}$$

Problem 10.5 The *radial Green's function* is defined by the equation

$$\frac{1}{r^2} \frac{d}{dr} \left[r^2 \frac{d}{dr} \mathcal{G}_{k,\ell}(r, r') \right] + \left[k^2 - \frac{\ell(\ell + 1)}{r^2} \right] \mathcal{G}_{k,\ell}(r, r') = \frac{1}{r^2} \delta(r - r')$$

(a) Verify the choice

$$\mathcal{G}_{k,\ell}^{(-)}(r, r') = C \left[\theta(r' - r) j_\ell(kr) h_\ell^{(-)}(kr') + \theta(r - r') j_\ell(kr') h_\ell^{(-)}(kr) \right]$$

by substituting in the above differential equation. Also find the normalization constant C.

(b) Show that the radial wave function satisfies the integral equation[6]

$$R_{k,\ell}(r) = j_\ell(kr) + \int_0^\infty dr' \, r'^2 U(r') \mathcal{G}_{k,\ell}(r, r') R_{k,\ell}(r')$$

Solution

(a) Using the fact that j_ℓ and $h_\ell^{(\pm)}$ are solutions of the free radial Schroedinger equation, as well as the identity $n_\ell'(x) j_\ell(x) - j_\ell'(x) n_\ell(x) = 1/x^2$, we can verify that the Green's function $\mathcal{G}_{k,\ell}(r, r')$ is a solution of the above equation. The normalization constant is $C = k$.

(b) Acting with the radial Schroedinger operator on both sides of the above integral equation, we get

$$U(r)R(r) = 0 + \int_0^\infty dr' r'^2 U(r') \frac{1}{r^2} \delta(r - r') R(r') = U(r)R(r)$$

Problem 10.6 Consider the double delta-shell potential $V(r) = (\hbar^2/2m)U(r)$, where

$$U(r) = -\lambda_1 \delta(r - a_1) - \lambda_2 \delta(r - a_2)$$

with $a_2 > a_1 > 0$, and calculate the phase shift δ_ℓ. In the case of s-waves ($\ell = 0$) investigate the existence of bound states and resonances.

Solution

Introducing the potential into the integral equation of the previous problem, we obtain the solution

$$R_{k,\ell}(r) = j_\ell(kr) - \lambda_1 a_1^2 \mathcal{G}_{k,\ell}(r, a_1) R_{k,\ell}(a_1) - \lambda_2 a_2^2 \mathcal{G}_{k,\ell}(r, a_2) R_{k,\ell}(a_2)$$

In the three different regions, we have the following expressions for the Green's function:

For $r \geq a_2 > a_1$,

$$\mathcal{G}_{k,\ell}(r, a_{1,2}) = k j_\ell(ka_{1,2}) h_\ell^{(-)}(kr)$$

For $a_1 \leq r \leq a_2$,

$$\mathcal{G}_{k,\ell}(r, a_1) = k j_\ell(ka_1) h_\ell^{(-)}(kr)$$

and

$$\mathcal{G}_{k,\ell}(r, a_2) = k j_\ell(kr) h_\ell^{(-)}(ka_2)$$

For $r \leq a_1 < a_2$,

$$\mathcal{G}_{k,\ell}(r, a_{1,2}) = k j_\ell(kr) h_\ell^{(-)}(ka_{1,2})$$

---

[6] $U = (2m/\hbar^2)V$.

Introducing

$$g_i \equiv ka_i^2\lambda_i, \qquad R_{k,\ell}(a_i) \equiv R_i$$
$$j_\ell(ka_i) \equiv j_i, \qquad h_\ell^{(-)}(ka_i) \equiv h_i$$

where $i = 1, 2$, we get

$$R_{k,\ell}(r) = \begin{cases} j_\ell(kr) - (g_1 R_1 j_1 + g_2 R_2 j_2) h_\ell^{(-)}(kr), & r \geq a_2 > a_1 \\ (1 - g_2 R_2 h_2) j_\ell(kr) - g_1 j_1 R_1 h_\ell^{(-)}(kr), & a_1 \leq r \leq a_2 \\ j_\ell(kr) (1 - g_1 R_1 h_1 - g_2 R_2 h_2), & r \leq a_1 < a_2 \end{cases}$$

These expressions carry R_1 and R_2 as unknown parameters. These can be determined from the system of the two equations that we get by considering the top expression at $r = a_2$ and the bottom at $r = a_1$, namely

$$(g_1 j_1 h_2) R_1 + (1 + g_2 j_2 h_2) R_2 = j_2$$
$$(1 + g_1 h_1 j_1) R_1 + (g_2 h_2 j_1) R_2 = j_1$$

Since we are interested in the phase shift, it suffices to consider the external wave function, which has the form

$$R_{k,\ell}(r) = j_\ell(kr) - (g_1 R_1 j_1 + g_2 R_2 j_2) h_\ell^{(-)}(kr) = j_\ell(kr) - A h_\ell^{(-)}(kr)$$

and in which a particular combination of R_1 and R_2 appears. Solving for this combination, we get

$$A = \frac{g_i j_1^2 + g_2 j_2^2 + g_1 g_2 j_1 j_2 (j_2 h_1 - j_1 h_2)}{1 + g_1 j_1 h_1 + g_2 h_2 j_2 + g_1 g_2 j_1 h_2 (j_2 h_1 - j_1 h_2)}$$

and, thus,

$$R_{k,\ell}(r) = j_\ell(kr) - \frac{g_i j_1^2 + g_2 j_2^2 + g_1 g_2 j_1 j_2 (j_2 h_1 - j_1 h_2)}{1 + g_1 j_1 h_1 + g_2 h_2 j_2 + g_1 g_2 j_1 h_2 (j_2 h_1 - j_1 h_2)} h_\ell^{(-)}(kr)$$

for $r > a_2 > a_1$.

Setting

$$\alpha_\ell \equiv 1 + g_2 j_2 n_2 + g_1 j_1 n_1 + g_1 g_2 n_2 j_1 (j_2 n_1 - j_1 n_2)$$
$$\beta_\ell \equiv g_1 j_1^2 + g_2 j_2^2 + g_1 g_2 j_1 j_2 (j_2 n_1 - j_1 n_2)$$

we can write

$$A = \frac{\beta_\ell}{\alpha_\ell + i\beta_\ell}$$

which implies that the external radial wave function can be given as

$$R_{k,\ell}(r) = \frac{1}{\alpha_\ell + i\beta_\ell} [\alpha_\ell \, j_\ell(kr) - \beta_\ell \, n_\ell(kr)]$$

$$= \exp\left[-i \tan^{-1}\left(\frac{\beta_\ell}{\alpha_\ell}\right)\right] [j_\ell(kr) \cos \delta_\ell - n_\ell(kr) \sin \delta_\ell]$$

Thus, the final expression for the phase shift is

$$\tan \delta_\ell = \frac{\beta_\ell}{\alpha_\ell} = \frac{g_1 j_1^2 + g_2 j_2^2 + g_1 g_2 j_1 j_2 (j_2 n_1 - j_1 n_2)}{1 + g_2 j_2 n_2 + g_1 j_1 n_1 + g_1 g_2 n_2 j_1 (j_2 n_1 - j_1 n_2)}$$

For s-waves ($\ell = 0$) the phase shift is just

$$\tan \delta_0 = \frac{\beta_0}{\alpha_0}$$

where α_0 and β_0 are obtained by substituting $j_0(ka_i) = (\sin ka_i)/ka_i$ and $n_0(ka_i) = -(\cos ka_i)/ka_i$. The explicit expressions are

$$\alpha_0 = 1 - \frac{\lambda_1}{2k} \sin 2ka_1 - \frac{\lambda_2}{2k} \sin 2ka_2$$

$$+ \frac{\lambda_1 \lambda_2}{4k^2} \{-1 + \cos[2k(a_1 - a_2)] + \cos 2ka_1 - \cos 2ka_2\}$$

and

$$\beta_0 = \frac{\lambda_1}{2k} + \frac{\lambda_2}{2k} - \frac{\lambda_1}{2k} \cos 2ka_1 - \frac{\lambda_2}{2k} \cos 2ka_2$$

$$+ \frac{\lambda_1 \lambda_2}{4k^2} \{\sin[2k(a_1 - a_2)] + \sin 2ka_2 - \sin 2ka_1\}$$

The partial scattering amplitude is

$$S_0 = e^{2i\delta_0} - 1 = -\frac{2 \tan \delta_0}{i + \tan \delta_0} = \frac{2i\beta_0}{\alpha_0 - i\beta_0}$$

Bound states correspond to

$$\alpha_0 = i\beta_0 \qquad (k = i\kappa)$$

and *resonances* correspond to positive energy values that make the partial cross section

$$\tfrac{1}{4}|S_0|^2 = \frac{\beta_0^2}{\beta_0^2 + \alpha_0^2}$$

maximal, i.e.

$$\alpha_0(k) = 0$$

The explicit condition for bound states is

$$
1 - \frac{\lambda_1}{2\kappa} - \frac{\lambda_2}{2\kappa} - \frac{\lambda_1}{2\kappa} \sinh 2\kappa a_1 - \frac{\lambda_2}{2\kappa} \sinh 2\kappa a_2 + \frac{\lambda_1}{2\kappa} \cosh 2\kappa a_1 + \frac{\lambda_2}{2\kappa} \cosh 2\kappa a_2
$$

$$
= \frac{\lambda_1 \lambda_2}{4\kappa^2} \{ -1 + \cosh[2\kappa(a_1 - a_2)] + \cosh 2\kappa a_1 - \cosh 2\kappa a_2
$$

$$
+ \sinh[2\kappa(a_1 - a_2)] + \sinh 2\kappa a_2 - \sinh 2\kappa a_1 \}
$$

Introducing the dimensionless numbers

$$
\nu \equiv \frac{a_2}{a_1}, \qquad \gamma_i \equiv \lambda_i a_i, \qquad \xi \equiv 2\kappa a_1
$$

we can write this condition as

$$
f(\xi) = \nu \xi^2 - \gamma_1 \nu \xi - \gamma_2 \xi + \gamma_1 \nu \xi e^{-\xi} + \gamma_2 \xi e^{-\nu \xi}
$$

$$
+ \gamma_1 \gamma_2 \left[1 - e^{-(\nu-1)\xi} - e^{-\xi} + e^{-\nu \xi} \right] = 0
$$

At the origin, $\xi = 0$, we have

$$
f(0) = f'(0) = 0
$$

and

$$
f''(0) = 2[\nu - \nu(\gamma_1 + \gamma_2) + \gamma_1 \gamma_2 (\nu - 1)]
$$

For very large values $\xi \to \infty$, we get $f(\xi) \sim \nu \xi^2$. Thus, there will be one zero of $f(\xi)$ corresponding to *one bound state*, provided that

$$
\gamma_1 + \gamma_2 > \gamma_1 \gamma_2 \frac{\nu - 1}{\nu} + 1
$$

This is equivalent to

$$
\lambda_1 a_1 + \lambda_2 a_2 > 1 + \lambda_1 \lambda_2 a_1 (a_2 - a_1)
$$

As an example, we consider the special case $\lambda_1 = \lambda_2 = a_1^{-1}$, $a_2 = 2a_1$. Plotting $f(\xi)$ in this case, we get the graph shown in Fig. 43. By inspection of the graph we can safely conclude that in the $\ell = 0$ case there will be one bound state.

The condition for resonance takes the explicit form

$$
g(\zeta) = \nu \zeta^2 - \gamma_1 \nu \zeta \sin \zeta - \gamma_2 \zeta \sin \nu \zeta
$$

$$
+ \gamma_1 \gamma_2 \{ -1 + \cos[(\nu - 1)\zeta] + \cos \zeta - \cos \nu \zeta \} = 0
$$

with $\zeta \equiv 2ka_1$. For $\zeta \to \infty$ this function goes to $+\infty$, while at the origin we have

$$
g(0) = g'(0) = 0
$$

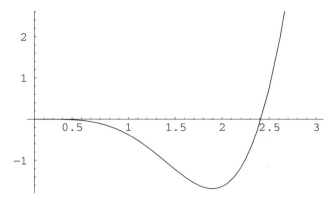

Fig. 43 Plot of $f(\xi)$.

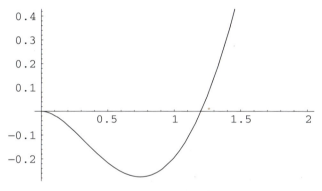

Fig. 44 Plot of $g(\zeta)$.

and

$$g''(0) = 2v(1 - \gamma_1 - \gamma_2) + g_1 g_2(v - 1)(2 - v)$$

There will be a zero and, therefore, resonance, provided that

$$\gamma_1 + \gamma_2 > 1 + \gamma_1 \gamma_2 \frac{(v - 1)(v - 2)}{2v}$$

or, equivalently, that

$$\lambda_1 a_1 + \lambda_2 a_2 > 1 + \frac{\lambda_1 \lambda_2}{2}(a_2 - a_1)(a_2 - 2a_1)$$

For the special value of the couplings and radii considered previously, $g(\zeta)$ is as shown in Fig. 44.

Problem 10.7 Consider the one-dimensional delta function potential

$$V(x) = \frac{\hbar^2 \lambda}{2m} \delta(x)$$

(a) Solve the energy eigenvalue problem for both signs of the coupling λ. Verify the ortho-normality and completeness of the eigenfunctions. In the case of the continuum (scattering states), write the eigenfunctions in terms of the scattering amplitude. Examine the analytic properties of the scattering amplitude in the k-plane. Are there poles? What do they correspond to?

(b) Determine the one-dimensional Green's function

$$\left(\frac{d^2}{dx^2} + k^2 \right) G_k(x, x') = -4\pi \delta(x - x')$$

and solve the above eigenvalue problem with the help of the *scattering integral equation*[7]

$$\psi_k(x) = \psi_k^{(0)}(x) - \frac{1}{4\pi} \int dx' \, G_k(x, x') U(x') \psi_k(x')$$

Solution

(a) For the continuum ($0 < E = \hbar^2 k^2 / 2m$) the energy eigenfunctions are

$$\psi_k(x) = \begin{cases} x < 0, & \frac{1}{\sqrt{2\pi}} \left(e^{ikx} + f e^{-ikx} \right) \\ x > 0, & \frac{1}{\sqrt{2\pi}} g e^{ikx} \end{cases}$$

From the continuity of the wave functions at the origin, we get

$$g = 1 + f$$

and from the discontinuity of the derivative

$$\psi'(+0) - \psi'(-0) = \lambda \psi(0)$$

or

$$2ikf = \lambda(1 + f)$$

These lead to the form

$$\psi_k(x) = \frac{1}{\sqrt{2\pi}} \left[e^{ikx} + f(k) e^{ik|x|} \right]$$

with

$$f(k) = \frac{1}{-1 + 2ik/\lambda}$$

Note however that the eigenfunctions

$$\tilde{\psi}_k(x) = \frac{1}{\sqrt{2\pi}} \left[e^{-ikx} + f(k) e^{ik|x|} \right]$$

[7] $U \equiv 2mV/\hbar^2$.

correspond to the same energy. The latter describe a particle incident on the potential from the right, while the former describe one incident from the left.

In the case of an attractive potential ($\lambda < 0$) there is also a single *bound state* with wave function

$$\psi_\Delta(x) = \sqrt{\kappa}\, e^{-\kappa|x|} = \sqrt{|\lambda|/2}\, e^{-|\lambda|\,|x|/2}$$

corresponding to the energy

$$E_\Delta = -\frac{\hbar^2 \kappa^2}{2m} = -\frac{\hbar^2 \lambda^2}{8m}$$

The proof of orthonormality is straightforward:

$$\int_{-\infty}^{+\infty} dx\, \psi_k^*(x)\psi_{k'}(x)$$

$$= \int_{-\infty}^{+\infty} \frac{dx}{2\pi} \left[e^{-ikx} + f(-k)e^{-ik|x|} \right] \left[e^{ik'x} + f(k')e^{ik'|x|} \right]$$

$$= \delta(k-k') + \int_{-\infty}^{+\infty} \frac{dx}{2\pi} \left[e^{-ikx+ik'|x|} f(k') + e^{ik'x-ik|x|} f(-k) \right.$$

$$\left. + f(-k)f(k')e^{-i(k-k')|x|} \right]$$

$$= \delta(k-k') - \left(\frac{i}{\pi}\right) \frac{1}{k^2 - k'^2} \left[k' f(k') + kf(-k) + (k+k')f(-k)f(k') \right]$$

$$= \delta(k-k')$$

Similarly, for the orthogonality of the continuum states to the discrete state, we have

$$\int_{-\infty}^{+\infty} dx\, \psi_\Delta(x)\psi_k(x) = \sqrt{\frac{\kappa}{2\pi}} \int_{-\infty}^{+\infty} dx\, e^{-\kappa|x|} \left[e^{ikx} + f(k)e^{ik|x|} \right] = 0$$

taking into account that $f^{-1}(k) = -1 - ik/\kappa$.

Completeness corresponds to

$$\int_0^{+\infty} dk\, \psi_k(x)\psi_k^*(0) + \int_0^{+\infty} dk\, \tilde{\psi}_k(x)\tilde{\psi}_k^*(0) + \psi_\Delta(x)\psi_\Delta(0) = \delta(x)$$

Using the fact that

$$f + f^* = -2|f|^2$$

and that

$$f^*(k) = f(-k)$$

we can write the continuum contribution above as

$$\int_0^{+\infty} \frac{dk}{2\pi} \left[e^{ikx} + f(k)e^{ik|x|} \right] \left[1 + f^*(k) \right]$$

$$+ \int_{-\infty}^0 \frac{dk}{2\pi} \left[e^{ikx} + f^*(k)e^{-ik|x|} \right] \left[1 + f(k) \right]$$

Thus, we arrive at the following form for the left-hand side of the completeness relation:

$$\delta(x) + \int_0^{+\infty} \frac{dk}{2\pi} \left[e^{ikx} f^* + e^{-ikx} f^* + (f - f^*)e^{ik|x|} \right] + \Theta(-\lambda)\kappa e^{-\kappa|x|}$$

$$= \delta(x) + \int_{-\infty}^{+\infty} \frac{dk}{2\pi} f(k)e^{ik|x|} + \Theta(-\lambda)\kappa e^{-\kappa|x|} = \delta(x)$$

The integration has been performed in the upper complex k-plane and it gives a non-vanishing result only when the pole of the amplitude $(-1 + 2ik/\lambda)^{-1}$ is there, i.e. when $\lambda < 0$.

The scattering amplitude is

$$f(k) = \frac{1}{-1 + 2ik/\lambda}$$

and it has a pole for

$$k = i\kappa_\Delta = -\frac{i\lambda}{2}$$

This pole corresponds to a bound state ($\lambda < 0$).

(b) Solving the Green's function equation through a Fourier transform,

$$G_k(x - x') = \int \frac{dq}{\sqrt{2\pi}} \mathcal{G}_k(q)e^{iq(x-x')} = 2 \int dq \, \frac{e^{iq(x-x')}}{q^2 - k^2 + i\epsilon}$$

we obtain

$$G_k(x - x') = \frac{2\pi i}{k} e^{ik|x-x'|}$$

Substituting this Green's function into the integral equation we get exactly the same scattering solutions as those we examined above.

Problem 10.8 Prove the formula

$$e^{i\delta_\ell} \sin \delta_\ell = -k \int_0^\infty dr \, r^2 \, U(r) j_\ell(kr) R_{k,\ell}(r)$$

and get from it a closed expression for the phase shift in the Born approximation. Apply this last expression for the potential

$$V(r) = g^2 e^{-\mu r}$$

with $\ell = 0$.

Solution

Using the integral equation

$$R_{k,\ell}(r) = j_\ell(kr) + \int_0^\infty dr'\, r'^2 U(r')\mathcal{G}_{k,\ell}(r, r')R_{k,\ell}(r')$$

and substituting the Green's function in the asymptotic region $r \to \infty$, we obtain

$$R_{k,\ell}(r) = j_\ell(kr) + k h_\ell^{(-)}(kr) \int_0^\infty dr'\, r'^2 U(r')j_\ell(kr')R_{k,\ell}(r')$$

This is of the form

$$R_{k,\ell}(r) = j_\ell(kr) + X h_\ell^{(-)}(kr)$$

with

$$X \equiv k \int_0^\infty dr'\, r'^2 U(r')\, j_\ell(kr')R_{k,\ell}(r')$$

However, from the asymptotic behaviour of $R_{k,\ell}(r)$,

$$R_{k,\ell}(r) \sim \frac{1}{kr} \sin\left(kr - \frac{\ell\pi}{2} + \delta_\ell \right) = j_\ell(kr) + \frac{i}{2}(e^{2i\delta_\ell} - 1)\, h_\ell^{(-)}(kr)$$

we obtain that

$$\frac{i}{2}(e^{2i\delta_\ell} - 1) = k \int_0^\infty dr'\, r'^2 U(r')j_\ell(kr')R_{k,\ell}(r')$$

or

$$e^{i\delta_\ell} \sin\delta_\ell = -k \int_0^\infty dr'\, r'^2 U(r')j_\ell(kr')R_{k,\ell}(r')$$

In the Born approximation, we can assume that δ_ℓ is small and also replace the radial eigenfunction in the integrand with $j_\ell(kr)$. Then we get

$$\delta_\ell \approx -k \int_0^\infty dr'\, r'^2 U(r')\, j_\ell^2(kr')$$

For the particular potential

$$U(r) = \frac{2mg^2}{\hbar^2} e^{-\mu r}$$

we obtain

$$\delta_\ell = -\frac{2mg^2 k}{\hbar^2} \int_0^\infty dr \, r^2 e^{-\mu r} j_\ell^2(kr) = -\frac{2mg^2}{k^2 \hbar^2} \int_0^\infty dx \, x^2 \, e^{-\mu x/k} \, j_\ell^2(x)$$

For s-waves, we get

$$\delta_0 = -\frac{mg^2}{k^2 \hbar^2} \int_0^\infty dx \, e^{-\mu x/k} \, (1 - \cos 2x)$$

$$= -\frac{mg^2}{\mu k \hbar^2} \frac{1}{1 + (\mu/2k)^2}$$

This is negative for our repulsive potential.

Problem 10.9

(a) Prove the identity

$$G^{(\pm)}(E) = G_0^{(\pm)}(E) \big[1 + V G^{(\pm)}(E) \big]$$

with

$$G_0^{(\pm)}(E) \equiv (E \pm i\epsilon - H_0)^{-1}, \qquad G^{(\pm)}(E) \equiv (E \pm i\epsilon - H)^{-1}$$

(b) From the equation

$$|\psi_{\mathbf{k}}^{(\pm)}\rangle = |\mathbf{k}\rangle + G^{(\pm)}(E) V |\mathbf{k}\rangle$$

derive the operator scattering equation (the *Schwinger–Lipmann equation*)

$$|\psi_{\mathbf{k}}^{(\pm)}\rangle = |\mathbf{k}\rangle + G_0^{(\pm)}(E) V |\psi_{\mathbf{k}}^{(\pm)}\rangle$$

(c) Establish the orthonormality property

$$\langle \psi_{\mathbf{k}'}^{(\pm)} | \psi_{\mathbf{k}}^{(\pm)} \rangle = \langle \mathbf{k}' | \mathbf{k} \rangle = \delta(\mathbf{k} - \mathbf{k}')$$

(d) If $k' = k$, prove the relation

$$\langle \psi_{\mathbf{k}'}^{(-)} | V | \mathbf{k} \rangle = \langle \mathbf{k}' | V | \psi_{\mathbf{k}}^{(+)} \rangle$$

(e) Introduce the operator

$$T(E) \equiv V + V G^{(+)}(E) V$$

and show that

$$T(E) - T^\dagger(E) = -2\pi i \, V \delta(E - H) V$$

Solution

(a) Multiplying by G_0^{-1} from the left and G^{-1} from the right, we arrive at the identity

$$G_0^{-1} = G^{-1} + V \qquad \leftrightarrow \qquad E - H_0 = E - H + V$$

(b) We have

$$|\mathbf{k}\rangle = (1 + GV)^{-1}|\psi_\mathbf{k}\rangle$$

This, substituted back, gives

$$|\psi_\mathbf{k}\rangle = |\mathbf{k}\rangle + GV(1 + GV)^{-1}|\psi_\mathbf{k}\rangle$$

This would be the Schwinger–Lipmann equation if

$$GV(1 + GV)^{-1} = G_0 V$$

which is equivalent to

$$GV = G_0 V(1 + GV)$$

or

$$GV = G_0(1 + VG)V$$

This, thanks to the identity proved in (a), is always true.

(c) We have

$$\begin{aligned}
\langle\psi_{\mathbf{k}'}|\psi_\mathbf{k}\rangle &= \langle\psi_{\mathbf{k}'}|\,[1 + G(E)V]\,|\mathbf{k}\rangle = \langle\psi_{\mathbf{k}'}|\mathbf{k}\rangle + \langle\psi_{\mathbf{k}'}|(E - H)^{-1}V|\mathbf{k}\rangle \\
&= \langle\psi_{\mathbf{k}'}|\mathbf{k}\rangle + \langle\psi_{\mathbf{k}'}|(E - E')^{-1}V|\mathbf{k}\rangle = \langle\psi_{\mathbf{k}'}|\mathbf{k}\rangle - \langle\psi_{\mathbf{k}'}|V(E' - H_0)^{-1}|\mathbf{k}\rangle \\
&= \langle\psi_{\mathbf{k}'}|\mathbf{k}\rangle - \langle\psi_{\mathbf{k}'}|VG_0(E')|\mathbf{k}\rangle = \langle\psi_{\mathbf{k}'}|\,[1 - VG_0(E')]\,|\mathbf{k}\rangle \\
&= \langle\mathbf{k}'|\mathbf{k}\rangle = \delta(\mathbf{k} - \mathbf{k}')
\end{aligned}$$

(d)

$$\begin{aligned}
\langle\psi_{\mathbf{k}'}^{(-)}|V|\mathbf{k}\rangle &= \langle\mathbf{k}'|\,[1 + G^{(-)}(E)V]^\dagger\, V|\mathbf{k}\rangle = \langle\mathbf{k}'|\,[1 + VG^{(+)}(E)]\, V|\mathbf{k}\rangle \\
&= \langle\mathbf{k}'|V\,[1 + G^{(+)}(E)V]\,|\mathbf{k}\rangle = \langle\mathbf{k}'|V|\psi_\mathbf{k}^{(+)}\rangle
\end{aligned}$$

(e)

$$\begin{aligned}
T(E) - T^\dagger(E) &= V\left(\frac{1}{E + i\epsilon - H} - \frac{1}{E - i\epsilon - H}\right)V \\
&= V\frac{-2i\epsilon}{(E - H)^2 + \epsilon^2}V = -2\pi i V\delta(E - H)V
\end{aligned}$$

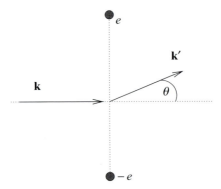

Fig. 45 Scattering from a dipole.

Problem 10.10 Consider an electric dipole consisting of two electric charges e and $-e$ at a mutual distance $2a$. Consider also a particle of charge e and mass m with an incident wave vector $\mathbf{k}$ perpendicular to the direction of the dipole; see Fig. 45.

(a) Calculate the scattering amplitude in the Born approximation. Find the directions at which the differential cross section is maximal.

(b) Consider a different system with a target consisting of two arbitrary charges q_1 and q_2 similarly placed. Calculate again the scattering amplitude and the directions of maximal scattering.

Solution

(a) The potential created by the dipole is

$$V(\mathbf{r}) = -\frac{e^2}{|\mathbf{r} + \mathbf{a}|} + \frac{e^2}{|\mathbf{r} - \mathbf{a}|}$$

We have placed the charge $-e$ at $-\mathbf{a}$ and the charge e at $\mathbf{a}$. The scattering amplitude in the Born approximation is

$$f_{\mathbf{k}}(\mathbf{k}') = -\frac{4\pi^2 m}{\hbar^2} \langle \mathbf{k}' | V | \mathbf{k} \rangle = -\frac{4\pi^2 m e^2}{\hbar^2} \int \frac{d^3 r}{(2\pi)^3} e^{i\mathbf{r}\cdot\mathbf{q}} \left(-\frac{1}{|\mathbf{r} + \mathbf{a}|} + \frac{1}{|\mathbf{r} - \mathbf{a}|} \right)$$

$$= -\frac{4\pi^2 m e^2}{\hbar^2} \int \frac{d^3 r}{(2\pi)^3} e^{i\mathbf{r}\cdot\mathbf{q}} \frac{1}{r} \left(-e^{-i\mathbf{q}\cdot\mathbf{a}} + e^{i\mathbf{q}\cdot\mathbf{a}} \right)$$

We have denoted $\mathbf{q} \equiv \mathbf{k} - \mathbf{k}'$. The integral involved is

$$\int d^3 r \, e^{i\mathbf{r}\cdot\mathbf{q}} \frac{1}{r} = 2\pi \int_0^\infty dr \, r \int_{-1}^1 (d\cos\theta) \, e^{iqr\cos\theta} = \frac{4\pi}{q^2}$$

Thus, the scattering amplitude is

$$f_{\mathbf{k}}(\mathbf{k}') = -\frac{2m e^2}{\hbar^2} \frac{1}{q^2} \left(-e^{-i\mathbf{q}\cdot\mathbf{a}} + e^{i\mathbf{q}\cdot\mathbf{a}} \right) = -\frac{4i m e^2}{\hbar^2} \left(\frac{\sin \mathbf{q}\cdot\mathbf{a}}{q^2} \right)$$

Taking $\mathbf{k} = k\hat{\mathbf{z}}$ and $\mathbf{a} = a\hat{\mathbf{x}}$, we have $\mathbf{k}' = k\sin\theta\,\hat{\mathbf{x}} + k\cos\theta\,\hat{\mathbf{z}}$, $q^2 = 2k^2(1 - \cos\theta)$ and $\mathbf{q} \cdot \mathbf{a} = -ka\sin\theta$. Thus

$$\left|f_{\mathbf{k}}(\mathbf{k}')\right|^2 = \frac{4m^2e^4}{(\hbar k)^4}\frac{\sin^2(ka\sin\theta)}{(1 - \cos\theta)^2}$$

The cross section becomes maximal when

$$2a\sin\theta = \frac{\pi}{k}(2n + 1) = \frac{\lambda}{2}(2n + 1)$$

(b) The scattering amplitude will be

$$\begin{aligned}
f_{\mathbf{k}}(\mathbf{k}') &= -\frac{4\pi^2 me}{\hbar^2}\int\frac{d^3r}{(2\pi)^3}\,e^{i\mathbf{r}\cdot\mathbf{q}}\frac{1}{r}\left(q_2 e^{-i\mathbf{q}\cdot\mathbf{a}} + q_1 e^{i\mathbf{q}\cdot\mathbf{a}}\right) \\
&= -\frac{2me}{\hbar^2}\frac{1}{q^2}\left(q_2 e^{-i\mathbf{q}\cdot\mathbf{a}} + q_1 e^{i\mathbf{q}\cdot\mathbf{a}}\right) \\
&= -\frac{me}{(\hbar k)^2}\frac{(q_1 + q_2)\cos(ka\sin\theta) + i(q_1 - q_2)\sin(ka\sin\theta)}{1 - \cos\theta}
\end{aligned}$$

The differential cross section is

$$\begin{aligned}
\left|f_{\mathbf{k}}(\mathbf{k}')\right|^2 &= \frac{m^2e^2}{(\hbar k)^4}\frac{(q_1 + q_2)^2\cos^2(ka\sin\theta) + (q_1 - q_2)^2\sin^2(ka\sin\theta)}{(1 - \cos\theta)^2} \\
&= \frac{m^2e^2}{(\hbar k)^4}\frac{q_1^2 + q_2^2 + 2q_1q_2\cos(2ka\sin\theta)}{(1 - \cos\theta)^2}
\end{aligned}$$

For $q_1q_2 > 0$, maximal cross section is achieved at

$$2a\sin\theta = n\frac{2\pi}{k} = n\lambda$$

while if $q_1q_2 < 0$ it is achieved at

$$2a\sin\theta = (2n + 1)\frac{\pi}{k} = (2n + 1)\frac{\lambda}{2}$$

Problem 10.11 Consider the scattering of particles of mass m from an attractive potential that has a constant strength $-V_0$ within a sphere of radius R but vanishes elsewhere. Calculate the differential and the total cross section for $kR \ll 1$ (i.e. small energies).

Solution

The phase shift is given by the formula

$$\tan\delta_\ell = \frac{j'_\ell(kR) - \alpha_\ell\,j_\ell(kR)}{n'_\ell(kR) - \alpha_\ell\,n_\ell(kR)}$$

where α_ℓ is the logarithmic derivative[8] at $r = R$. In the limit $kR \ll 1$, using the small-argument behaviour of the spherical Bessel functions, we get

$$\tan \delta_\ell \approx -\frac{(2\ell + 1)}{[(2\ell + 1)!!]^2} (kR)^{2\ell+1} \frac{\alpha_\ell R - \ell}{\alpha_\ell R + \ell + 1}$$

These phase shifts are small; the scattering will be dominated by small values of ℓ. The differential cross section is

$$\frac{d\sigma}{d\Omega} \sim \frac{1}{k^2} |\delta_0 + 3\delta_1 P_1(\cos\theta) + \cdots|^2 \sim C_0 + C_1 \cos\theta + \cdots$$

where

$$C_0 = \frac{\delta_0^2}{k^2} = R^2 \frac{R^2 \alpha_0^2}{(1 + R\alpha_0)^2}$$

$$C_1 = \frac{6\delta_0\delta_1}{k^2} = 2R^2(kR)^2 \frac{\alpha_0 R(-1 + R\alpha_1)}{(1 + R\alpha_0)(2 + R\alpha_1)}$$

and

$$\frac{C_1}{C_0} = 2(kR)^2 \frac{(-1 + R\alpha_1)(1 + R\alpha_0)}{R\alpha_0(2 + R\alpha_1)}$$

It is clear that $C_1/C_0 \ll 1$ in the limit of low energies.

The total cross section will be

$$\sigma = \int d\Omega \, (C_0 + C_1 \cos\theta + \cdots) \approx 4\pi C_0 = 4\pi R^2 \left(\frac{\alpha_0 R}{1 + \alpha_0 R}\right)^2$$

$$= 4\pi R^2 \left(1 - \frac{\tan qR}{qR}\right)^2$$

Problem 10.12 The scattering amplitude of a particle of mass m in a potential $V(r)$ can be written as

$$f_{\mathbf{k}}(\hat{\mathbf{r}}) = -\frac{4m\pi^2}{\hbar^2} \langle \mathbf{k} | V | \psi_{\mathbf{k}}^{(+)} \rangle$$

where $\psi_{\mathbf{k}}^{(+)}$ is the scattering wave function, which satisfies the integral equation[9]

$$|\psi_{\mathbf{k}}^{(+)}\rangle = |\mathbf{k}\rangle + G_0^{(+)}(E) \, V \, |\psi_{\mathbf{k}}^{(+)}\rangle$$

Write down the *Born expansion* of the scattering amplitude. Using the optical

---

[8] The logarithmic derivative at $r = R$ is $\alpha_\ell = j'(qR)/j_\ell(qR)$. The wave number q is defined as $q \equiv \sqrt{2m(E + V_0)/\hbar^2}$.

[9] The Green's function operator is defined as

$$G^{(+)}(E) = \frac{1}{E + i\epsilon - H}$$

theorem calculate the total cross section for the potential $V(r) = g^2 e^{-\mu r}$ to the lowest non-trivial order.

Solution

The scattering amplitude in the forward direction is

$$f_{\mathbf{k}}(\mathbf{k}) = -\frac{4\pi^2 m}{\hbar^2} \langle \mathbf{k}|V|\psi_{\mathbf{k}}^{(+)}\rangle$$

$$= -\frac{4\pi^2 m}{\hbar^2} \left[\langle \mathbf{k}|V|\mathbf{k}\rangle + \langle \mathbf{k}|V G_0^{(+)}(E)V|\mathbf{k}\rangle + \cdots \right]$$

Its imaginary part will be

$$\text{Im}\,[f_{\mathbf{k}}(\mathbf{k})] = -\frac{4\pi^2 m}{\hbar^2} \left\{ \text{Im}\,[\langle \mathbf{k}|V|\mathbf{k}\rangle] + \text{Im}\left[\langle \mathbf{k}|V G_0^{(+)}(E)V|\mathbf{k}\rangle\right] + \cdots \right\}$$

$$= -\frac{2\pi^2 m}{i\hbar^2} \left\langle \mathbf{k}\left|V\left(\frac{1}{E + i\epsilon - H_0} - \frac{1}{E - i\epsilon - H_0}\right)V\right|\mathbf{k}\right\rangle$$

$$= -\frac{2\pi^2 m}{i\hbar^2}(-2i)\left\langle \mathbf{k}\left|V\frac{\epsilon}{(E - H_0)^2 + \epsilon^2}V\right|\mathbf{k}\right\rangle$$

$$= \frac{4\pi^3 m}{\hbar^2}\langle \mathbf{k}|V\delta(E - H_0)V|\mathbf{k}\rangle = \frac{4\pi^3 m}{\hbar^2}\langle \mathbf{k}|V\int d^3q\,|\mathbf{q}\rangle\langle \mathbf{q}|\delta(E - H_0)V|\mathbf{k}\rangle$$

$$= \frac{4\pi^3 m}{\hbar^2}\int d^3q\,\langle \mathbf{k}|V|\mathbf{q}\rangle\,\langle \mathbf{q}|V|\mathbf{k}\rangle\delta\,(E(k) - E(q))$$

$$= \frac{4\pi^3 m}{\hbar^2}\int d^3q\,|\langle \mathbf{k}|V|\mathbf{q}\rangle|^2\,\delta\,(E(k) - E(q))$$

$$= \frac{8\pi^3 m^2}{\hbar^4}\int d^3q\,|\langle \mathbf{k}|V|\mathbf{q}\rangle|^2\,\delta(q^2 - k^2)$$

Let us now calculate the potential matrix element, first setting $\mathbf{Q} \equiv \mathbf{k} - \mathbf{q}$. We have

$$\langle \mathbf{k}|V|\mathbf{q}\rangle = g^2\int \frac{d^3r}{(2\pi)^3}\,e^{-i\mathbf{Q}\cdot\mathbf{r} - \mu r}$$

$$= \frac{g^2}{4\pi^2}\int_0^\infty dr\,r^2 e^{-\mu r}\int_{-1}^1 (d\cos\theta)\,e^{-iQr\cos\theta}$$

$$= \frac{ig^2}{4\pi^2 Q}\int_0^\infty dr\,r\left[e^{-(\mu + iQ)r} - e^{-(\mu - iQ)r}\right]$$

$$= -\frac{ig^2}{4\pi^2 Q}\frac{\partial}{\partial\mu}\int_0^\infty dr\left[e^{-(\mu + iQ)r} - e^{-(\mu - iQ)r}\right]$$

$$= -\frac{ig^2}{4\pi^2 Q}\frac{\partial}{\partial\mu}\left(\frac{1}{\mu + iQ} - \frac{1}{\mu - iQ}\right)$$

$$= \frac{ig^2}{4\pi^2 Q}\left(\frac{1}{(\mu + iQ)^2} - \frac{1}{(\mu - iQ)^2}\right) = \frac{\mu g^2}{\pi^2}\frac{1}{(\mu^2 + Q^2)^2}$$

For $q = k$, we may write

$$Q^2 = 2k^2(1 - \cos \theta)$$

Thus, the imaginary part of the scattering amplitude is

$$
\begin{aligned}
\mathrm{Im}\,[f_{\mathbf{k}}(\mathbf{k})] &= \frac{8g^4\mu^2 m^2\pi}{\hbar^4} \int d^3q\, \frac{1}{(\mu^2 + Q^2)^4}\,\delta(q^2 - k^2) \\
&= \frac{8g^4\mu^2 m^2\pi^2}{\hbar^4} \int_0^\infty dq\, q\delta(q - k) \int_{-1}^1 (d\cos\theta)\,\frac{1}{(\mu^2 + Q^2)^4} \\
&= \frac{8g^4 k\mu^2 m^2\pi^2}{\hbar^4} \int_{-1}^1 (d\cos\theta)\,\frac{1}{(\mu^2 + 2k^2 - 2k^2\cos\theta)^4} \\
&= \frac{g^4\mu^2 m^2\pi^2}{2k^7\hbar^4} \int_{-1}^1 (d\cos\theta)\,\frac{1}{(-\mu^2/2k^2 - 1 + \cos\theta)^4} \\
&= \frac{16}{3}\left(\frac{g^4 m^2 k\pi^2}{\hbar^4\mu^4}\right) \frac{3\mu^2(\mu^2 + 4k^2) + 16k^4}{(\mu^2 + 4k^2)^3}
\end{aligned}
$$

The total cross section is

$$\sigma = \left(\frac{64\pi^3 g^4 m^2}{3\hbar^4\mu^4}\right) \frac{3\mu^2(\mu^2 + 4k^2) + 16k^4}{(\mu^2 + 4k^2)^3}$$

Problem 10.13 Consider a one-dimensional potential that vanishes beyond some point, i.e. $V(x) = 0$ for $|x| \geq a > 0$.

(a) The scattering wave functions satisfy the following integral equation[10]

$$\psi_k^{(+)}(x) = \phi_k(x) + \int_{-\infty}^\infty dx'\, G_0^{(+)}(x - x')V(x')\psi_k^{(+)}(x')$$

where $\phi_k(x) = e^{ikx}/\sqrt{2\pi}$. Determine the Green's function $G_0^{(+)}(x - x')$. Show that it can be written as the position matrix element of an operator.

(b) Show that in the asymptotic region $|x| \gg a$, we can write

$$\psi_k(x) \sim \phi_k(x) + \frac{e^{ik|x|}}{\sqrt{2\pi}}\, f(k, k')$$

Determine the *scattering amplitude* $f(k, k')$ in terms of V and ψ.

(c) What is the connection of the scattering amplitude $f(k, k')$ to the reflection and transmission coefficients, familiar from standard one-dimensional problems? Prove the relation

$$\mathrm{Re}\,[f(k, k)] = -\tfrac{1}{2}\left[|f(k, k)|^2 + |f(k, -k)|^2\right]$$

---

[10] The wave number corresponds to the energy in the standard way, $E = \hbar^2 k^2/2m$.

(d) Consider the exactly soluble problem for which $V(x) = g\delta(x)$. Calculate the scattering amplitude in the Born approximation and compare with the known exact answer. Show that, in the attractive case, the Born approximation is valid only when the energy is large in comparison to the bound-state energy.

Solution

(a) Acting on both sides of the integral equation with the operator $H_0 - E$, we obtain

$$[H_0(x) - E]\,\psi_k^{(+)}(x)$$
$$= [H_0(x) - E]\,\phi_k(x) + \int_{-\infty}^{\infty} dx'\,[H_0(x) - E]\,G_0^{(+)}(x - x')V(x')\psi_k^{(+)}(x')$$

or

$$-V(x)\psi_k^{(+)}(x) = \int_{-\infty}^{\infty} dx'\,[H_0(x) - E]\,G_0^{(+)}(x - x')V(x')\psi_k^{(+)}(x')$$

which implies that

$$[E + i\epsilon - H_0(x)]\,G_0^{(+)}(x - x') = \delta(x - x')$$

We have added the term $+i\epsilon$ to the energy in order to fix the boundary conditions at infinity. Fourier transforming, we obtain

$$\int \frac{dq}{2\pi} e^{iq(x-x')} \left(E + i\epsilon - \frac{\hbar^2 q^2}{2m}\right) \tilde{G}_0^{(+)}(q) = \int \frac{dq}{2\pi} e^{iq(x-x')}$$

and

$$\tilde{G}_0^{(+)}(q) = \left(\frac{\hbar^2 k^2}{2m} - \frac{\hbar^2 q^2}{2m} + i\epsilon\right)^{-1}$$

Thus finally we get

$$G_0^{(+)}(x - x') = \int_{-\infty}^{\infty} \frac{dq}{2\pi}\, e^{iq(x-x')} \left(\frac{\hbar^2 k^2}{2m} - \frac{\hbar^2 q^2}{2m} + i\epsilon\right)^{-1}$$

$$= \frac{2im}{\hbar^2}\left[\Theta(x - x')\frac{e^{ik(x-x')}}{2k} - \Theta(x' - x)\frac{e^{ik(x'-x)}}{(-2k)}\right] = -\frac{im}{\hbar^2 k}\, e^{ik|x-x'|}$$

It is easy to see that the position matrix elements of the operator

$$G_0^{(+)}(E) = \frac{1}{E + i\epsilon - H_0}$$

give exactly the Green's function

$$\langle x|G_0^{(+)}(E)|x'\rangle = \int dq\,\langle x|q\rangle\,\langle q|G_0^{(+)}(E)|x'\rangle$$

$$= \int \frac{dq}{2\pi}\, e^{i(x-x')q} \left(E + i\epsilon - \frac{\hbar^2 q^2}{2m}\right)^{-1}$$

(b) In the asymptotic region we can make the approximation

$$|x - x'| = \sqrt{(x - x')^2} = \sqrt{x^2 + (x')^2 - 2xx'} \approx \sqrt{x^2 - 2xx'}$$
$$\approx |x| \left(1 - \frac{x'}{x}\right) = |x| - x' \frac{|x|}{x}$$

Thus, defining $k' \equiv k|x|/x$, we approximate the Green's function as follows:

$$G_0^{(+)}(x - x') = -\frac{im}{\hbar^2 k} e^{ik|x - x'|} \approx -\frac{im}{\hbar^2 k} e^{ik|x| - ik'x'}$$

Then we have

$$\psi_k^{(+)}(x) \approx \frac{1}{\sqrt{2\pi}} \left[e^{ikx} - \frac{i\sqrt{2\pi}\, m}{\hbar^2 k} e^{ik|x|} \int dx'\, e^{-ik'x'} V(x') \psi_k^{(+)}(x') \right]$$

The scattering amplitude is

$$f(k, k') = -\frac{mi\sqrt{2\pi}}{\hbar^2 k} \int dx'\, e^{-ik'x'} V(x') \psi_k^{(+)}(x')$$

(c) In the far positive region, $x \gg a$, the wave function is

$$\frac{e^{ikx}}{\sqrt{2\pi}} + f(k, k) \frac{e^{ikx}}{\sqrt{2\pi}}$$

In the far negative region, $x \ll -a$, it is

$$\frac{e^{ikx}}{\sqrt{2\pi}} + f(k, -k) \frac{e^{-ikx}}{\sqrt{2\pi}}$$

From these expressions we can recognize immediately

(1) the *incident wave and current*

$$\frac{e^{ikx}}{\sqrt{2\pi}}, \qquad J_i = \frac{\hbar k}{m(2\pi)}$$

(2) the *reflected wave and current*

$$f(k, -k) \frac{e^{-ikx}}{\sqrt{2\pi}}, \qquad J_r = -\frac{\hbar k}{m(2\pi)} |f(k, -k)|^2$$

(3) the *transmitted wave and current*

$$[1 + f(k, k)] \frac{e^{-ikx}}{\sqrt{2\pi}}, \qquad J_t = \frac{\hbar k}{m(2\pi)} |1 + f(k, k)|^2$$

Thus, the reflection and transmission coefficients are

$$\mathcal{T} = J_t/J_i = |1 + f(k, k)|^2, \qquad \mathcal{R} = |J_r|/J_i = |f(k, -k)|^2$$

Probability conservation dictates that

$$\mathcal{R} + \mathcal{T} = 1$$

which is equivalent to

$$|1 + f(k, k)|^2 + |f(k, k)|^2 = 1$$
$$\implies \quad |f(k, -k)|^2 + |f(k, k)|^2 = -2\,\mathrm{Re}[f(k, k)]$$

(d) The exact answer for the scattering amplitude of the delta function potential is

$$f = \frac{-ig}{ig + \hbar^2 k/m}$$

and the bound-state energy is

$$E_b = -\frac{mg^2}{2\hbar^2}$$

On the other hand, the Born approximation to the scattering amplitude gives

$$f_B \approx -i\frac{gm}{\hbar^2 k}$$

Notice that the exact answer can be rewritten as ($g < 0$)

$$f = \frac{-i}{i - \sqrt{E/|E_b|}}$$

In the high-energy limit $E \gg |E_b|$, this is approximated by

$$f \approx i\sqrt{\frac{|E_b|}{E}} = -i\frac{mg}{\hbar^2 k} = f_B$$

which coincides with the Born approximation result.

Problem 10.14 The neutron–proton scattering amplitude is of the form

$$f = \chi_f^\dagger \left(a + 4\hbar^{-2}b\,\mathbf{S}_n \cdot \mathbf{S}_p\right)\chi_i$$

where χ_i and χ_f are the initial and final states. Calculate the cross section for neutron–proton scattering when the initial and final proton spins are not measured.

Solution

The operator appearing in the amplitude is

$$\mathbf{S}_n \cdot \mathbf{S}_p = \tfrac{1}{2}\left\{ S_p^{(+)}S_n^{(-)} + S_p^{(-)}S_n^{(+)} + 2S_{pz}S_{nz} \right\}$$

The cross section corresponding to the scattering of neutrons from protons when the target spin is not measured will be

$$\sigma_{m_n'}^{m_n'} = \sum_{m_p, m_p'} \left| f_{m_p, m_n}^{m_p', m_n'} \right|^2$$

In particular, we have

$$\sigma_+ \equiv \sigma_+^+ = \sum_{m_p, m_p'} \left| f_{m_p, +}^{m_p', +} \right|^2 \qquad \text{and} \qquad \sigma_- \equiv \sigma_+^- = \sum_{m_p, m_p'} \left| f_{m_p, +}^{m_p', -} \right|^2$$

The first relevant amplitude is

$$f_{m_p, +}^{m_p', +} = \left\langle m_p \left| \left\langle m_n = \tfrac{1}{2} \right| \left(a + 4\hbar^{-2} b\, \mathbf{S}_n \cdot \mathbf{S}_p \right) \right| m_n' = \tfrac{1}{2} \right\rangle \right| m_p' \right\rangle$$
$$= \left(a + 2bm_p \right) \delta_{m_p m_p'}$$

and leads to the cross section

$$\sigma_+ = \sum_{m_p, m_p'} \left| a + 2bm_p \right|^2 \delta_{m_p m_p'} = \sum_{m_p} \left| a + 2bm_p \right|^2$$
$$= \left| a + b \right|^2 + \left| a - b \right|^2 = 2|a|^2 + 2|b|^2$$

Similarly, we have for the other amplitude

$$f_{m_p, +}^{m_p', -} = \left\langle m_p \left| \left\langle m_n = \tfrac{1}{2} \right| \left(a + 4\hbar^{-2} b\, \mathbf{S}_n \cdot \mathbf{S}_p \right) \right| m_n' = -\tfrac{1}{2} \right\rangle \right| m_p' \right\rangle$$
$$= 2b\delta_{m_p', 1/2}\delta_{m_p, -1/2}$$

and for the corresponding cross section

$$\sigma_- = \sum_{m_p, m_p'} 4|b|^2 \delta_{m_p', 1/2}\delta_{m_p, -1/2} = 4|b|^2$$

The *polarization* ϖ will be

$$\varpi = \frac{\sigma_+ - \sigma_-}{\sigma_+ + \sigma_-} = \frac{|a|^2 - |b|^2}{|a|^2 + 3|b|^2}$$

Problem 10.15 Consider the scattering of a particle by a distribution of scattering centres. Each scatterer is located at a point $\mathbf{r}_i$ and scatters with a given potential $V_0(|\mathbf{r} - \mathbf{r}_i|)$. Write down the scattering amplitude in the Born approximation.

(a) Consider the case of a cube of side a with the scatterers placed at its eight vertices.

(b) Do the same for an infinite cubic lattice of lattice spacing a.

Solution

The Born-approximation scattering amplitude is ($\mathbf{q} \equiv \mathbf{k} - \mathbf{k}'$)

$$f_{\mathbf{k}}(\hat{\mathbf{r}}) = -\frac{m}{2\pi\hbar^2} \int d^3r' \, e^{i\mathbf{q}\cdot\mathbf{r}'} \sum_i V_0(|\mathbf{r}' - \mathbf{r}_i|)$$

$$= -\frac{m}{2\pi\hbar^2} \sum_i e^{i\mathbf{q}\cdot\mathbf{r}_i} \int d^3\rho \, e^{i\mathbf{q}\cdot\boldsymbol{\rho}} V_0(\rho) = -\frac{m\sqrt{2\pi}}{\hbar^2} \tilde{V}_0(q) \sum_i e^{i\mathbf{q}\cdot\mathbf{r}_i}$$

where $\tilde{V}_0$ is the Fourier transform of the given potential V_0.

(a) In the case of the cube, the sum is

$$8\cos(aq_x/2)\cos(aq_y/2)\cos(aq_z/2)$$

The cross section will be

$$\frac{d\sigma}{d\Omega} = \frac{64m^2(2\pi)}{\hbar^4}|\tilde{V}_0|^2 \cos^2\frac{q_x a}{2} \cos^2\frac{q_y a}{2} \cos^2\frac{q_z a}{2}$$

Maximal value is achieved when all $q_i = 2n_i\pi/a$.

(b) In the case of an infinite lattice, the sum is

$$\sum_{n_x=0}^{\infty} e^{iaq_x n_x} \sum_{n_y=0}^{\infty} e^{iaq_y n_y} \sum_{n_z=0}^{\infty} e^{iaq_z n_z}$$

$$= (1 - e^{iaq_x})^{-1}(1 - e^{iaq_y})^{-1}(1 - e^{iaq_z})^{-1}$$

$$= -8i \, e^{-i(q_x+q_y+q_z)a/2} \left(\sin\frac{q_x a}{2} \sin\frac{q_y a}{2} \sin\frac{q_z a}{2} \right)^{-1}$$

The cross section will be

$$\frac{d\sigma}{d\Omega} = \frac{64m(2\pi)}{\hbar^4}|\tilde{V}_0|^2 \left(\sin^2\frac{q_x a}{2} \sin^2\frac{q_y a}{2} \sin^2\frac{q_z a}{2} \right)^{-1}$$

Maximal (infinite) cross section corresponds to any of the momentum transfers

$$q_x = \frac{2n_x\pi}{a}, \qquad q_y = \frac{2n_y\pi}{a}, \qquad q_z = \frac{2n_z\pi}{a} \qquad (n_x, n_y, n_z = 1, 2, \ldots)$$

Bibliography

G. Baym. *Lectures in Quantum Mechanics*, New York: W. A. Benjamin, 1969.

J. S. Bell. *Speakable and Unspeakable in Quantum Mechanics*, Cambridge University Press, 1993.

A. Capri. *Problems and Solutions in Non-relativistic Quantum Mechanics*, World Scientific, 2001.

C. Cohen-Tannoudji *et al. Quantum Mechanics*, vols. I and II, Wiley, 1977.

F. Constantinescu and E. Magyari. *Problems in Quantum Mechanics*, Pergamon Press, 1971.

P. A. M. Dirac. *Quantum Mechanics*, 4th edn, London: Oxford University Press, 1958.

S. Flügge. *Practical Quantum Mechanics*, Springer-Verlag, 1971.

S. Gasiorowitcz. *Quantum Physics*, New York: Wiley, 1996.

I. I. Goldman and V. D. Krivchenkov. *Problems in Quantum Mechanics*, New York: Dover Publications, 1993.

K. Gottfried. *Quantum Mechanics*, vol. 1, New York: W. A. Benjamin, 1966.

W. Greiner. *Quantum Mechanics: An Introduction*, Springer-Verlag, 1989.

L. Landau and E. M. Lifshitz. *Quantum Mechanics*, Reading MA: Addison-Wesley, 1965.

F. Mandl. *Quantum Mechanics*, London: Butterworths Scientific Publications, 1957.

A. Messiah. *Quantum Mechanics*, vols. I and II, North Holland, 1970.

E. Merzbacher. *Quantum Mechanics*, Wiley, 1970.

J. J. Sakurai. *Modern Quantum Mechanics*, Reading MA: Addison-Wesley, 1995.

L. Schiff. *Quantum Mechanics*, New York, MacGraw-Hill, 1968.

G. L. Squires. *Problems in Quantum Mechanics*, Cambridge University Press, 1995.

Yung-Kuo Lim (ed.). *Problems and Solutions on Quantum Mechanics*, World Scientific, 1998.

Index